Agent Systems in Electronic Business

Eldon Y. Li
National Chengchi University, Taiwan

Soe–Tsyr Yuan
National Chengchi University, Taiwan

Information Science
REFERENCE

INFORMATION SCIENCE REFERENCE

Hershey · New York

Acquisitions Editor: Kristin Klinger
Development Editor: Kristin Roth
Assistant Development Editor: Meg Stocking
Editorial Assistant: Jessica Thompson
Senior Managing Editor: Jennifer Neidig
Managing Editor: Sara Reed
Copy Editor: April Schmidt
Typesetter: Jamie Snavely
Cover Design: Lisa Tosheff
Printed at: Yurchak Printing Inc.

Published in the United States of America by
      Information Science Reference (an imprint of IGI Global)
      701 E. Chocolate Avenue, Suite 200
      Hershey PA 17033
      Tel: 717-533-8845
      Fax: 717-533-8661
      E-mail: cust@igi-global.com
      Web site: http://www.igi-global.com

and in the United Kingdom by
      Information Science Reference (an imprint of IGI Global)
      3 Henrietta Street
      Covent Garden
      London WC2E 8LU
      Tel: 44 20 7240 0856
      Fax: 44 20 7379 0609
      Web site: http://www.eurospanonline.com

Library of Congress Cataloging-in-Publication Data

Agent systems in electronic business / Eldon Y. Li & Soe-Tsyr Yuan, editors.

      p. cm.

   Summary: "This book delivers definitive research on the use of agent technologies to advance the practice of electronic business in today's organizations, targeting the needs of enterprises in open and dynamic business opportunities to incorporate skilled use of multiple independent information systems. It clearly articulates the stages involved in developing agent-based e-business systems"--Provided by publisher.

   Includes bibliographical references and index.

   ISBN-13: 978-1-59904-588-7 (hbk.)

   ISBN-13: 978-1-59904-590-0 (ebook)

   1. Electronic commerce--Technological innovations. 2. Intelligent agents (Computer software) I. Li, Eldon Yu-zen, 1952- II. Yuan, Soe-Tsyr.

   HF5548.32.A35 2008

   621.381068'8--dc22

                        2007024486

British Cataloguing in Publication Data
A Cataloguing in Publication record for this book is available from the British Library.

Advances in Electronic Business Series (AEBUS)
ISBN: Pending

Editor-in-Chief: Eldon Li, PhD, The Chinese University of Hong Kong, Hong Kong

Agent Systems in Electronic Business

Edited By: Eldon Li, National Chengchi University, Taiwan & Soe-Tsyr Yuan, National Chengchi University, Taiwan

Information Science Reference • copyright 2008 • 300pp • H/C (ISBN: 978-1-59904-588-7) • $180.00 (list price)

Agent technologies are believed to be one of the most promising tools to conduct business via networks and the Web in an autonomous, intelligent, and efficient way. The ever-expanding application of business automation necessitates clarification of the methods and techniques of agent-based electronic business systems. Agent Systems in Electronic Business delivers definitive research to academics and practitioners in the field of business automation on the use of agent technologies to advance the practice of electronic business in today's organizations, targeting the needs of enterprises in open and dynamic business opportunities to incorporate skilled use of multiple independent information systems. This comprehensive resource clearly articulates the stages (electronic commerce transactions, business processes and e-business infrastructure) involved in developing agent-based e-business systems.

Advances in Electronic Business, Volume 2

Edited By: Eldon Li; National Chengchi University, Taiwan & California Polytechnic State University, USA

CyberTech Publishing • copyright 2007 • 332 pp • H/C (ISBN: 1-59140-678-1) • US $85.46 (our price) • E-Book (ISBN: 1-59140-680-3) • US $63.96 (our price)

The Internet has generated a large amount of information that is created and shared between individuals and organizations. Because of the amount of information flying through cyberspace, the time to locate and digest the information increases exponentially, but the question of what information can be shared and how to share it remains unsolved. Advances in Electronic Business, Volume 2 explores the semantic web and intelligent web services, two methods created to help solidify the meaning and relationship of data, and explains how they relate to business processes. Professionals, policy-makers, academics, researchers, and managers in IT, business, and commerce will find this book useful in understanding the semantic web and intelligent web services impact on e-commerce.

Advances in Electronic Business, Volume 1

Eldon Li , Timon C. Du; uan Ze University, Taiwan; The Chinese University of Hong Kong, Hong Kong

IGI Publishing • copyright 2005 • 356 pp • H/C (ISBN: 1-59140-381-2) • US $76.46 (our price) • E-Book (ISBN: 1-59140-383-9) • US $55.96 (our price)

Advances in Electronic Business advances the understanding of management methods, information technology, and their joint application in business processes. The applications of electronic commerce draw great attention of the practitioners in applying digital technologies to the buy-and-sell activities. This timely book addresses the importance of management and technology issues in electronic business, including collaborative design, collaborative engineering, collaborative decision making, electronic collaboration, communication and cooperation, workflow collaboration, knowledge networking, collaborative e-learning, costs and benefits analysis of collaboration, collaborative transportation and ethics.

The Advances in Electronic Business (AEBUS) Book Series advances the understanding of management methods, information technology, and their joint application in business processes. The applications of electronic commerce draw great attention of the practitioners in applying digital technologies to the buy-and-sell activities. This book series addresses the importance of management and technology issues in electronic business, including collaborative design, collaborative engineering, collaborative decision making, electronic collaboration, communication and cooperation, workflow collaboration, knowledge networking, collaborative e-learning, costs and benefits analysis of collaboration, collaborative transportation and ethics. The Internet has generated a large amount of information that is created and shared between individuals and organizations. Because of the amount of information flying through cyberspace, the time to locate and digest the information increases exponentially, but the question of what information can be shared and how to share it remains unsolved. AEBUS explores the semantic web and intelligent web services, two methods created to help solidify the meaning and relationship of data, and explains how they relate to business processes. Professionals, policy-makers, academics, researchers, and managers in IT, business, and commerce will find this book useful in understanding the semantic web and intelligent web services impact on e-commerce.

DISSEMINATOR of KNOWLEDGE

Hershey • New York

Order online at www.igi-global.com or call 717-533-8845 x10 –
Mon-Fri 8:30 am - 5:00 pm (est) or fax 24 hours a day 717-533-8661

Table of Contents

Section III
Agent Technologies in Electronic Business Processes

Section IV
Agent Technologies in E-Business Infrastructure

Section V
Cross-Fertilized Techniques in Business Antomation

Detailed Table of Contents

Section I
Overview

Agents are rapidly emerging as a new paradigm for developing software applications. They are being used in an increasing variety of applications, ranging from relatively small systems such as assistants to large, open, mission-critical systems like electronic marketplaces. One of the most promising areas of applications for agent technology is e-business. In this chapter, we describe a group of architectural patterns for agent-based e-business systems. These patterns relate to front-end e-business activities that involve interaction with the user, and delegation of user tasks to agents. Patterns capture well-proven, common solutions, and guide developers through the process of designing systems. This chapter should be of interest to designers of e-business systems using agent technology. The description of the patterns is followed by the case study of an online auction system to which the patterns have been applied.

Section II
Agent Technologies in Electronic Commerce Transactions

Comparison-shopping agents became the important link in the B2C e-commerce domain since late 1990s. Since its emergence in 1995, the evolution of comparison-shopping agents experienced a few ups and downs. This chapter covers key events and issues of comparison-shopping agent evolution in three intertwined threads: the emergence of representative agents, the evolution of comparison-shopping agent technology, and the evolution of their business models.

The Internet and World Wide Web are becoming more and more dynamic in terms of their contents and usage. Agent-based shopping support (ASS) aims at keeping up with this dynamic environment by mimicking shoppers' purchasing behavior in the electronic commerce transaction process in the sense of matching the profiles of Web sites and shoppers. Evolutionary agent-based shopping supports are emerging as intelligent shopping support. This chapter contains the earliest attempt to gather and investigate the nature of current research. The idea of applying concepts of product characteristics from the matrix of Internet marketing strategies is introduced for solving problems of natural language information search. The process of focus-group research methodology is applied in acquiring the essential knowledge for examining shopper's knowledge of search. An architecture of ASS in the case of outbound group package tour in Taiwan is presented. This work demonstrates the process of knowledge acquisition to tackle the problem of ineffective online information search by a customer-centric method.

The theme of this chapter includes topics of matching, auction, and negotiation. We have shown that the knowledge of game theory is very important when designing an agent-based matching or negotiation system. The problem of bounded rationality in multi-agent systems is also discussed; we put forward the mechanism design and heuristic methods as solutions. A real negotiation scenario is presented to demonstrate our proposed solutions. In addition, we discuss the future trends of the agent technology in e-commerce systems.

Market mechanism or auction design research is playing an important role in computational economics for resolving multi-agent allocation problems. In this chapter, we review relevant background of trading agents, and market designs by evolutionary computing methods. In particular, a genetic algorithm (GA) can be used to design auction mechanisms in order to automatically generate a desired market mechanism for electronic markets populated with trading agents. In previous research, an auction space model was studied, in which the probability that buyers and sellers are able to quote on a given time step is optimized by a simple GA in order to maximize the market efficiency in terms of Smith's coefficient of convergence. In this chapter, we also show some new results based on experiments with homogeneous and heterogeneous agents in a more realistic auction space model. This research provides a way of designing efficient auctions by evolutionary computing approaches.

Section III
Agent Technologies in Electronic Business Processes

Chapter VI

It is observed that agent (or software agent) based systems largely imitate organizations of human actors. Thus, the nature of agent-based systems can be better understood by first studying the ordinary human actors or organizations that own the agent-based systems. In this chapter, we first study agent systems and discuss characteristics of software agents, then we introduce a generic pattern of agents interaction derived from the communication patterns of human actors. Agent-based systems are studied in the context of inter-organizational business process using diagrams and notations adapted by the authors. The methods and concepts used in this chapter are based on the semiotics approach and the language action perspective. For the illustration of our concept of agent-based systems, we discuss a case study conducted based on a real life business.

Chapter VII

This chapter explores the utilization of a multi-agent system in the field of supply chain management for electronic business. It investigates the coordination and cooperation processes, and proposes and discusses a newly developed model for an enhanced and effective cooperation process for e-business. The contribution made by this research provides a theoretical solution and model for agents that adopt the enhanced strategy for e-business. Both large organizations and SMEs will benefit by increasing and expanding their businesses globally, and by participating and sharing with business partners to achieve common goals. As a consequence, the organizations involved will each earn more profit.

Chapter VIII

Emergent processes are business processes whose execution is determined by the prior knowledge of the agents involved and by the knowledge that emerges during a process instance. The amount of process knowledge that is relevant to a knowledge-driven process can be enormous and may include common sense knowledge. If a process' knowledge can not be represented feasibly then that process can not be managed; although its execution may be partially supported. In an e-market domain, the majority of transactions, including trading orders, requests for advice and information, are knowledge-driven processes for which the knowledge base is the Internet, and so representing the knowledge is not at issue. Multiagent systems are an established platform for managing complex business processes. What is needed for emergent process management is an intelligent agent that is driven not by a process goal, but by an in-flow of knowledge, where each chunk of knowledge may be uncertain. These agents should assess the extent to which it chooses to believe that the information is correct, and so they require an inference mechanism that can cope with information of differing integrity.

Chapter IX

This chapter presents a new concept for supporting electronic collaboration, operations, and relationships among trading partners in the value chain without hindering human autonomy. Although autonomous intelligent-agents, or electronic robots (e-bots), can be used to inform this endeavor, the chapter advocates the development of e-sensors, i.e., software based units with capabilities beyond intelligent-agent's functionality. E-sensors are hardware-software capable of perceiving, reacting and learning from its interactive experience thorough the supply chain, rather than just searching for data and information through the network and reacting to it. E-sensors can help avoid the 'bullwhip' effect. The chapter briefly reviews the related intelligent-agent and supply-chain literature and the technological gap between fields. It articulates a demand-driven, sense-and-response system for sustaining e-collaboration and e-business operations as well as monitoring products and processes. As a proof of concept, this research aimed a test solution at a single supply-chain partner within one stage of the process.

Section IV
Agent Technologies in E-Business Infrastructure

One consequence of market globalization has been the growing incidence of collaborative ventures among companies from different countries. Small and large, experienced and novice, companies increasingly are choosing partnerships as a way to compete in the global marketplace. International joint ventures have emerged as the dominant form of partnership in light of intense global competition and the need for strategic organizational viability. The success of international joint ventures depend on many factors, but the most critical is vendors selection from among many suppliers based on their ability to meet the quantity requirements, delivery schedule, and the price limitation. The supplier selection negotiation mechanism is often the most complex, since it requires evaluation and decision-making under uncertainty, based on multiple attributes (criteria) of quantitative and qualitative nature, involving temporal and resource constraints, risk and commitment problems, varying tactics and strategies, domain specific knowledge and information asymmetries, and so forth. In this chapter, we propose a negotiation mechanism employing fuzzy logic to evaluate different quantitative/qualitative scale of each attribute, generating similarity matching with bilateral alternatives offered by buyer agent and seller agents, and then modeling some constraints-based rules for sellers when receiving counter-proposal by buyer, consequently proceeding to trade-off mechanism between both sides to gain an agreement. The negotiation mechanism is mainly classified in five parts. We first defined negotiation parameters set and iso-curve computation in the preliminary setting. Second, a negotiation alternative processing service will be proposed to select buyer's alternative based on iso-curve. After selecting a negotiation alternative, the buyer agent will send its alternative (counter-proposal) to seller agents to determine if it satisfies seller's constraints or not then decide on iso-curve relaxation which is buyer's subjective behavior.

Consequently, we use trade-off to find out buyer's partner and determine which attributes need to change along with the iso-curve. An example application to negotiate a supplier selection among agents is to demonstrate for the agents how to adjust its negotiation attribute parameters and reach its agreement. Last, in the postnegotiation analysis, we will compare two results from the preceding trade-off strategies (i.e., risk-seeking and risk-averse) to let decision-making know how to make the most benefit decision for company. In our proposed negotiation mechanism scheme, agents autonomously negotiate multi-attribute fuzzy values of trade-off in an international joint ventures selection tested with a notebook computer manufacturing company scenario.

Chapter XI

Service-oriented computing promises an effective approach to seamless integration and orchestration of distributed resources for dynamic business processes along the supply chains. In this chapter, the integration and adaptation needs of next generation e-logistics, which motivates the concept of a middleware integration framework, are first explained. Then, an overview of a service oriented intelligent middleware service framework for fulfilling the needs is presented with details regarding how one can embed the autonomy oriented computing (AOC) paradigm in the framework to enable autonomous service brokering and composition for highly dynamic integration among heterogeneous middleware systems. The authors hope that this chapter can provide not only a comprehensive overview on technical research issues in the e-logistics field but also a guideline of technology innovations which are vital for next generation on-demand e-logistics applications.

Chapter XII

This chapter proposes a new type of multi-agent mobile negotiation support system named MAM-NSS in which both buyers and sellers are seeking the best deal given limited resources. Mobile commerce, or m-commerce, is now on the verge of explosion in many countries, triggering the need to develop more effective decision support systems capable of suggesting timely and relevant action strategies for both buyers and sellers. To fulfill research purpose like this, two AI methods such as CBR (case-based reasoning) and FCM (fuzzy cognitive map) are integrated, and named MAM-NSS. The primary advantage of the proposed approach is that those decision makers involved in m-commerce regardless of buyers and sellers can benefit from the negotiation support functions that are derived from referring to past instances via CBR and investigating inter-related factors simultaneously through FCM. To prove the validity of the proposed approach, a hypothetical m-commerce problem is developed in which theaters (sellers) seek to maximize profit by selling its vacant seats to potential customers (buyers) walking around within reasonable distance. For experimental design and implementation, a multi-agent environment Netlogo is adopted. Simulation reveals that the proposed MAM-NSS could produce more robust and promising results that fit the characteristics of m-commerce.

This chapter focuses on the issue of malicious agents in open multi-agent systems (MAS) with discussions in relation to the existing crime study, that is, supply-demand analysis and deterrence theory. Our work highlights the importance of mechanisms to make intervention into the MAS, in an attempt to deter malicious agents from maximizing their utilities through illegal actions. Indeed, in market-oriented MAS, human interventions sometimes would be necessary, given the condition that these interventions would not destroy the ecological stability of the agent society. However, an automatic intervention mechanism seems to be the winning card in the end, among them, the reputation mechanism, agent coalition formation, and so forth. It is our hope that our work can shed light on these issues as well as their deployment in MAS.

Section V
Cross-Fertilized Techniques in Business Antomation

There are certain features that distinguish killer apps from other ordinary applications. This chapter examines those features in the context of the Semantic Web, in the hope that a better understanding of the characteristics of killer apps might encourage their consideration when developing Semantic Web applications. Killer apps are highly transformative technologies that create new e-commerce venues and widespread patterns of behavior. Information technology generally, and the Web in particular, have benefited from killer apps to create new networks of users and increase its value. The Semantic Web community, on the other hand, is still awaiting a killer app that proves the superiority of its technologies. The authors hope that this chapter will help to highlight some of the common ingredients of killer apps in e-commerce, and discuss how such applications might emerge in the Semantic Web.

Chapter XV

A location model represents the inclusive objects and their relationships in a space, and helps engender the values of location based services (LBS). Nevertheless, LBS for enterprise decision support are rare due to the common use of static location models. This chapter presents for enterprises a framework of dynamic semantic location modeling that is novel in three ways: (1) It profoundly brings location models into enterprise business models. (2) With a novel method of dynamic semantic location modeling, enterprises effectively recognize the needs of the clients and the partners scattered in different locations, advancing existing business relationships by exerting appropriate service strategies through their mobile workforces. (3) Through the Location Model Platform of information sharing, enterprises are empowered to discover the potential business partners and predict the values of their cooperation, gaining competitive advantages when appropriate partnership deals are made by enterprise mobile workforces. This proposed framework has been implemented with the J2EE technology and attained the positive evidence of its claimed values.

Chapter XVI

This chapter describes and classifies service composition approaches according to ubiquitous and pervasive computing requirements. More precisely, because of the tremendous amount of research in this area, we present the state of the art in service composition and identify key issues related to the efficient implementation of service composition platforms in ubiquitous and pervasive computing environments.

Preface

INTRODUCTION

The growth, integration, and sophistication of information technology and communications are changing our society and economy. Consumers now routinely use a variety of networks and the Web to identify sellers, evaluate products and services, compare prices, and exert market leverage. Businesses use the networks and the Web even more extensively to conduct and re-engineer production processes, streamline procurement processes, reach new customers, and manage internal operations. However, the Web was not meant for commercial purposes, and thus a number of issues (e.g., trust, privacy) limit its use for this commercial purpose. On the other hand, the infrastructure for e-business needs to accommodate business that takes place in large-scale, open, and dynamic network environments blending multiple independent systems. One way to cope with the limitation and complexity is to design them as agent-based systems. This book is written to address the importance of agent systems in electronic business.

Agent technologies offer high-level abstractions and mechanisms that address issues such as reasoning, communication, coordination, cooperation, commitments, goals, beliefs, intentions, and so forth. Agent technologies are believed to be one of the most promising tools to conduct business via the networks and the Web in an autonomous, intelligent, and efficient way. On the other hand, electronic business concerns three primary dimensions: electronic business processes (how business is conducted), electronic commerce transactions (buying and selling), and e-business infrastructure (what infrastructures are used to support electronic business processes and conduct electronic commerce transactions). Accordingly, agent technologies could be unfolded along these three dimensions in unveiling business automation.

Meanwhile, electronic business increasingly uses agents and the other emergent technologies acting on behalf of human buyers, sellers, and business. These emergent technologies include Web services, mobile computing, P2P technologies, the Semantic Web, and so forth. Such cross-fertilized techniques have created all sorts of challenging yet interesting agent-based systems in the electronic business area.

This book is organized into five parts: the first part gives an overview chapter of agent-based e-business systems; the second part includes four chapters addressing agent technologies for electronic commerce transactions; the third part provides four chapters discussing agent technologies for electronic business processes; the fourth part gives four chapters presenting agent technologies for e-business infrastructure; finally, the fifth part includes three chapters examining some cross-fertilized techniques in business automation.

BOOK ORGANIZATION

Section I: Overview

Chapter I (*Patterns for Designing Agent-Based E-Business Systems*) describes a group of architectural patterns for agent-based e-business systems. These patterns relate to front-end e-business activities that involve interaction with the user and delegation of user tasks to agents. Patterns capture well-proven, common solutions, and guide developers through the process of designing systems. These patterns are just the beginning of a pattern language for agent-based e-business system design. As the use of agent technology in e-business matures, this language will evolve as well. Finally, a number of examples then illustrate the application of these patterns.

Section II : Agent Technologies in Electronic Commerce Transactions

Chapter II (*The Evolution of Comparison-Shopping Agents*) discusses comparison-shopping agents and addresses key events and issues of comparison-shopping agent evolution in three intertwined threads (the emergence of representative agents, the evolution of comparison-shopping agent technology, and the evolution of their business models). These identified key issues include the complexity of product information dimension, consumer expectations, attitudes of online retailer and service providers, and competition intensity in the comparison-shopping markets. These issues intertwine together and shape the development of comparison-shopping agent systems and their activity environments.

Chapter III (*A Study of Intelligent Shopping Support: A Case Study of Outbound Group Package-Tour Products in Taiwan*) proposes a customer-centric solution to improve the search effectiveness of agent-based shopping support in terms of developing a knowledge base. This knowledge base consists of representation of the product characteristics in the form of ontology as those concepts affect shoppers' decision-making behavior. A case of outbound group-package tour in Taiwan is investigated and a focus-group interview is applied in order to validate the ontological framework of the knowledge base. In other words, this chapter demonstrates a process of knowledge acquisition to tackle the problem of ineffective online information search by a customer-centric method.

Chapter IV (*Agent-Based Matching of Demands and Supplies in Business Transaction Formation*) identifies certain problems in conducting electronic business and supply chain management and defines the expected benefits for supply chains with agents working together in coordination and cooperative processes. Moreover, architecture of effective cooperative processing for agents is also proposed.

Chapter V (*Evolutionary Auction Design for Agent-Based Marketplaces*) designs a new way of efficient auctions by evolutionary computing approaches for its important role in computational economics for resolving multi-agent allocation problems. In the design, a genetic algorithm is used to design auction mechanisms in order to automatically generate a desired market mechanism for electronic markets populated with trading agents. Moreover, this new design also shows that the optimal auction for heterogeneous agents could be a human-designed one while the optimal auction for homogeneous agents is a hybrid one, given the same supply-demand schedule.

Section III: Agent Technologies in Electronic Business Processes

Chapter VI (*An Inter-Organizational Business Process Study from Agents Interaction Perspective*) introduces a generic pattern of agents interaction derived from the communication patterns of human actors. Two theoretical concepts (the semiotics approach and the language action perspective) are then employed for agent based e-commerce and e-business systems study in the context of inter-organizational business process using diagrams and notations. This chapter demonstrates that the second concept is very useful (as it is diagrammatically rich and easy to communicate to nontechnical users, and provides sufficient notations to model complex hierarchical agent based e-business systems). This chapter also gives a case study conducted based on a real life business for the illustration of the concept of agent-based systems.

Chapter VII (*Applications of Agent-Based Technology as Coordination and Cooperation in the Supply Chain Based E-Business*) proposes a newly developed model for an enhanced and effective cooperation process in the field of supply chain management for electronic business by a utilization of a multi-agent system. This chapter also provides a theoretical solution and model for agents that adopt the enhanced strategy for e-business. Both large organizations and SMEs will benefit as the proposed model will enhance their global business by participating and sharing with other businesses to achieve common goals.

Chapter VIII (*An Agent-Based Framework for Emergent Process Management*, reprint from Debenham, J. (2006), *International Journal of Intelligent Information Technologies, 2*(2), 30-48) provides a definition for emergent process and presents an inference mechanism to manage emergent business processes (whose execution is determined by the prior knowledge of the agents involved and by the knowledge that emerges during a business process instance) that contribute to the establishment of dynamic business relationships.

Chapter IX (*Beyond Intelligent Agents: E-Sensors for Supporting Supply Chain Collaboration and Preventing the Bullwhip Effect*, reprint from Rodriguez, W., Zalewski, J., & Kirche, E. (2007), *International Journal of e-Collaboration, 3*(2), 1-15) presents a new concept of e-sensors for supporting electronic collaboration, operations, and relationships among trading partners in the value chain without hindering human autonomy. E-sensors are next-generation hardware-software agents that are capable of perceiving, reacting and learning from its interactive experience through the supply chain, rather than just searching for data and information through the network and reacting to it.

Section IV: Agent Technologies in E-Business Infrastructure

Chapter X (*An Automated Negotiation Mechanism for Agents Based on International Joint Ventures*) develops an automated one-to-many multi-attribute negotiation mechanism for facilitating multi-agent technology approaches in scenarios of international joint ventures (emerged as the dominant form of partnership in light of intense global competition and the need for strategic organizational viability), securing positive impact on supplier selection, and partners' profit. This negotiation mechanism is contributable in terms of a practical approach utilizing the fuzzy logic to represent the attributes and jointed buyer's behavior within negotiation in order to negotiate in a real world.

Chapter XI (*An Agent-Mediated Middleware Service Framework for E-Logistics*) presents a service oriented e-logistics middleware service framework for e-logistics infrastructure and network integration and development. This middleware aims to enable more responsive supply chains and better planning and management of complex inter-related systems, such as materials planning, inventory management, capacity planning, logistics, and production systems. Furthermore, this chapter also contributes to the understanding of how logistics middleware systems can seemingly be integrated and self-adapt for optimizing the global supply network.

Chapter XII (A *Multi-Agent System Approach to Mobile Negotiation Support Mechanism by Integrating Case-Based Reasoning and Fuzzy Cognitive Map*) proposes a new type of multi-agent mobile negotiation support system in which both buyers and sellers are seeking the best deal given limited resources. Two AI methods such as CBR (case-based reasoning) and FCM (fuzzy cognitive map) are used and integrated in the system in order for the decision makers involved in m-commerce regardless of buyers and sellers to benefit from the negotiation support functions that are derived from referring to past instances via CBR and investigating inter-related factors simultaneously through FCM.

Chapter XIII (*A Study of Malicious Agents in Open Multi-Agent Systems: The Economic Perspective and Simulation*) discusses the issue of malicious agents in open multi-agent systems with several examples in relation to the existing crime study. In addition, various governance schemes are proposed from the perspective of both the system designers and users, justified by an empirical analysis of the governance strategies by means of deterrence theory. Meanwhile, this chapter highlights the importance of mechanisms to make intervention into the multi-agent systems, in an attempt to deter malicious agents from maximizing their utilities through the illegal actions.

Section V: Cross-Fertilized Techniques in Business Automation

Chapter XIV (*Features for Killer Apps from a Semantic Web Perspective*) describes certain features that distinguish killer apps from other ordinary applications in the context of the Semantic Web. By examining the common ingredients of killer apps in e-commerce, the chapter discusses how such applications might emerge in the Semantic Web. This discussion is important because the Semantic Web community is still awaiting a killer app that can prove the superiority of its technologies; on the other hand, the Semantic Web provides a context for killer app development and a context based on the ability to integrate information from a wide variety of sources and interrogate it.

Chapter XV (*Semantic Location Modeling for Mobile Enterprises*) presents a novel framework of location modeling in integrating enterprise business models with location models. This framework enables enterprises to effectively recognize the needs of the clients and the partners scattering in different locations, advancing existing business relationships by exerting appropriate service strategies through their mobile workforces. Moreover, enterprises are also empowered to discover the potential business partners and predict the values of their cooperation, gaining competitive advantages when appropriate partnership deals are made by enterprise mobile workforces. In other words, this framework would bestow enterprise mobile workforce location-sensitive decision information so as to properly serve the clients or justifiably negotiate contracts with other enterprises.

Chapter XVI (*Service Composition Approaches for Ubiquitous and Pervasive Computing: A Survey*) describes and classifies service composition approaches according to ubiquitous and pervasive computing requirements. Moreover, this chapter identifies the key issues related to the efficient implementation of service composition platforms in ubiquitous and pervasive computing environments (characterized with self-adaptation, self-organization, and emergence in systems that operate in an open and dynamic environment). These issues can be classified into two major directions: the languages to specify and describe services and the architectures to enable scalable, fault tolerance, and adaptive service composition.

Eldon Y. Li
National Chengchi University, Taiwan and Cal Poly – San Luis Obispo, USA

Soe-Tsyr Yuan
National Chengchi University, Taiwan

Acknowledgment

This book has been prepared in close cooperation with experts in the area of software agents and collaborative commerce from 16 universities and four companies over eight countries or regions. The editors would like to thank all the chapter authors, in particular Michael Weiss, Yun Wan, Wen-Shan Lin, Tim Chou, Raymund J. Lin, Zengchang Qin, Joseph Barjis, Samuel Chong, Golenur Begum Huq, Robyn Lawson, Yee Ming Chen, Pei-Ni Huang, Z. Luo, M. Wang, W.K. Cheung, J. Liu, F. Tong, C.J. Tan, Kun Chang Lee, Namho Lee, Pinata Winoto, Tiffany Y. Tang, Kieron O'Hara, Harith Alani, Yannis Kalfoglou, Nigel Shadbolt, Soe-Tsyr Yuan, Pei-Hung Hsieh, M. Bakhouya, and J. Gaber, for their great effort in preparing the manuscripts and their insights and excellent contributions to this book.

Our thanks also go to all reviewers who provided constructive and comprehensive reviews for their assistance in improving the quality of this book. Special thanks to IGI Global, who published the book with clear guidelines over the whole editorial process. Last but not the least, we would like to thank Kristin Roth, Meg Stocking, and Jessica Thompson who provided all the help during the editorial process in the past year.

In closing, we wish to thank all of the readers for their support to this book. We hope they find this book informative and useful.

Eldon Y. Li
National Chengchi University, Taiwan and Cal Poly – San Luis Obispo, USA

Soe-Tsyr Yuan
National Chengchi University, Taiwan

List of Reviewers

Michael Weiss, Carleton University, Canada

Yun Wan, University of Houston, USA

Wen-Shan Lin, National Chiayi University, Taiwan

Raymund J. Lin, Institute for Information Industry, Taiwan

Zengchang Qin, University of California, Berkeley, USA

Joseph Barjis, Georgia Southern University, USA

Golenur Begum Huq, University of Western Sydney, Australia

Robyn Lawson, University of Western Sydney, Australia

Yee Ming Chen, Yuan-Ze University, Republic of China

W.K. Cheung, Hong Kong Baptist University, Hong Kong

Kun Chang Lee, Sungkyunkwan University, Korea

Tiffany Y. Tang, The Hong Kong Polytechnic University, Hong Kong

Harith Alani, University of Southampton, UK

Soe-Tsyr Yuan, National Chengchi University, Taiwan

J. Gaber, Université de Technologie de Belfort-Montbéliard, France

HsiuJu Yen, National Central University, Taiwan

Jason Chen, Yuan Ze University, Taiwan

Timon C. Du, Chinese University of Hong Kong, Hong Kong

Houn-Gee Chen, National Taiwan University, Taiwan

Jen-Ho Wu, National Sun Yat-sen University, Taiwan

Chaochange Chiu, Yuan Ze University, Taiwan

Cheng-Yuan Ku, National Chung Chen University, Taiwan

David C. Yen, Miami University, USA

Paul T.Y. Tseng, Tatung University, Taiwan

Wei-Lung Chang, National Chengchi University, Taiwan

Wei-Feng Tung, National Chengchi University, Taiwan

Yuan-Chu Hwang, National Chengchi University, Taiwan

Section I
Overview

Chapter I
Patterns for Designing Agent-Based E-Business Systems

Michael Weiss
Carleton University, Canada

ABSTRACT

Agents are rapidly emerging as a new paradigm for developing software applications. They are being used in an increasing variety of applications, ranging from relatively small systems such as assistants to large, open, mission-critical systems like electronic marketplaces. One of the most promising areas of applications for agent technology is e-business. In this chapter, we describe a group of architectural patterns for agent-based e-business systems. These patterns relate to front-end e-business activities that involve interaction with the user, and delegation of user tasks to agents. Patterns capture well-proven, common solutions, and guide developers through the process of designing systems. This chapter should be of interest to designers of e-business systems using agent technology. The description of the patterns is followed by the case study of an online auction system to which the patterns have been applied.

INTRODUCTION

Agents are rapidly emerging as a new paradigm for developing software applications. They are being used in an increasing variety of applications, ranging from relatively small systems such as assistants to large, open, mission-critical systems like electronic marketplaces. One of the most promising areas of applications for agent technology is e-business (Papazoglou, 2001). In this chapter, we describe a group of architectural patterns for agent-based e-business systems. These patterns relate to front-end e-business activities that involve interaction with the user, and delegation of user tasks to agents.

The chapter is structured as follows. First, we provide a background on patterns and their application to the design of agent systems. Then, we discuss the forces or design constraints that need to be considered during the design of agents for e-business systems. This is followed by a description of the agent patterns for e-business. A number of examples illustrate the application of these patterns. Finally, we discuss current trends and opportunities for future research and offer concluding remarks.

BACKGROUND

Patterns are reusable solutions to recurring design problems and provide a vocabulary for communicating these solutions to others. The documentation of a pattern goes beyond documenting a problem and its solution. It also describes the forces or design constraints that give rise to the proposed solution (Alexander, 1979). These are the undocumented and generally misunderstood features of a design. Forces can be thought of as pushing or pulling the problem towards different solutions. A good pattern balances these forces. A set of patterns, where one pattern leads to other patterns that refine or are used by it, is known as a pattern language. A pattern language can be likened to a process: it guides designers who wants to use those patterns through their application in an organic manner. As each pattern of the pattern language is applied, some of the forces affecting the design will be resolved, while new unresolved forces will arise as a consequence. The process of using a pattern language in a design is complete when all forces have been resolved.

There is by now a growing literature on using patterns to capture common design practices for agent systems. Aridor and Lange (1998) describe domain-independent patterns for the design of mobile agent systems. They classify mobile agent patterns into traveling, task, and interaction patterns. Kendall, Murali Krishna, Pathak, et al. (1998) use patterns to capture common building blocks for the architecture of agents. They integrate these patterns into the layered agent pattern, which serves as a starting point for a pattern language for agent systems based on the

strong notion of agency. Schelfthout, Coninx, et al. (2002), on the other hand, document agent implementation patterns suitable for developing weak agents.

Deugo, Weiss, and Kendall (2001) identify a set of patterns for agent coordination, which are, again, domain-independent. They classify agent patterns into architectural, communication, traveling, and coordination patterns. They also describe an initial set of global forces that push and pull solutions for coordination. Kolp, Giorgini, and Mylopoulos (2001) document domain-independent organizational styles for multi-agent systems using the Tropos methodology. Weiss (2004) motivates the use of agents through a set of patterns that document the forces involved in agent-based design and key agent concepts.

On the other hand, Kendall (1999) reports on work on a domain-specific pattern catalog developed at BT Exact. Several of these patterns are documented using role models in a description of the ZEUS agent building kit (Collis & Ndumu, 1999). Shu and Norrie (1999) and the author in a precursor to this chapter have also documented domain-specific patterns, respectively, for agent-based manufacturing and electronic commerce. However, unlike most other authors, they present the patterns in the form of a pattern language. This means that the relationships between the patterns are made explicit in such a way that they guide a developer through the process of designing a system.

Lind (2002) and Mouratidis, Weiss, and Giorgini (2006) suggest that we can benefit from integrating patterns with a development process, while Tahara, Oshuga, and Hiniden (1999) and Weiss (2003) propose pattern-driven development processes. Lind (2002) suggests a view-based categorization scheme for patterns based on the MASSIVE methodology. Mouratidis et al. (2006) document a pattern language for secure agent systems that uses the modeling concepts of the Tropos methodology. Tahara et al. (1999) propose a development method based on agent patterns and distinguish between macro and micro architecture patterns. Weiss (2003) documents a process for mining and applying agent patterns.

FORCES

The design of agent-based systems in the e-business domain is driven by a number *forces*, including autonomy, the need to interact, information overload, multiple interface, ensuring quality, adaptability, privacy concerns, search costs, and the need to track identity. Not all of these forces can be equally satisfied by a given design, and trade-offs need to be made. The patterns described in this chapter help with making informed trade-offs.

AUTONOMY

The currently dominant metaphor for interacting with computers is direct manipulation. Direct manipulation requires the user to initiate all tasks explicitly and to monitor all events. For example, a user searches the Web for an auction that offers the desired item for sale, and subsequently monitors the state of the auction. The obvious drawback of this approach is that most of the time the user is occupied in tasks that are peripheral to the primary objectives. The user's ability to find the best deal available at any of the many online auctions in operation is also greatly limited.

Agents can be used to implement a complementary interaction style, in which users *delegate* some of their tasks to software agents which then perform them autonomously on their behalf. This indirect manipulation style engages the user in a cooperative process in which human and software agents both initiate communication, monitor events, and perform tasks. Autonomy is the capability of an agent to follow its goals without interactions or commands from the user or another agent. An autonomous agent does not require the

user's approval at every step of executing its task, but is able to act on its own.

With agents performing autononmous actions, users are now facing issues of trust and control over their agents. The issue of *trust* is that by engaging an agent to perform tasks (such as selecting a seller), the user must be able to trust the agent to do so in an informed and unbiased manner. The agent should not, for example, have entered contracts with sellers to favor them in return for a cut on their proceeds to the developer of the agent or the server that hosts and executes the agent. The user would also like to specify the *degree of autonomy* of the agent. For example, the user may not want to delegate decisions to the agent that have legal or financial consequences, although a buyer agent is capable of not only finding the cheapest seller, but also placing a purchase order.

NEED TO INTERACT

Agents typically only have *a partial representation* of their environment, and are thus *limited* in their ability—in terms of their expertise, access to resources, location, and so forth—to interact with it. Thus, they rely on other agents to achieve goals that are outside their scope or reach. They also need to coordinate their activities with those of other agents to ensure that their goals can be met, avoiding interference with one another. The behavior of an individual agent is thus often not comprehensible outside its social structure—its relationships with other agents. For example, the behavior of a buyer agent in an auction cannot be fully explained outside the context of the auction itself, and of the conventions that govern it (for example, in which order—ascending or descending—bids must be made, and how many rounds of bidding there are in the auction).

An important issue in designing systems of interacting agents is dealing with *openness*. The Internet and e-business applications over the Internet are both examples of open systems.

Open systems pose unique challenges in that their components are not known in advance; they can change unexpectedly, and they are composed of heterogeneous agents implemented by different developers at different times with different tools and methodologies. Similarly, as we do not control all the agents, one can also not assume that the agents are cooperative. Some agents may be benevolent and agree on some protocol of interaction, but others will be *self-interested* and follow their own best interests. For example, in an electronic marketplace, buyer and seller agents are pursuing their own best interests (making profit) and need to be constrained by conventions.

INFORMATION OVERLOAD

People and organizations wish to find relevant information and offerings to make good deals and generate profit. However, the large set of sellers, in conjunction with the *multiple interfaces* they use, makes it difficult for a human to overview the market. One solution has been to provide *portals* or common entry points to the Web. These portals periodically collect information from a multitude of information sources and condense them to a format that users find easier to process, typically taking the form of a hierarchical index. One disadvantage of this solution is that the categories of the index will be the same for every user. Individual preferences are not taken into account when compiling the information, and niche interests may not be represented at all.

MULTIPLE INTERFACES

One of the difficulties in finding information (e.g., when comparing the offerings of different sellers) is the large number of different interfaces used to present the information. Not only are store fronts organized differently, sellers do not follow the same conventions when describing their

products and terms of sale. For instance, some sellers include the shipping costs in the posted price; others will advertise one price, but add a handling charge to each order. A solution is to agree on *common vocabularies*, but these must also be widely adopted. With the introduction of the extensible markup language (XML) for associating metacontent with data and current developments in the Semantic Web such as ontology representation languages (OWL), this is slowly becoming a reality. For example, a price in a catalog can be marked up with its currency whether it already includes the shipping cost. However, the difficulty with any standard format is that it takes a considerable amount of time to find agreement among the interested parties. One also needs to allay the fear of sellers in losing business to competitors once their product information becomes easily accessible.

ENSURING QUALITY

Shopping online lacks the immediate mechanisms for establishing trustworthiness. How can you trust a seller, with whom you have had no previous encounter, whether the order you placed will be fulfilled satisfactorily? For example, any seller in an online auction could claim that the item offered for sale is in superior condition, when the buyer cannot physically verify that claim. One solution is to solicit *feedback* about the performance of a seller (respectively, buyer) from buyers (respectively, sellers) after order fulfillment. For example, the online auction site eBay keeps records of how a seller was rated by other buyers. Potential new buyers will take the ratings from previous buyers into account before considering buying from a seller. However, eBay's solution falls short in two ways. Old low ratings are not discarded or discounted when more recent ratings are higher. Also, if a seller gets a low overall rating, it is easy for the seller to assume a new identity and start afresh with a new rating. A mechanism for ensuring quality must avoid this, as discussed in the context of reputation system design by Zacharia, Moukas, and Maes (1999).

ADAPTABILITY

Users differ in their status, level of expertise, needs, and preferences. The issue of adaptability is that of tailoring information to the features of the user, for example, by selecting the products most suitable for the user from a catalog, or adapting the presentation style during the interaction with the user. Any approach to tailoring information involves creating and maintaining a user model. When creating a user model, two cases need to be distinguished (Ardissono, Barbero, et al., 1999): for first time visitors, no information about them is available, and the user characteristics must be recognized during the interaction; on subsequent visits, a detailed user model about a visitor is already available and can be used to tailor the information.

Several design considerations for user modeling are detailed in Ardissono et al. (1999). Users need to register permanently with the system to have their data stored; otherwise user models will only be maintained during a single interaction. In the context of online shopping, a system must also deal with direct and indirect users. A customer of a Web store may browse for products for himself, as well as for other people (indirect users), who have different needs and preferences. Finally, the user model must be able to reflect changes in interest over time. One approach to collecting user information is to ask the user to provide the information explicitly, for example, by filling out a form. This allows one to create a profile of the user that is potentially very accurate, and to provide personalized service to the user from the beginning. However, there are at least two problems with this solution. First, by requiring the user to provide this information upfront; the threshold for the user to do so is very high.

Only very advanced users will want to tune their own profiles. Second, when the user's interests change, this will not be reflected in the profile, unless the user keeps updating her profile. Again, in practice, users do not update their profiles after installation.

PRIVACY CONCERNS

Personalization requires the collection and release of personal information to the agent providing the personalized service. One way of personalizing interactions between buyers and sellers is for the seller to collect information about a buyer from the buyer's behavior (e.g., their clickstream). The buyer may not be aware of the information collected, nor does she always have control over what information is gathered about her. Although effective from the seller's perspective, this is not a desirable situation from the perspective of the buyer. Users are typically not willing to allow just anyone to examine their preferences and usage patterns, in particular without their knowledge or consent. They want to remain in control, and decide on an interaction-by-interaction basis which information is conveyed to the seller. A solution that addresses the force of privacy concerns must put the user in charge of which information is collected and who it is made available to. An additional complexity results from the desire of some buyers to remain anonymous. If a buyer remains anonymous, a seller cannot provide personalized service. Thus, generally, users are willing to share personal information with sellers, if the expected gains outweigh the possible threats for their privacy.

SEARCH COSTS

It can be expensive for buyers and sellers to find each other. In a static marketplace, each buyer can store a contact list of sellers for each product, and then quickly locate an appropriate seller when a particular product is needed. However, an electronic marketplace is dynamic. Buyers and sellers can join and leave the marketplace, and change their requirements and offerings qualitatively and quantitatively at any point in time. It, therefore, becomes impossible for a market participant to maintain an up-do-date list of contacts. Another problem is that of restricting the buyer's options. If each buyer maintains its own list of contacts, they run the risk of not being aware of better deals available elsewhere. One possible solution to these problems is to use a mediator which can match potential trading partners in the market. With the introduction of mediators, buyers and sellers no longer maintain their own lists of contacts, or need to contact a large number of alternative trading partners to find the optimal one. One trade-off of this solution is, however, that individual preferences or history of interaction with a particular trading partner cannot be accounted for by a mediator. Thus, it is reasonable to maintain individual lists of trading partners that one has dealt with in the past, keeping track of the quality provided and using this personalized ranking of partners to filter the list of contacts provided by a mediator.

IDENTITY

For various reasons, buyers and sellers need to be represented by unique identities. The most important reasons are authentication, nonrepudiation, and tracking. One way of assigning a unique identity to trading partners is to use one of the many unique labels which are readily available on the Internet, for example, an e-mail address, or a Yahoo! account name. A problem with this approach is that it is also very easy to obtain a new identity, thus making authentication, nonrepudiation, or tracking schemes that rely on such identities impractical. Similarly, a user could obtain multiple identities and pretend to represent

multiple different parties, where instead there is only one. A solution that remedies this situation must make it advantageous for individuals to keep their identities over those users who change them often (Zacharia et al., 1999).

PATTERNS

The patterns we identified and their relationships are shown in Figure 1. The arrows indicate refinement links between the patterns. Each arrow in the diagram points in the direction from a "larger" pattern to a "smaller" pattern. The starting point for the language is the AGENT SOCIETY pattern, which motivates the use of agents for building the application. At the next level of refinement, the diagram leads the designer to consider the patterns agent as DELEGATE, AGENT AS MEDIATOR, AND COMMON VOCABULARY.[1]

agent as delegate and the patterns it links to deal with the design of agents that act on behalf of a single user. The agent as mediator pattern guides the designer through the design of agents that facilitate between a group of agents and their users. COMMON VOCABULARY provides guidelines for defining exchange formats between agents. The rest of Figure 1 shows refinements of the AGENT AS DELEGATE pattern. For example, the USER AGENT pattern prescribes to use a single locus of interaction with the user and represent the concurrent transactions a user participates in as buyer and seller agents. User interaction also includes profiling the user (USER PROFILING) and subscribing to information (e.g., the status of an auction) relevant to the user (NOTIFICATION).

In the following, each pattern is represented by its context, the problem it addresses, a discussion of the forces, its solution, and a resulting context. The context is represented by the dependencies between the patterns. The problem is a succinct statement on what problem the pattern addresses. The solution takes the form of a role diagram. These roles will be filled by agents. For example,

Figure 1. Patterns for e-business agents and their dependencies (arrows indicate dependencies and dashed lines patterns that are not described here)

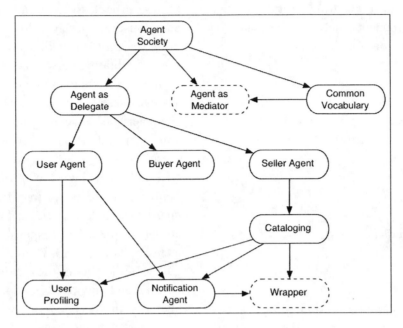

Figure 2. Role diagram for the USER AGENT *pattern*

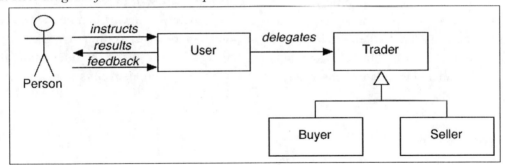

consider the USER AGENT pattern. It is applied after AGENT AS DELEGATE, and, in turn, refined by USER PROFILING and NOTIFICATION. The it addresses the problem of how users instruct agents to act on their behalf (as buyers and sellers) and how they keep in control over what the agent does (e.g., does it have authority to complete a trade?). The role diagram for the USER AGENT pattern is shown in Figure 2. Role diagrams and their semantics are discussed further in AGENT SOCIETY. The resulting

context points to related patterns in this pattern language.

AGENT SOCIETY

Context

Your application domain satisfies at least one of the following criteria: your domain data, control, knowledge, or resources are decentralized; your application can be naturally thought of as a system of autonomous cooperating entities, or you have legacy components that must be made to interoperate with new applications.

Problem

How do you model systems of autonomous cooperating entities in software?

Forces

- Autonomy
- Need to interact

Solution

Model your application as a society of agents. Agents are autonomous computational entities (autonomy), which interact with their environment (reactivity) and other agents (social ability) in order to achieve their own goals (proactiveness). Often, agents will be able to adapt to their environment and have some degree of intelligence, although these are not considered mandatory characteris-

tics. These computational entities act on behalf of users or groups of users (Maes, 1994). Thus, agents can be classified as *delegates*, representing a single user and acting on his behalf, or *mediators* or *intermediaries* acting on behalf of a group of users, facilitating between them.

It is important to point out that objects cannot achieve these goals directly. A differentiating characteristic between agents and objects is their *autonomy*. Autonomy is here used in an extended sense. It not only comprises the notion that agents operate in their own thread of control, but also implies that agents are long-lived (they execute unattended for long periods), take initiative (they do not simply act in response to their environment), react to stimuli from the environment as guided by their goals (the *receiving* agent decides whether and how to respond to a stimulus), and interact with other agents to leverage their abilities in support of their own as well as collective goals. *Active objects*, on the other hand, are autonomous only in the first of these senses.

A society of agents can be viewed from two dual perspectives: either a society of agents emerges as a result of the interaction of agents, or the society imposes constraints and policies on

Figure 3. Micro-macro view of an agent society

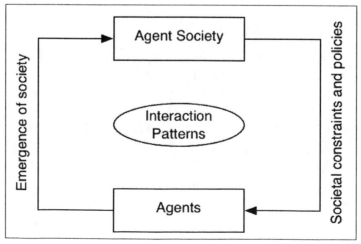

its component agents. Both perspectives, which we can refer to as micro and macro view of the society, respectively, mutually reinforce each other, as shown in Figure 3. For example, emergent behaviors such as specialization of agents leads to the notion of roles that agents play, which allow us to codify expectations about an agent. Roles, in turn, impose restrictions on the possible behaviors of agents (Ferber, 1999).

This suggests two approaches to systematically designing agent societies. In the first approach, we identify top-level goals for the system and decompose them recursively, until we can assign them to individual agents. In the second approach, we construct an agent society incrementally from a catalog of interaction patterns (Kendall et al., 1999). These interaction patterns are described in terms of roles that agents can play and their interactions and may also specify any societal constraints or policies that need to be satisfied. Roles are *abstract* loci of control (Ferber, 1999; Kendall, 1999; Riehle & Gross, 1998). Protocols (or patterns of interaction) describe the way the roles interact. Policies define constraints imposed by the society on these roles (Weiss, Gray, & Diaz, 1997; Zambonelli, Jennings, et al., 2001). As an example of a policy, consider an agent-mediated auction, which specifies conventions specific to its auction type (for example, regarding the order of bids; ascending in an English auction,

descending in a Dutch auction) that participating agents must comply with in order for the auction to function correctly.

Many types of applications, including electronic commerce systems, can be modeled using User, Task, Service, and Resource roles, and their subtypes. Figure 4 depicts these roles and their subtypes used in this chapter in a role diagram, using the notation introduced in Kendall (1999). *Role diagrams* are more abstract than class diagrams. Each role in the diagram defines a position and a set of responsibilities. A role has collaborators—other roles that it interacts with. Arrows between roles indicate dynamic interactions between roles; their direction represents the direction in which messages are sent. The triangle indicates a subtyping relationship between roles; subtypes inherit the responsibilities of their parent roles and add responsibilities of their own (e.g., Buyer from Trader).

The User role encapsulates the behavior of managing a user's task agents, providing a presentation interface to the user and collecting profile information about the user, among other responsibilities. A Task role represents users in a specific task, typically a long-lived, rather than one-shot transaction. In this context, we only consider Trader agents, which can either represent the user as a Buyer or Seller. The common functionality of both roles (for example,

Figure 4. Top-level roles and their subtypes in this pattern language

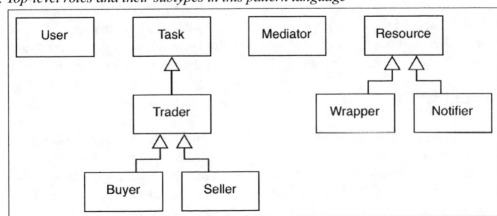

they represent the negotiation strategies of their users) is represented in the trader role. Another example of a Task role would be a search role that encapsulates the behavior of making queries to the Web on the user's behalf and filtering the results. One could also model transactions as a separate Transaction role as proposed by Ulmer (2004), representing the transactions a Task agent engages in.

A Mediator role typically provides a service to a group of users. It mediates the interaction between two or more agents through this service. One example is a directory agent that provides a reference to an agent given its name (white pages agent), or references to agents that can provide a certain product or service (yellow pages agent). More advanced services facilitate more complex interactions between agents, for example, enforcing an auction protocol. For example, see the recommender agents described in Weiss (2004) or the direction and auction agents in Fonseca, Griss, and Letsinger (2001).

The Resource role abstracts information sources. These can be legacy data sources wrapped by "glue" code that converts generic requests from other agents to the API of the data source. These can be agents that process more complex queries by first breaking them down into subqueries and then collating the results. Resource agents are a form of Indirection Layer (Avgeriou & Zdun, 2005). Notifier agents also belong in this category; they can be instructed to generate an alert when a certain condition in a data source is being met. For example, the Amazon Deliver's agent sends e-mails to users, when a book meeting user-specified criteria (such as author) is added to a catalog.

Resulting Context

For members of an agent society to understand each other, they need to agree on common exchange formats, as described in COMMON VOCABULARY. For agents that act on behalf of a single user, which is what e-business front-end system are concerned with, consult AGENT AS DELEGATE. For agents that facilitate between a group of users and their agents, refer to AGENT AS MEDIATOR described in more detail in Weiss (2004).

AGENT AS DELEGATE

Context

You are designing your system as a society of autonomous agents (AGENT SOCIETY) and want to delegate a user's time-consuming, peripheral tasks to software assistants.

Problem

How do you instruct assistants? How much discretion should you give to assistants? How do you manage the assistants that perform tasks on your behalf?

Forces

- Information overload
- Search costs

Solution

Use agents to act *on behalf* of the user. They are the system's interface to the user and manage task-specific agents such as buyer or seller agents on the user's behalf. User agents can also learn about the needs of the user by building a user profile from a history of interactions. This user profile allows sellers to customize their offerings to specific user tastes, while the user agents control what part of a profile sellers can access.

The structure of this pattern is shown in Figure 5. This role diagram shows the roles agents need to fill in the pattern. Note that we introduced the role of a Trader agent, as discussed in AGENT SOCIETY, which contains the beliefs and behavior common to both Buyer and Seller agents. For

Figure 5. Role diagram for AGENT AS DELEGATE

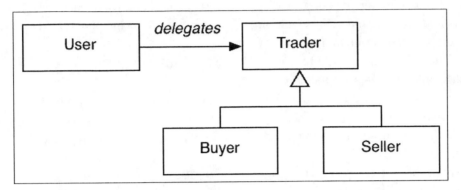

example, both Buyer and Seller agents have a belief for the desired price and make offers and counter-offers to other agents.

Resulting Context

For the interaction with the user to gather their requirements and feedback on the quality of the service received by a seller and for managing Task agents, consult USER AGENT. Also consult USER AGENT for measures to control access to user profiles. For the design of Trader agents, see BUYER AGENT and SELLER AGENT.

COMMON VOCABULARY

Context

When agents in an AGENT SOCIETY interact, they need to agree on common exchange formats. One scenario is that you are using agents to represent customers and sellers in individual transactions—BUYER AGENT and SELLER AGENT. Buyer and seller agents need to understand each other in order to exchange messages with one other. Another scenario is that you need to derive a user profile from user interactions—USER PROFILING—and you require your information sources such as product catalogs to be well structured.

Problem

How do you enable agents (for example, buyer and seller agents) to interact? How can you make it easier for a comparison-shopping agent to extract data from a catalog?

Forces

• Multiple interfaces

Solution

In the example, for buyer and seller agents to understand each other, they need to agree on a common message format that is grounded in a common ontology. The ontology defines product-related concepts that each party must use during interaction, their attributes, and valid value ranges. On receiving a message, the seller agent translates the request to a seller-specific format and fetches the contents of the product database.

It is generally impractical to define a general-purpose ontology for agent interaction. These are unlikely to include the intricacies of all possible domains. Instead, the common ontology will be application-specific. Given such a shared ontology, the communicating agents need to map their internal representations to the shared ontology. Much progress has been made on XML-based ontologies for electronic commerce, for example, xCBL, cXML, and RosettaNet, in recent years. Given that buyer and seller agents adopt one of these standards as their internal representation of domain concepts, the problem remains of building inter-ontology translators. One pragmatic approach to ontology mapping using XSLT (XML stylesheet language transformation) is described in Carlson (2001). However, refer to the references in the Resulting Context for more complex mappings.

The structure of this pattern is shown in Figure 6. This figure was adapted from Collis and Lee (1999).

Resulting Context

This pattern does not itself lead to other patterns in this pattern language. But patterns for ontologies have been documented elsewhere (Aranguren, 2005; Reich, 2000). A current overview on techniques for ontology mapping, which could be mined for further ontology mapping patterns, can be found in Staab and Stuckenschmidt (2006).

USER AGENT

Context

You are delegating tasks to an agent—AGENT AS DELEGATE—which autonomously performs the task on behalf of the user. Now you want to create a locus of control through which the users can interact with the agents created on their behalf.

Problem

How do you manage the user's profile and control access to it and the various (concurrent) transactions in which the user participates?

Forces

* Privacy concerns
* Ensuring quality

Solution

User agents form the *interface* between the user and the system receive the user's queries and feedback, and present information *tailored* to users. A user agent delegates the user's queries or orders (offers to buy/sell an item) to a trader agent and manages the various buyer and seller agents on behalf of its user. The user agent manages the user's profile and controls who can access it. The Open Profiling Standard (W3C, 1997) was an early example of employing user agents in this way. In this proposal, profile information for different sellers was stored in a personal profile repository. Sections of the profile could be restricted to a subset of the sellers. In the SOaP system (Voss &

Figure 6. Role diagram for COMMON VOCABULARY

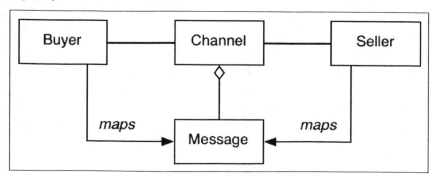

Kreifelts, 1997), user agents track all task agents for tasks in which a user is engaged.

The structure of this pattern is shown in Figure 7.

Resulting Context

A user can subscribe to the progress of a query or an order by instructing a NOTIFICATION AGENT to monitor its status. For more on modeling the user that requires explicit user input is kept at a minimum; consult USER PROFILING.

BUYER AGENT

Context

You are delegating tasks to an agent—AGENT AS DELEGATE—and provide a locus of control where users can interact with their agents—USER AGENT. You need an agent to represent a user as a customer in a transaction with sellers in an electronic marketplace.

Problem

How do you communicate your goals and buying strategy?

Forces

- Autonomy

Solution

A buyer agent can *tailor* the selection and presentation of products to the needs of its user. It *locates* seller agents that offer the requested product or service and negotiates with them about price and other terms of sale (e.g., shipping). Kasbah and Tete-a-Tete (Guttman & Maes, 1999) are systems that use this approach. In these systems, each buyer agent is given such information as the desired price, the maximum price to pay, and the time by which a purchase needs to be completed. In Tete-a-Tete, the agents are negotiating about other criteria than just price such as shipping terms or extended warranties. The virtual marketplace architecture in Greengrass, Sud, and Moore (1999) uses mobile buyer agents to represent the user in a particular transaction. A further consideration is whether a buyer agent needs final approval to complete a transaction.

The buyer role is defined as a subrole of the trader role in Figure 7 above.

Resulting Context

For buyer and seller agents to understand each other, they need to agree on a COMMON VOCABULARY

Figure 7. Role diagram for USER AGENT

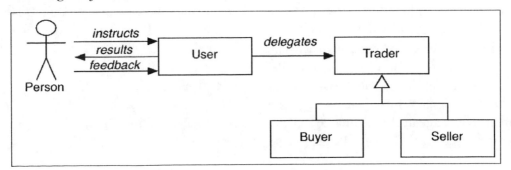

for exchanging messages during the negotiation. Buyer and seller agents find each other through AGENTS AS MEDIATORS (Weiss, 2004). In Greengrass et al. (1999) and Fonseca et al. (2001), a directory facilitator plays the role of a mediator.

SELLER AGENT

Context

You are delegating tasks to an agent—AGENT AS DELEGATE—and provide a locus of control where users can interact with their agents—USER AGENT. You now need an agent to represent a seller in a transaction with customers in an electronic marketplace.

Problem

How do you communicate your selling strategy to an agent?

Forces

• Autonomy

Solution

A seller agent *offers* products and services to buyer agents (directly or through a market). It answers queries for information about its owner's products or services, responds to RFPs, and enters into negotiation with buyer agents. A seller agent encapsulates its *execution state* and the seller's selling *strategy* (e.g., a function for setting bids). In Kasbah (Guttman & Maes, 1999) seller agents periodically send offers to buyer agents, each time adjusting their offers according to a concession strategy. In the virtual marketplace by Greengrass et al. (1999) stationary seller agents negotiate with mobile buyer agents. In the mobile shopper scenario in Fonseca et al. (2001), seller agents bid

in competition with one another to satisfy a buyer agent's call for proposal.

The seller role is defined as a subrole of the trader role in Figure 7.

Resulting Context

For buyer and seller agents to understand each other, they need to agree on a common exchange format, as described in a COMMON VOCABULARY. Buyer and seller agents find each other through AGENTS AS MEDIATORS (Weiss, 2004). If you wish to make different offers to different buyers, creating a custom catalog for each buyer by tailoring the contents of a generic catalog to the buyer's profile, consult CATALOGING.

USER PROFILING

Context

You provide a locus of control—USER AGENT—where users can interact with their agents. You receive instructions and feedback from the user. You want to derive a profile that describes the user from these interactions. Once you have deduced a user profile, you can provide it to sellers who want to personalize the customer's experience, for example, by creating a custom catalog based on the user's preferences and buying pattern.

Problem

How can you capture the users' profiles without requiring too much explicit input? This can become involved, if you want to track users' profiles as their interest changes.

Forces

• Adaptability

Solution

During the interaction with a user, the system builds a user profile by static user modeling, classifying the user into predefined user classes (stereotypes), or dynamic user modeling (monitoring user activity and changing the profile dynamically). For example, in Ardissono et al. (1999), user profiles contain a collection of interests in addition to generic information about the user. Interests contain descriptions of the topics the user is interested in. Interests may also contain weights (e.g., fuzzy logic quantifiers) that define the degree of interest the user has in a topic. The agent may also tailor its interaction style, providing more or less detail depending on the user's receptivity.

It is desirable to limit the amount of explicit input required from the user, for example, in the form of rules. Although manual setting of the user profile results in a very precise profile, this is difficult to perform correctly except for an advanced user (Brenner et al., 1998). On the other hand, it is advantageous to know as much as possible about the user to start with in order to keep the learning phase short. The user agent can request the user to provide demographic information such as sex, age, and profession, which allow us to group the user into a class. The interests of the user are then initialized to those typical for this class of users.

From the point of view of limiting user input, initializing the user profile by observing the user's behavior over time is most desirable (Brenner, Zarnekow, & Wittig, 1998). The user's personal preferences can then be deduced using cluster analysis. This has the added advantage that it is easy to adapt the user profile to changes in their interests. Once the user profiles have been initialized, the user agent can update the profile by soliciting feedback on query results from the user, or to rate a seller on her quality of service after the order has been fulfilled. A typical example for learning a user profile from user feedback is the Letizia user agent (Lieberman, 1998). The Sporas system for reputation management (Zacharia et al., 1999) ranks sellers on their past performance as assessed by user feedback.

In Figure 7, the arrow labeled "feedback" indicates how the user agent receives feedback from the user after evaluating the quality of the result returned.

Resulting Context

To construct complex user profiles (e.g., based on user navigation patterns), the information sources such as product catalogs must be well structured—COMMON VOCABULARY—and comparable on the basis of their content. If the items in an information source cannot be compared in terms of attributes, collaborative filtering can still help you create useful simpler user profiles—RECOMMENDER (Weiss, 2004).

CATALOGING

Context

You are deriving a user profile from user interactions—USER PROFILING. You are using an agent to represent a seller in a transaction in a marketplace—SELLER AGENT. You wish to tailor the contents of your catalog to individual buyers based on their user profiles.

Problem

How do you customize a catalog?

Forces

- Information overload
- Multiple interfaces

Solution

A product extractor dynamically creates a customized view on the catalog. It selects the products that best match the customer's preferences by comparing the product features with those contained in the profile. A user profile contains predictive information about the user's preferences toward product features as defined by the product categories. The view may also be customized regarding generic traits of the user (e.g., his level of expertise, job, or age).

Tailoring the product catalog to the user resolves the force of information overload. A large electronics parts catalog, for example, may contain 50,000 parts, of which only a small subset are relevant to the needs of an individual customer. By extracting a customized view of the catalog we show only those parts matching features specified in the user profile. An example of using this pattern is the personalized Web store architecture presented in Ardissono et al. (1999).

Existing product catalogs can be integrated with the agent-based subsystem through catalog agents. Each catalog agent is a WRAPPER (Weiss, 2004) that "agentifies" an existing catalog, converting a catalog-specific schema to a common schema used by the product extractor. In addition to filtering the contents of the catalog against the buyer's profile, discounts can also be offered to individual customers to entice existing customers to remain loyal to the business, as well as to attract new customers.

The structure of the pattern is shown in Figure 8.

Resulting Context

To control what information is disclosed to the seller, the user profile should be maintained by the user agent, and supplied to the seller agent on request. For addressing privacy concerns when accessing user profiles, consult USER AGENT. For a discussion on obtaining the user profiles for tailoring the content of a generic catalog to a buyer's profile, consult USER PROFILING. Consult NOTIFICATION AGENT, if you want to notify the customer about changes to the catalog (e.g., the addition of a new widget that matches the user's profile). Each catalog agent is a wrapper, that agentifies an existing catalog, converting a catalog-specific schema to a common schema—WRAPPER (Weiss, 2004).

Figure 8. Role diagram for CATALOGING

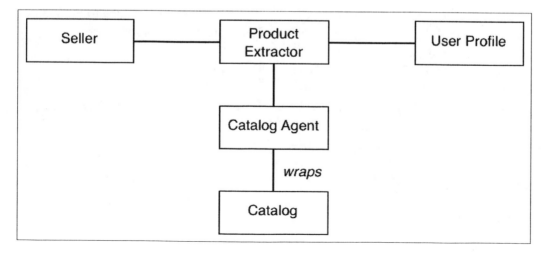

NOTIFICATION AGENT

Context

You provide a locus of control—USER AGENT—where users can interact with their agents, and want to subscribe to the progress of a transaction. You want to notify the customer about changes to the catalog—CATALOGING. In both cases, you need to monitor a data source (e.g., a product catalog) for changes or the progress of a transaction.

Problem

How to keep users informed about changes of interest to them.

Forces

- Information overload

Solution

The solution involves a notification agent that resides close to the information source. It monitors the information source for changes of interest by registering with the information source or by continuously polling it. Notification agents can be considered a special type of mobile agent that only move to a single remote location. A notification agent is created with a condition, typically comprising an event (such as the posting of a product for sale) and a boolean expression on the event data (such as the name of a product). Whenever it detects an event of the specified type at the remote location, it evaluates this condition. If the condition is satisfied, the agent notifies the user agent.

The benefits of using this pattern are similar to those of sending a mobile agent to another host and letting it interact remotely. It reduces the number of messages between the user agent and the information source. The example of a meta-auction engine such as AuctionWatcher (2002) makes this clear: it creates an agent that monitors the progress of an auction on the user's behalf and notifies the user about events of relevance, for example, when the user has been outbid and when the user should increase her maximum bid. Without a meta-auction engine, the user agent would have to poll the state of the auction periodically.

The structure of this pattern is shown in Figure 9.

Resulting Context

You want to use the same notification agent with multiple sellers. Therefore, you decide to employ WRAPPER (Weiss, 2004) agents that translate from the format of seller-specific information sources to that understood by the notification agent.

Figure 9. Role diagram for NOTIFICATION AGENT

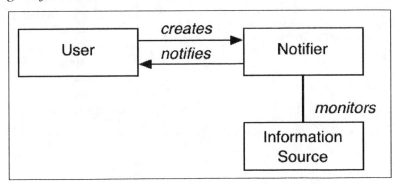

Examples

The following examples illustrate some uses of the patterns described in this chapter. The first example, an online auction, demonstrates how agents can be used in the negotiation phase (Guttman & Maes, 1999) of an e-business transaction, in which buyers and sellers negotiate the terms of a transaction (such as price and warranty). The second and third examples show agents in operation during the information phase (Guttman & Maes, 1999), in which buyers navigate product catalogs and use mediators to locate sellers that offer a certain product or establish their actual needs.

Online Auction

In this example, in an online auction, users post items for sale, while other users place bids on those items. A user may be represented—AGENT AS DELEGATE—by multiple BUYER and SELLER AGENTS in concurrent transactions. These agents are all managed by a USER AGENT that provides a single interface to the user. In our example, the user Bob is represented in the auction by a seller agent, while the users Alice and Mary act as buyers via their buyer agents, as shown in Figure 10. The auction mechanism is implemented by a mediator agent—AGENT AS MEDIATOR. This agent provides a MEETING PLACE (Deugo et al., 2001) for buyers and sellers. It maintains a catalog of items for sale and a list of auctions with administrative information for each one (highest bid, reserve price, and remaining duration of the auction). The mediator receives requests from sellers to create an auction, bids from buyers, and, if the type of auction allows it (e.g., in an English auction), informs each buyer agent about a new bid, as well as the outcome.

Buyer agents implement the bidding strategies of their users (characterized by parameters such as starting price, maximum price, and rate of increase). They also inform them about important events regarding the current auction (e.g., if they are winning or losing the auction). While buyer agents only notify a user about the current auction, notifier agents—NOTIFICATION AGENT—can be used to monitor other auctions on a user's behalf. The user creates a notifier agent by initializing it with a specification of the product or service she is looking for. On receiving a NOTIFICATION about an auction, a user can decide to create a new buyer agent to join it.

Figure 10. Collaboration diagram for the agent-based online auction example

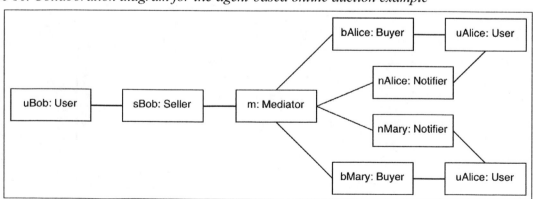

Locating Sellers by Product or Service

This example switches the positions of buyer and seller agents in the online auction example. Bob wants to buy a certain product or service and, via his USER AGENT, creates a BUYER AGENT and initializes it with a specification of the product or service. This buyer agent queries a MEDIATOR AGENT—in this case, a search agent or a directory agent—for the locations of sellers for the product or service in question. At this architectural level, we could use static or mobile agents to represent buyer agents. Greengrass et al. (1999) describe a mobile agent scenario in which a buyer agent moves to the known site of a directory agent, queries this directory agent, and obtains an itinerary of the locations of SELLER AGENTS to visit. Fonseca et al. (2001) document a mobile shopping system, where an agent representing a shopper in a mall contacts a directory facilitator agent on entry to the mall, and obtains directions to stores which carry products of interest to the user, provides those directions to the user, and allows him to check the prices of other stores. For further details on this case study, and an illustration of how the patterns can be used as part of a pattern-driven design process, see Weiss (2003).

Customizing Navigation

Another example is using agents to assist users in navigating through an (online) product catalog. From a front-end perspective, a USER AGENT can be used to collect profile information about a user and control how this information is made available to seller agents. The profile information for different sellers is stored in a personal profile repository in the user agent—USER PROFILING. Sections of the profile can be restricted to a subset of the sellers only. Using the profile information, the seller agent can classify the user into predefined stereotypes and thus infer a profile of product pref-erences—USER PROFILING. With these preferences, a product extractor agent—CATALOGING—can tailor the contents of the product catalog, such as in the personalized Web store in Ardissono et al. (1999). In addition to filtering the catalog contents against the buyer's profile, the product extractor can propose discounts to retain existing or attract new customers.

Although we have focused on using user profiles in the front-end, this is not to say that they cannot equally be used in the back-end. However, the basic trade-off in this case is that users will have less control over profiles collected and stored by sellers. Customers may decide to do business with sellers who do not collect profiles in this way, or make the process of collecting profiles transparent and the profiles themselves accessible.

CONCLUSION

In this chapter, we described a group of architectural patterns for designing agent-based e-business systems and gives several examples illustrating their use. These patterns relate to front-end e-business activities that involve interaction with the user and delegation of tasks from users to agents. Future work will describe patterns for back-end e-business activities that do not involve direct interaction with the user, but rather depict mechanisms for mediating between agents representing users. These patterns will expand on the description of mediator agents in this chapter. Together, these patterns are just the beginnings of a pattern language for agent-based e-business system design, based on our current understanding of the technology. As the use of agent technology in e-business matures, this language will evolve as well. It is our hope that the proposed set of patterns and the pattern format may provide a starting point for future effort in this direction.

FUTURE RESEARCH DIRECTIONS

To date, only a subset of the patterns of agent-based e-business systems have been documented. This provides many opportunities for future research. These opportunities can be grouped into two areas. On one hand, a better understanding of using agent patterns as part of a pattern-driven design process (Weiss, 2003) is required; we need to be able to reason about design trade-offs between patterns; see Mussbacher, Amyot, and Weiss (2007) for a typical approach and survey of related efforts, and capture pattern knowledge in pattern repositories (Knublauch & Rose, 2002) that can be consulted by developers to address specific design problems during the development of an agent-based system. Finally, we want to generate code for specific agent frameworks from those patterns.

On the other hand, many more agent patterns in the e-business domain remain to be mined and documented. Areas of particular interest are patterns where agents are used as mediators, ranging from directory agents to sophisticated broker and market maker agents (Deugo et al., 2001), and patterns for agents that provide mobile, context-aware services. One example of current research in this area is a set of auction patterns documented by Jureta, Faulkner, and Kolp (2005). However, this would also include patterns for search, reputation management, and integration. A taxonomy of e-business agents, which could provide a suitable starting point for a systematic mining effort is given in Papazoglou (2001).

Conceptual frameworks for these activities are also required. Weiss (2003) proposes a pattern-driven design framework for harvesting recurring design solutions and documenting them as patterns, and guiding the designer through the selection of patterns appropriate to their specific design context. The approach suggests a five step process, the first three related to mining patterns: identify the forces in a domain, document the roles of pattern participants, and document patterns and their dependencies. The last two apply to patterns: identify the overall design goals and select patterns.

Finally, we need to document patterns for nonfunctional design issues such as deployment, scalability, and security of agent-based e-business systems, and we must gain a better understanding of how to integrate such patterns with the current agent development processes. Some progress along these lines has been made in the area of agent security patterns. A good starting point is Mouratidis et al. (2006).

ACKNOWLEDGMENT

An earlier version of this chapter was accepted for PLoP 2001 (Weiss, 2001). Thanks go to my shepherd Dirk Riehle for his valuable comments. The material in this chapter has benefited from feedback in several tutorials and classes held since on this material and was expanded to reflect the recent progress in the area of agent patterns.

REFERENCES

Alexander, C. (1979). *The timeless way of building.* Oxford University Press.

Aranguren, M. (2005). *Ontology design patterns for the formalisation of biological ontologies.* Master's Thesis, Manchester University.

Ardissono, L., Barbero, C., et al. (1999). An agent architecture for personalized Web stores. In *Proceedings of the Conference on Autonomous Agents* (pp. 182-189). ACM.

Aridor, Y., & Lange, D. (1998). Agent design patterns: Elements of agent application design. In *Proceedings of the Conference on Autonomous Agents* (pp. 108-115). ACM.

AuctionWatcher (2002). Retrieved August 14, 2007, from http://www.auctionswatch.info

Avgeriou, P., & Zdun, U. (2005). Architectural patterns revisited: A pattern language. In *Proceedings of the European Conference on Pattern Languages of Programs*.

Brenner, W., Zarnekow, R., & Wittig, H. (1998). *Intelligent software agents: Foundations and applications*. Springer.

Carlson, D. (2001). *Modeling XML applications with UML: Practical e-business applications*. Addison-Wesley.

Collis, J., & Lee, L. (1999). Building electronic marketplaces with the ZEUS agent tool-kit. *Lecture Notes in Artificial Intelligence, 1571*, 1-24.

Deugo, D., Weiss, M., & Kendall, E. (2001). Reusable patterns for agent coordination. In A. Omicini et al. (Eds.), *Coordination of Internet Agents* (pp. 347-368). Springer.

Ferber, J. (1999). *Multi-agent systems: An introduction to distributed srtificial intelligence* (pp. 13-16). Addison-Wesley.

Fonseca, S., Griss, M., & Letsinger, R. (2001). *An agent-mediated e-commerce environment for the mobile shopper* (HP Labs Tech. Rep. No. HPL-2001-157).

Greengrass, E., Sud, J., & Moore, D. (1999, June). Agents in the virtual marketplace: Component dtrategies. In *Proceedings of SIGS* (pp. 42-52).

Griss, M. (1999). *My agent will call your agent ... but will it respond?* (HP Labs Tech. Rep. No. HPL-1999-159).

Guttman, R., & Maes, P. (1999). Agent-mediated integrative negotiation for retail electronic commerce. *Lecture Notes in Artificial Intelligence, 1571*, 70-90.

Jureta, I., Faulkner, S., & Kolp, M. (2005). Best practices agent patterns for on-line auctions. In *Proceedings of the International Conference on Enterprise Information Systems* (pp. 814-822).

Kendall, E. (1999). Role models: Patterns of agent system analysis and design. In *Proceedings of the Agent Systems and Applications/Mobile Agents*. ACM.

Kendall, E., Murali Krishna, P., Pathak, C., et al (1998). Patterns of intelligent and mobile agents. In *Proceedings of the Conference on Autonomous Agents*. ACM.

Kolp, M., Giorgini, P., & Mylopoulos, J. (2001). A goal-based organizational perspective on multi-agent architectures. In *Proceedings of the Workshop on Agent Theories, Architectures, and Languages* (LNCS 2333, pp. 128-140). Springer.

Lieberman, H. (1998). *Integrating user interface agents with conventional applications*. Paper presented at the Conference on Intelligent User Interfaces. ACM.

Lind, J. (2002). Patterns in agent-oriented software engineering. In *Proceedings of the Workshop on Agent-Oriented Software Engineering* (LNCS 2585, pp. 45-58). Springer.

Maes, P. (1994, July). Agents that reduce work and information overload. *Communications of the ACM*, pp. 31-41.

Mouratidis, H., Weiss, M., & Giorgini, P. (2006). Modelling secure systems using an agent-oriented approach and security patterns. *International Journal on Software Engineering and Knowledge Engineering, 16*(3), 471-498.

Mussbacher, G., Amyot, D., & Weiss, M. (2007). Formalizing patterns with the user requirements notation. In T. Taibi (Ed.), *Design Pattern Formalization Techniques*. Hershey, PA: IGI Global.

Noriega, P., & Sierra, C. (1999). Agent-mediated electronic commerce. *Lecture Notes in Artificial Intelligence, 1571*. Springer.

Papazoglou, M. (2001, April). Agent-oriented technology in support of e-business. *Communications of the ACM*, pp. 71-77.

Reich, J. (2000). *Ontological design patterns: Modelling the metadata of molecular biological ontologies, information and knowledge*. Paper presented at the Conference on Database and Expert System Applications.

Riehle, D., & Gross, T. (1998). Role model based framework design and integration. In *Proceedings of the Conference on Object-Oriented Programs, Systems, Languages, and Applications* (pp. 117-133). ACM.

Schelfthout, K., Coninx, T., et al (2002). *Agent implementation patterns*. Paper presented at the OOPSLA Workshop on Agent-Oriented Methodologies.

Shu, S., & Norrie, D. (1999). *Patterns for adaptive multi-agent systems in intelligent manufacturing*. Paper presented at the International Workshop on Intelligent Manufacturing Systems.

Staab, S., & Stuckenschmidt, H. (Eds.). (2006). *Semantic Web and peer-to-peer*. Springer.

Steinmetz, E., Collins, J., Jamieson, S., et al. (1999). Bid evaluation and selection in the MAGNET automated contracting system. *Lecture Notes in Artificial Intelligence, 1571*, 105-125.

Tahara, Y., Oshuga, A., & Hiniden, S. (1999). *Agent system development method based on agent patterns*. Paper presented at the International Conference on Software Engineering. ACM.

Ulmer, D. (2004). *Architectural solutions to agent-enabling e-commerce portals with pull/push abilities*. Unpublished doctoral thesis, Pace University.

Voss, A., & Kreifelts, T. (1997). *SOaP: Social agents providing people with useful information*. Paper presented at the Conference on Supporting Groupwork. ACM.

W3C. (1997). *Open profiling standard* (version 1.0, note 1997/6).

Weiss, M. (2001). *Patterns for e-commerce agent architectures: Agents as delegates*. Paper presented at the Conference on Pattern Languages of Programs.

Weiss, M. (2003). Pattern-driven design of agent systems: Approach and case study. In *Proceedings of the Conference on Advanced Information System Engineering* (pp. 711-723). Springer.

Weiss, M. (2004). A pattern language for motivating the use of agents. In *Agent-Oriented Information Systems: Revised Selected Papers* (LNAI 3030, pp. 142-157). Springer

Weiss, M., Gray, T., & Diaz, A. (1997). Experiences with a service environment for distributed multimedia applications. feature interactions in telecommunications and distributed systems, IOS, 242-253.

Wooldridge, M., & Jennings, N. (1995). Intelligent agents: Theory and practice. *The Knowledge Engineering Review, 10*(2), 115-152.

Zacharia, G., Moukas, A., & Maes, P. (1999). *Collaborative reputation mechanisms in electronic marketplaces*. Paper presented at the Hawaii International Conference On System Science. IEEE.

Zambonelli, F., Jennings, N., et al. (2001). Agent-oriented software engineering for Internet applications. In A. Omicini, et al. (Eds.), *Coordination of Internet Agents* (pp. 326-346). Springer.

ADDITIONAL READING

Deugo, D., Kendall, E., & Weiss, M. (1999). Agent pPatterns. Tutorial at the *International Symposium on Agent Systems and Applications/Mobile Agents*.

Retrieved August 15, 2007, from http://www.scs. carleton.ca/~deugo/Patterns/Agent/Presentations/ AgentPatterns

Jureta, I., Faulkner, S., & Kolp, M. (2005). Best practices agent patterns for on-line auctions. *International Conference on Enterprise Information Systems* (pp. 814-822).

Knublauch, H., & Rose, T. (2002). Tool-supported process analysis and design for the development of multiagent systems. *Workshop on Agent-Oriented Software Engineering* (LNCS 2585, pp. 186-197). Springer.

Mussbacher, G., Amyot, D., & Weiss, M. (2007). Formalizing patterns with the user requirements notation. In T. Taibi (Ed.), *Design Pattern Formalization Techniques*. Hershey, PA: IGI Global.

ENDNOTE

[1]   It is customary to indicate references to patterns through a SMALL CAPS font.

Section II
Agent Technologies in Electronic Commerce Transactions

Chapter II
The Evolution of Comparison–Shopping Agents

Yun Wan
University of Houston at Victoria, USA

ABSTRACT

Comparison-shopping agents became the important link in the business-to-consumer (B2C) e-commerce domain since the late 1990s. Since its emergence in 1995, the evolution of comparison-shopping agents experienced a few ups and downs. This chapter covers key events and issues of comparison-shopping agent evolution in three intertwined threads: the emergence of representative agents, the evolution of comparison-shopping agent technology, and the evolution of their business models.

INTRODUCTION

Generally speaking, comparison-shopping agents are Web-based business-to-consumer (B2C) information searching services that could retrieve product information from multiple online retailers, aggregate them, and then present them to online shoppers to assist shopping decision making. There are several synonyms for comparison-shopping agent, such as *shopbot, product comparison agent, recommendation agent, buyer's agent* as well as *aggregator* (Bakos, 1998; Brynjolfsson & Smith, 2000; Crowston, 1997; Green, 1998; Kuttner, 1998; Madnick & Siegel, 2001; Sinha, 2000). Since the popularity of the World Wide Web (WWW) in 1994, electronic decision aids or Web-based business-to-consumer agents became crucial to Web users for their information acquisition and processing support. Among these Web-based B2C agent systems, comparison-shopping agents are the most visible and systematically developed agent group.

BACKGROUND

The popularity of comparison-shopping agents and the formation of this industry originated from sheer consumer needs when they have to navigate in the World Wide Web to look for product information especially the price.

The emergence and commercialization of Internet provides a completely new channel for consumer shopping—online shopping. Compared with traditional channels, this online channel has two important features: low entry barrier and almost unlimited shelf space.

As described in the seminal book *Information Rules* (Shapiro & Varian, 1998), online channel has unprecedented low entry cost for potential retailers. Nowadays, any individual can launch an e-commerce site by uploading the product data to a template provided by Internet service protocols (ISPs) with slight customization of the interface. As a result, we were observing an increasing number of micro-online retailers that operated by only one or two individual.

This new channel also has a unique feature of almost unlimited shelf space compared with physical channels. For example, Wal-Mart, the biggest brick-and-mortar chain store, at any time has 100,000 items available on the shelf in a typical supercenter. However, Amazon.com, the biggest online store can already offer as many as 18 million unique items even without inclusion of third-party vendors who utilize the platform provided by Amazon to sell to its customer base.

These two features led to an exponential increase of product information online in the past 10 years, which have been enriching our shopping experience ever since. However, with so many products available online, finding them was not as easy as expected unless one could remember all those domain names beyond a few established portals. Thus, this posed a practical problem for online shoppers: *how to find the relevant product information especially price information from those unknown stores?*

Even when searching established portals from memory, experience told us that retrieving the same product price from each site for comparison-shopping was exceedingly time-consuming. So can we find all these relevant information with one click instead of several clicks per site?

These questions are directly related to the research topic of online information searching and processing behavior.

To make a shopping decision, the consumer needs relevant information. However, there exists a cost to retrieve this information (Shugan, 1980). In a traditional environment, cost incurred in searching information is prohibitive if one wants to visit several physical stores to retrieve the price information for comparison-shopping, for example, time, transportation to each store, and so forth. In an online environment, shoppers can avoid costs such as transportation. But still there is limited time one could spend on searching the Web for a product with the preferred feature and price.

Online shopping, as an information search process, has two components: intersite and intrasite search (Hodkinson & Kiel, 2003). The intersite search mainly refers to use general purpose search engine to locate shopping information from multiple sites; intrasite search refers to search the specific product information in the site of one specific online retailer. Both components incur cost to online shoppers. The intersite search requires considerable time and domain knowledge (familiarity with search engines) for the online shopper to identify useful information; the intrasite search also requires the learning time for online shopper to familiar oneself with the specific site he searches.

We can expect that, when the volume of B2C electronic commerce increases further and reaches certain plateaus, the efficiency and effectiveness of manual comparison-shopping would be a major bottleneck in the effort to achieve customer satisfaction. Consumers would face an increasing amount of product information to sort through from general search engines. Meanwhile, when they conduct intrasite search, they will be limited by the time they could spend on shopping and most probably restricted to search only a few established portals though some unknown online retailer where only a click away could offer a more competitive price with reasonable service quality.

In summary, though there is much optimism about the Internet being a great equalizer for online shopping, to find the relevant product information and make a sound decision for online shopping, online shoppers still need assistance. Comparison-shopping agents emerged and came to rescue in this scenario. They provide a solution to greatly reduce both the searching and processing cost of online shopping.

THE DEVELOPMENT OF COMPARISON-SHOPPING AGENTS

The Emergence (1995)

The emergence of comparison-shopping agents was triggered by the popularity of World Wide Web. With the release of the first multimedia Web browser, Mosaic, in 1995, the online population reaches the critical mass to serve as a large enough customer base for B2C e-commerce. Subsequently, we see a Cambrian age of B2C e-commerce models including comparison-shopping agents. Two of the earliest comparison-shopping agents known to the public are BargainFinder.com and Pricewatch.com.

The BargainFinder Experiment

BargainFinder is an agent designed for a phenomenal experiment conducted by Andersen Consulting and Smart Store Center in 1995. Interesting enough, the intent of this experiment was not testing whether a comparison-shopping agent could assist consumer in online shopping.

Instead, it was designed to test the impact on online retailers by the price arbitrage behavior of online shoppers.

They suspected that online shoppers may take advantage of the rich product information provided by premium online stores like Amazon and then shop the actual product from another retailer charging a lower price. This behavior may lead to a situation that online retailers competing solely on price and compromising online shopping experience. Solely competing on price would also lead Cournot equilibrium (Cournot, 1838), a ruinous outcome for online retailers.

To test reactions of consumers and merchants to such pricing information, BargainFinder was built and deployed on the Web for public access. The basic interaction mode between Bargain-Finder and online shopper defined the style for all subsequent comparison-shopping agents:

[BargainFinder] takes the name of a particular record album, searches for it at nine Internet stores, and returns to the user a list of the prices found. After the search, the user can select one of the stores and be taken electronically into it and directly to the album. He then has the option of getting more information, looking for other albums, or buying the product. (Krulwich, 1996)

It turns out that even though BargainFinder was a very primitive agent, most online shoppers would like to use it at least occasionally according to the survey response conducted by the team. Also, as expected, 90% of the users of Bargain-Finder clicked on the cheapest price in the list. The responses from online retailers diversified. Some refuse contact with the team or even blocked the access of BargainFinder while others seek the collaboration with BargainFinder and hope the agent could search their sites too.

These initial reactions of both online shoppers and online retailers represented typical behaviors later experienced by new agents.

Pricewatch: The Online Catalog Advertiser

Pricewatch.com emerged the same year as the BargainFinder experiment, though with much less publicity. Actually, Pricewatch.com deliberately keeps a very low profile. Even nowadays, there was no introduction information about the company on its Web site. Both retailers and online shoppers can only contact the company via Web-based e-mail.

Instead of launching as an experiment, Pricewatch.com was launched as an electronic catalog for multiple online retailers. Any online retailer could upload its product information to Pricewatch and the latter charged a commission fee for traffic it generated for these retailers.

Different from BargainFinder's music CD information, the product information domain of Pricewatch.com was mainly computer components and accessories, ranging from CPU, memory, storage, I/O, to networking gears, and so forth. These products are relatively more complex than books and CDs in terms of product description; for example, to identify a CPU, one needs at least a brand and a model number compared with an ISBN for a book. Because the relative complexity of product information for computer-related items, different from BargainFinder, Pricewatch.com needed more cooperation from online retailers in both product data provision and real time access.

Strictly speaking, neither BargainFinder nor Pricewatch were created as comparison-shopping agents as indicated in previous sections. But the outcome from a public perspective made them the first pair of comparison-shopping agents in the B2C e-commerce domain. Actually, they represented two distinctive technology features and potential business models later adopted by their successors and evolved into the much more sophisticated versions we experience today.

The Early Boom (1996-1998)

Since the BargainFinder experiment, the commercial potential of comparison-shopping agents attracts a lot of attention from entrepreneurs. Hence, a large number of comparison-shopping agents emerged between 1996 and 1998 and they triggered the early booming age of comparison-shopping agents. Here we discuss five most representatives ones: Pricescan, Jango, Junglee, ComparisonNet, and mySimon.

Pricescan.com

Pricescan.com was launched in 1997. Based on the concept of BargainFinder, Pricescan can not only aggregate price information from multiple online retailers for computer products but also provide nifty features like displaying high, low, and average price trends over the past several weeks for each product. Pricescan.com emphasized its proconsumer position in providing price comparison information. According to David Cost, the cofounder, Pricescan did not charge online retailers to be listed in its database. In addition, to bring consumers the best prices, it obtained pricing information not only from vendor Web

Table 1. Comparison of BargainFinder and Pricewatch in 1995

	Technology	Potential Business Model
BargainFinder	Data Extraction	Advertisement
Pricewatch	Data Feeding	Commission-based

sites but also from off-line sources like magazine ads (PRWeb.com, 1998). Pricescan.com survives through the revenue generated from the banner advertisements on its Web site.

Jango.com

Jango.com or NETBot was based on the prototype of a comparison-shopping agent designed by three researchers in Washington State University (Doorenbos, Etzioni, & Weld, 1997). A notable feature of Jango was that it could automate the building of wrapper for a specific online vendor. Like BargainFinder, Jango was more an experiment than a real commercial agent services. Jango was soon acquired by Excite for $35 million in stock in October 1997.

Junglee the Virtual Database

Junglee was the nickname of a comparison-shopping technology called virtual database (VDB) created in 1996 by three Stanford graduate students. The core of Junglee is a improved wrapper technique that made it easier to search for complex product information online (Gupta, 1998). Instead of having its own Web presence, Junglee.com provides the search service to multiple online portals like Yahoo.com. Junglee.com was acquired by Amazon.com for $230 million in 1998.

CompareNet.com

CompareNet.com was founded by Trevor Traina and John Dunning in 1996 and backed by venture capitals like Media Technology and Intel. It provided comparison information on rather diverse categories like electronics, home office equipment, home appliances, automobiles, motorcycles, sporting goods, and software and computer peripherals. It was acquired by Microsoft in 1999.

mySimon.com

mySimon.com was founded in 1998 by two cofounders. It used its own proprietary wrapper technology ("Virtual Agent") to collect information from almost every online store. MySimon.com was noted for its easy-to-use interface and was acquired by CNET in 2000. Though being acquired, mySimon.com escaped the faceless destination of its peers and has become one of the remaining major comparison-shopping agents nowadays.

As indicated above, from 1996 to 1998, we experienced the first booming age of comparison-shopping agents. With improved technologies, the first generation of comparison-shopping agents emerged and achieved their respective star status by sophisticating the wrapping and information classification technologies in different ways.

However, many of them were subsequent acquired by major Web portals before they secured a large enough customer base. As a result, a majority of online shoppers had not experienced the comparison-shopping service they provided before many of them were either completely absorbed into other technology infrastructures or discarded. According to estimation (Baumohl, 2000a), there were only about 4 million online shoppers used comparison-shopping agents in October 2000, less than 1% of the Internet users in the United States at that time.

The Reshuffling and Resurgence (1999-2002)

The Reshuffling

By the end of 1999, the first booming of comparison-shopping agents came to its end due to the large scale acquisitions by established e-commerce portals. Another dampening factor from the acquisition is that many e-commerce portals could not strategically synthesize these agents into their existing infrastructures. As a result, many excellent agent technologies and burgeoning brand names were abandoned and became obsolete. The acquisition of Junglee by Amazon was one example. The crumbling of Excite@ Home in 2000 and 2001 also ended the further development of Jango.com.

Thus, the comparison-shopping category in B2C e-commerce market experienced its first reshuffling from 1999 to 2001. Meanwhile, the second generation of comparison-shopping agents emerged with an emphasis on improved business models and alternative information retrieval technologies pioneered by Pricewatch, the data feeding.

The Resurgence: The Top Three and Niche Players

Since 2000, a new generation of comparison-shopping agents emerged and became increasingly popular. If the technical innovation characterized the first generation comparison-shopping agents, the sophisticated business models distinguished the second generation agents. The three representative ones in online retailing are shopping.com (renamed from dealtime.com), PriceGrabber.com, and Shopzilla.com as well as leading agents in other categories as well as numerous niche B2C e-commerce domains.

Shopping.com was founded in 1997 as Dealtime.com. Together with CNET Networks' mySimon.com, Shopping.com was among the first group of comparison-shopping agents to use intensive marketing efforts to build the concepts of Web-based comparison-shopping among consumers (White, 2000). Within a short three years of strategic buildup, Shopping.com managed to rank fourth (behind eBay, Amazon, and Yahoo Shopping) among U.S. multicategory e-commerce sites in terms of unique monthly visitors (Baumohl, 2000; Census.gov, 2004).

PriceGrabber.com was another major comparison-shopping agent that emerged in 1999. It improved the service by providing more specific price information like tax and shipping cost as well as availability of the product from vendors.

Shopzilla.com was transformed from Bizrate. com in 2004. Bizrate.com was an online retailer rating services launched in 1996. Like other first generation e-commerce startups, Bizrate. com found the comparison-shopping service an attractive category and thus made a natural transformation because it already possessed an important information element of comparison-shopping, the rating of online retailers.

Sensing the challenges from new startups and observing the opportunities in this new category, established online Titans began to adding or transforming their shopping channel into a comparison-shopping style tool. By 2004, we already found comparison-shopping services like Froogle by Google, Yahoo! Shopping by Yahoo, and MSN Shopping by Microsoft, and so forth.

Besides online retailing, comparison shopping was well-established in online traveling due to its well-developed technology infrastructure dated back to the 1950s. In this category, we not only observed well-established services like Expedia. com, Orbitz.com, and Travelocity.com but also the agents of agents like Kayak.com, which aggregates and repackages information collected from existing comparison-shopping agents.

Comparison-shopping service was also a natural adaptation for finance and insurance businesses that are essentially broker-coordinated. In this category, most comparison-shopping agents served as an additional channel that is parallel to human brokers to interact with consumers thus increasing customer experience, for example, lendingtree.com.

Comparison-shopping services also emerged in numerous niche markets. There are comparison-shopping agents specialized from books and CDs to wine, ink cartridges, or whatever products that are hard for consumers to search and compare online.

Except some unique cases (like Pricescan.com), almost all comparison-shopping agents emerged in this period and adopted the cost-per-click (CPC) business model, the one evolved from Pricewatch.com. The CPC model allows startup agents like shopping.com and pricegrabber.com to become profitable without initial capital infusion of large venture capitals.

The Consolidation and Restructuring (2003 and After)

Starting from 2003, with the increasing sophistication and homogenization of existing comparison-shopping agent technologies and business models, the space of further development was shrinking gradually. The leader in comparison-shopping services began to expand their market share via merger and acquisition with compatible agent services. By such a strategy, they hope to create synergy and increase customer experience thus compete more effectively.

The most notable case is Dealtime's acquisition of resellerratings.com and epinion.com in February and March 2003, respectively. Dealtime.com was mainly focusing on price comparison. Resellerratings.com was one of the earliest agent services that focused on collecting ratings about online vendors. Epinion.com was specializing at collecting product review information. When these three services merged together, online shoppers became able to obtain almost all they want for making a shopping decision regarding a product on one page. Another case is expansion of services by Bizrate.com. Bizrate.com, like Resellerratings.com, was providing vendor service evaluation for comparison-shopping since 1996. However, facing the challenge from competitors, it expanded its services to price comparison. Because of its established relationship with online vendors, its new service, Shopzilla.com, became popular fairly quickly.

The popularity of comparison-shopping has also spread across national boundaries. In Europe, Kelkoo, which launched in the same year as Dealtime in 2000, experienced multiple mergers with other small Shopbots like Zoomit, Dondecomprar, and Shopgenie and now has become Europe's largest e-commerce Web site after Amazon and eBay and the largest e-commerce advertising platform both in the UK and Europe (Kelkoo.com, 2005).

In online traveling, Expedia.com, Travelocity.com, and Obitz.com became the "Big Three." They adopted the referral model and collaborate with airline, hotel, and car rental companies. In online personal finance, leading comparison-shopping agents like lendingtree.com and bankrate.com aggregate interests rate comparisons and online mortgage applications to provide one-stop service for personal finance needs.

As a result of these consolidations, a more matured market structure was formed.

The two themes accompanying the evolution of comparison-shopping agents are its technology and business models. In the next two sections, we will discuss each of them in detail.

THE EVOLUTION OF TECHNOLOGY

The Data Extraction Approach

The early development of the Internet ignored the commercial potential of the Web and chose the Web page design protocol as HTML, a subset of SGML. HTML is a semistructured language in a sense that it is more for description of display format for human view than for description of the semantic meaning of the content the page contains, which led to the difficulty for decision aid agents like comparison-shopping agents to analyze the text and extract relevant information. As a result, to retrieve product information from different online stores, comparison-shopping agents have to employ the "wrapper" technique (also called "screen scraping" in the popular press), a procedure that could recognize the pattern of the information layout on the page and retrieve information from a specific online store (Choi, 2001; Doorenbos et al., 1997; Madnick & Siegel, 2001). Actually, the first generation agents were mainly competing on wrapping technologies, and those top-rated agents were all being noticed because of their competitive advantage in wrapper technologies to extract data from online vendors. BargainFinder was the predecessor, so we use it as an example to illustrate the basic system of comparison-shopping agents.

The system architecture of BargainFinder is illustrated in Figure 1. In BargainFinder, the wrapper algorithm employed is relatively simple because it only had to deal with a few online music stores. With a wrapper, BargainFinder could translate an online shopper's query into the format that is understandable by the targeted online music stores and, at the same time, transform price information in different formats retrieved from these online music stores into a format that can be processed by its business logic layer.

The succeeding comparison-shopping agents emerged from 1996 to 1998 and were focused on optimizing these wrapper techniques like the virtual database technology patented by Junglee. However, most of these innovations were incremental instead of revolutionary. Largely due to a lack of breakthrough in natural language processing and the semistructured Web environment, the

Figure 1. The design of BargainFinder (Krulwich, 1996)

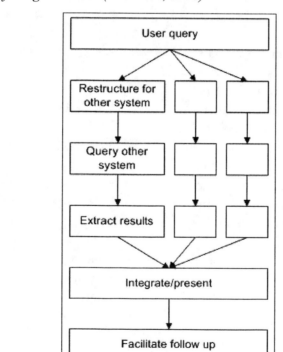

improvement of wrapper techniques reached a plateau after a few years of rapid progress.

The limitation to the data extraction approach makes the handling of more complex product information a challenge. Meanwhile, with a large number of online vendors wanting to participate in comparison shopping, it was impossible to manually generate a wrapper for each individual retailer site; thus, it was a natural move for agents to seek collaboration from online retailers. The retailers were initially expected to make the product information displayed on their site following a certain format and later were asked to feed the data into the agent by themselves.

The Data Feeding Approach

The accidental pioneer for the data feeding approach was probably Pricewatch.com. As an online catalog advertiser, it was considered a technology innovation for its proprietary "Info-Link" System that could automatically read the content of participating online retailers that put the information in a prespecified format. Of course, the degree of innovation of this method was lower than wrapper techniques.

This is also for the data feeding techniques most comparison-shopping agents employ nowadays. The data feeding approach is essentially a software system that allows online retailers to feed the product information into the database of comparison-shopping agents.

They were a rather conventional technology compared with wrapper and this approach depends on cooperation from online retailers.

However, this method turned out a necessary strategy because of at least two reasons.

First, to deal with more complex product information on the Web like those describing PC and its accessories, traveling information, finance and insurance information, and so forth, wrapper techniques are limited. Without cooperation from retailers or service providers, agents could not collect a complete record.

Second, the competition among agents also demanded comparison-shopping service providers to aggregate more product related information to customers like price updating, inventory information (is the product in stock or not?), tax and shipping cost, and so forth. Such information needs cooperation from online retailers, and the optimal solution comes from the feeding of this information by retailers themselves.

Another important fact related to the wide adoption of data feeding techniques were the popularity of comparison-shopping agents among online retailers. For major comparison-shopping agents, there were thousands of online retailers participating. Thus, the semiautomatic data extraction method could not work efficiently.

By shifting research and development (R&D) competition from data extraction, agents adopting the data feeding method could invest more R&D in improving the technical challenge of classifying and aggregating a large array of product information from different retailers across different categories. It became possible for one agent to provide a product comparison service across hundreds if not thousands of product categories. Thus, we observed the emergence of such mega-agents like shopping.com, pricegrabber.com, and shopzilla.com.

Without a steep entry barrier in back-end technology, both established online portals and new startups could enter into the comparison-shopping market with relatively little technical challenges; as a result, agents competed more on improving interface and business models.

Meanwhile, by division of labor, data feeding technology became more and more sophisticated and specialized so that many new startups propped up to provide data feeding services for online retailers and help them manage multiple accounts with different comparison-shopping agents.

THE EVOLUTION OF BUSINESS MODELS

The BargainFinder Model

As we discussed previously, there is a need by online shoppers in finding product information effectively, especially price information. Both BargainFinder and Pricewatch served this need though neither were created for this purpose. We have to wait until the late 1990s for some visionary technical entrepreneurs to create such dedicated agents with a viable business model.

The business model adopted by comparison-shopping agents experienced two stages.

The first business model, which can be characterized as the BargainFinder model, emphasizes the independence of agents from online retailers. This model was prevalent in first generations of comparison-shopping agents because most of them were developed by technology pioneers, and they came to the market with support from venture capitals thus having little concern about revenue generation in the short term. Because agents that adopted this model did not generate revenues from online retailers listed, they had to use other methods. Most of them used banner ads like Pricescan.com.

The second business model emphasizes the business partnership with online retailers. In this model, the agents only listed subscribed retailers. And online retailers listed had to pay a referral fee to the agents for the traffic it generated. Because Pricewatch.com was the first agent using this model, we call it the Pricewatch model.

The BargainFinder model emphasizes the impartial assistant's role of agents to consumers. At the time, when BargainFinder was released to the public, consumers were mainly focusing on killer apps that could enable them to reduce search cost in both inter and intrasearch. Thus, the design goal of BargainFinder was as consumer-oriented as possible. This principle also guided latter agent's

designs (Ashish & Knoblock, 1997; Choi, 2001; Firat, Madnick, & Siegel, 2000). This led to features like the ability to proactively search online stores, the indiscrimination of data sources, and the listing of comparison information according to the preference of consumers.

These features improve the fairness, freshness, and trustworthiness of the comparison information being provided.

The proconsumer principle of the BargainFinder model prevents agents from having partnerships with specific online retailers. Thus, agents have to retrieve comparison information proactively from all retailers for the consumers. Because they did not discriminate against data sources, the comparison information they collected tended to be fair information about the product instead of information prepared by online retailer for targeted market. Since comparison-shopping agents adopting BargainFinder model tend to retrieve price information from the data source directly, the product information they retrieved was in real time. This improved the freshness of comparison information. Because the BargainFinder model excluded business relationships between the product information listed and its online retailers, the comparison information displayed has no preference over special online vendors. This even-handedness increases the trustworthiness of information provided and minimizes the risk of making bad decisions from biased information.

The BargainFinder model became dominant because it was then a killer app that could enable online shoppers to search price information with nearly zero searching cost. The appeal of the BargainFinder experiment to online shoppers and the short time success of its successors like Pricescan.com established the consumer group of early adopters of comparison-shopping (Rogers, 2003), which did attract attention of small and micro-online retailers. However, this stage was far from smooth, and doubts still existed and blockades from online retailers happened all the time, sometimes even with legal threats (Plitch, 2002).

The Pay-Per-Click (PPC) Model

Different from BargainFinder, Pricewatch.com needed cooperation from online stores in both product data format and real time access. Consequently, a different business model for comparison-shopping agents emerged.

Back in 1995, when assembling a computer was still a hobby shared only by a small circle of highly technical-savvy individuals, small online retailers of computer parts and accessories need a more focused channel to advertise their product. Since Pricewatch.com could bring together these potential buyers of computer products, online retailers found it more cost-effective to advertise in Pricewatch.com though they have to pay the referral fee. This made the collaboration between Pricewatch.com and online retailers possible. However, to make it happen, Pricewatch.com had to position itself as an advertiser for online stores and cannot claim its neutral position in providing product information.

Though it was a relatively solid business model, the Pricewatch model was not popular before 2000 compared with BargainFinder model. This situation only changed when the comparison information provided by those Shopbots adopting the Pricewatch model far exceeded those provided by the BargainFinder model in terms of number of online retailers each side covered and comprehensiveness of comparison information each side provided. Also, many of those agents adopting the BargainFinder model were either acquired or went bankrupt.

Exponentially expanded B2C electronic market and competition among agents made more sophisticated comparison-shopping service the competitive advantage, which required the cooperation from online retailers a necessity.

Meanwhile, the number of small and micro-online retailers that wanted to subscribe to comparison shopping reached a critical mass, which made the marketing pull model, like being adopted by Pricewatch.com, viable. Thus, the business model adopted by Pricewatch.com eventually evolved into the dominant *Pay-Per-Click* (PPC) model adopted by almost all major comparison-shopping agents nowadays.

The PPC model has become increasingly sophisticated within the last five years. Some agents like Shopping.com adopt a nondiscriminative policy to all its subscribers, which is the classic method used by Pricewatch.com; while other agents like PriceGrabber.com differentiated the subscribed online retailer by charging different levels of merchant program fees and placed the information from premium subscribers in prominent positions on the search results page.

FUTURE TRENDS AND RESEARCH

With the consolidation of the comparison-shopping market by these established players, the competition in the comparison-shopping market is becoming more and more intense. However, the advance of Web-centric technologies further decreases the entry barrier and it actually became easier for new startups to compete with these full-fledged shopbots or find new niche markets.

There are at least three new trends and future research directions for the evolution of comparison-shopping agents.

First, new comparison-shopping agents collecting and aggregating product information from existing comparison-shopping agents emerged and became popular. One prominent example is Kayak.com. Instead of collecting traveling information from airline, hotel, and car rental companies, Kayak retrieves the information from the Web, especially from existing comparison-shopping agents. This gives Kayak an edge compared with existing agents in terms of coverage of information sources because it can conduct the metasearch.

We expect that such derivative agents would continue to emerge from other market sectors when the information infrastructure of the Web becomes more structured and standardized. It would be interesting to know which factors determined the emergence and further developments of such derivative agents in different product and service categories.

Second, since the competition on the price comparison-centric model has matured, we expect comparison-shopping agents will explore new competitive advantage on other product information. The aforementioned acquisition of epinion. com and resellerratings.com by shopping.com is one example. Another example is the recently launched become.com by cofounders of mySimon. com. Become.com integrates price information with other product information. It searches online discussion groups, review sites, and various online forums to collect product information. Online shoppers could not only search product price-related information but also *research* a product based on materials collected by the agent. Such a vertical search agent could be extremely useful when its technology is further improved. We expect more intensive research in this direction to explore the decision-making behavior of online shoppers when using such agents, thus improving decision-making efficacies.

Third, with the mobile access of Internet become prevalent, comparison-shopping agents are also finding new opportunities in this new market. A recently launched mobile comparison-shopping agent service called frucall.com allows a shopper in brick and mortar stores to key in the bar code of a product he found, and then the agent tells the shopper if the same product with lower price will be found in other stores in the same local area, say, within five miles. It is reasonable to expect that these kind of B2C agent services will emerge very soon when people are more familiar with mobile access of the Internet. More innovative business models would be studied and emerge with such ubiquitous comparison-shopping agents.

CONCLUSION

This chapter gave an overview of comparison-shopping agents, which included definition, origination, an illustration of evolution, and future trends.

We first illustrate the chronological development of comparison-shopping agent systems from 1995 to the current status. Then, we explained the technology and business model evolution of comparison-shopping agents during the same time period.

During our analysis, we identified key issues that influenced the development direction of comparison-shopping agents, which includes the complexity of product information dimension, consumer expectations, attitudes of online retailer and service providers, and competition intensity in the comparison-shopping markets. These issues intertwined together and shaped the development of comparison-shopping agent systems and their activity environments.

Comparison-shopping agent system is an important category of Web-based business-to-consumer agent systems, and they had become the essential component of B2C e-commerce domain. They have and will continue to generate great momentum to transform the electronic market in the foreseeable future. We expect more research in this area in the near future.

REFERENCES

Ashish, N., & Knoblock, C. A. (1997). *Semi-automatic wrapper generation for Internet information sources.* Paper presented at the Second IFCIS Conference on Cooperative Information Systems, Kiawah Island, SC.

Bakos, Y. (1998). The emerging role of electronic marketplaces on the internet. *Communications of the ACM.*

Baumohl, B. (2000, December 11). Can you really trust those bots? *TIME Magazine, 156.*

Brynjolfsson, E., & Smith, M. (2000). Frictionless commerce? A comparison of Internet and conventional retailers. *Management Science, 46*(4).

Census.gov. (2004). Retail e-commerce sales in second quarter 2004 were $15.7 billion. Retrieved August 15, 2007, from http://www.census.gov/mrts/www/current.html

Choi, J. (2001). A customized comparison-shopping agent. *IEICE TRANS. COMMUN., E84-B*(6), 1694-1696.

Cournot, A. (1838). *Researches into the mathematical principles of the theory of wealth.* New York: Macmillan.

Crowston, K. (1997). *Price behavior in a market with Internet buyer's agents.* Paper presented at the International Conference on Information Systems, Atlanta, GA.

Doorenbos, R. B., Etzioni, O., & Weld, D. S. (1997). *A scalable comparison-shopping agent for the World Wide Web.* Paper presented at the International Conference on Autonomous Agents, Marina del Rey, CA.

Firat, A., Madnick, S., & Siegel, M. (2000, December). *The camn Web wrapper engine.* Paper presented at the Workshop on Information Technology and Systems, Brisbane, Queensland, Australia.

Green, H. (1998, May 4). A cybershopper's best friend. *BusinessWeek.*

Gupta, A. (1998, February 23-27. *Junglee: Integrating data of all shapes and sizes.* Paper presented at the Fourteenth International Conference on Data Engineering, Orlando, FL.

Hodkinson, C. S., & Kiel, G. C. (2003). Understanding WWW information search behaviour: An exploratory model. *Journal of End User Computing, 15*(4), 27-48.

Kelkoo.com. (2005). *Company information of Kelkoo.com.* Retrieved August 15, 2007, from http://www.kelkoo.co.uk/b/a/co_4293_128501_corporate_information_company.html

Krulwich, B. (1996). The BargainFinder agent: Comparison price shopping on the Internet. In J. Williams (Ed.), *Bots, and Other Internet Beasties* (pp. 257-263). Indianapolis: Macmillan Computer Publishing.

Kuttner, R. (1998, May). The Net: A market too perfect for profits. *BusinessWeek.*

Madnick, S., & Siegel, M. (2001). Seizing the opportunity: Exploring Web aggregation. *MISQ Executive.*

Plitch, P. (2002). Are bots legal? *Wall Street Journal, 240*(54), R13.

PRWeb.com. (1998, February 5). Anneneberg, Bloomberg, and now David Cost and Jeff Trester's PriceScan, Proves that Price Information is POWER! Retrieved August 15, 2007, from http://ww1.prweb.com/releases/1998/2/prweb3455.php

Rogers, E. M. (2003). *Diffusion of Innovations* (5th ed.). New York: Free Press.

Shapiro, C., & Varian, H. R. (1998). *Information rules: A strategic guide to the network economy.* Cambridge, MA: Harvard Business School Press.

Shugan, S. M. (1980). The cost of thinking. *Journal of Consumer Research, 7,* 99-111.

Sinha, I. (2000, March-April). Cost transparency: The net's real threat to prices and brands. *Harvard Business Review,* pp. 43-50.

White, E. (2000, October 23). No comparison. *Wall Street Journal.*

Chapter III
A Study of Intelligent Shopping Support:
A Case Study of Outbound Group Package-Tour Products in Taiwan

Wen-Shan Lin
National Chiayi University, Taiwan

ABSTRACT

The Internet and World Wide Web are becoming more and more dynamic in terms of their contents and usage. Agent-based shopping support (ASS) aims at keeping up with this dynamic environment by mimicking shoppers' purchasing behavior in the electronic commerce transaction process in the sense of matching the profiles of Web sites and shoppers. Evolutionary agent-based shopping supports are emerging as intelligent shopping support. This chapter contains the earliest attempt to gather and investigate the nature of current research. The idea of applying concepts of product characteristics from the matrix of Internet marketing strategies is introduced for solving problems of natural language information search. The process of focus-group research methodology is applied in acquiring the essential knowledge for examining shopper's knowledge of search. An architecture of ASS in the case of outbound group package tour in Taiwan is presented. This work demonstrates the process of knowledge acquisition to tackle the problem of ineffective online information search by a customer-centric method.

INTRODUCTION

Commercial Web sites are regarded as the virtual channel to communicate with shoppers by transmitting marketing messages online. Evolutionary **agents** are emerging as bringing buying and selling together in the sense of mediating in the process of online business transactions. When **shoppers search** online product information, they place judgments on searching for the best place, that is, the Web site in this case and the best offer to shop. From shoppers' points of view, an intelligent shopping support should complete search tasks without human interventions and know well both the domain of electronic commerce and what shoppers want. By taking this notion into consideration, a comprehensive study of agent-based shopping supports (ASS) is required with the aim to improve the shopping supports in the subject of electronic business. And this kind of study is lacking in current literature. The research objectives are twofold: The investigating of relevant literature of intelligent shopping support is conducted on one hand; the development of a knowledge-based ASS is carried out on the other hand. A business case of group package tour in Taiwan is selected to explore and demonstrate the feasibility for improving the search effectiveness of ASS based on knowledge engineering method regarding transforming human knowledge of search into the ASS's **knowledge base**. This chapter presents an ongoing research project; therefore, it is included the current progresses for acquiring essential knowledge by conducting a **focus-group interview** and an architecture of ASS.

This chapter is arranged as follows. In the second section, the background information of this chapter regarding the problems of ASS is investigated and examined. The third section proposes a shopper-centric solution and details the process of knowledge acquisition by an experiment of focus group. The fourth section presents the architecture of the ASS and its ontological knowledge base with some snapshots proposed in this chapter. Finally, results, discussions, and future trends are stated.

LITERATURE REVIEW

ASS in Supporting Online Shoppers

It is proposed that agents offer ways to buy and sell in electronic marketplace (Amin & Ballard, 2000; Liu & You, 2003; Vahidov, 2005; Yuan, 2002). It is also addressed that by knowing what human wants in terms of their profiles and buyer behavior are essential for keeping the efficiency of ASS. This chapter takes this notion as central for further explorations in the subject of agent-based shopping support. In other words, in view of the fact that the shopper-centric solution is proposed as a promising direction for agent developments it is required a comprehensive examination on current literature in terms of its unsolved challenges toward this direction. In this section, related works about ASS are analyzed regarding advantages and disadvantages of the efficiency for assisting shoppers.

There are a number of successful or unsuccessful cases of ASS in the Internet shopping domain (Hostler, Yoon, & Guimaraes, 2005; Liu & You, 2003), including *AuctionBot* in a consumer-to-consumer (C2C) e-auction system, *BarginFinder* and *Jango* in a business-to-consumer (B2C) e-shopping system, and MAGNET in a business-to-consumer (B2B) e-supply chain system. These ASSs are developed for assisting shoppers in the decision-making process which is modeled as the six stages: (1) **consumer requirements definition**, (2) **product brokering**, (3) **merchant brokering**, (4) negotiation, (5) purchase and delivery, and (6) after-sale services and evaluation. ASSs are commonly supported the first three stages as human users are in need to handle the problems of information overload. **Comparative shopping** and negotiation capabilities have been proposed lately. It is suggested that the context-aware multi-agent intelligent architecture should be adopted for implementing autonomous negotiation between buyers and sellers. In addition, **case-based reasoning** and negotiation mechanisms are utilized in the process of matching the best deal for shoppers. However, it is revealed difficult to gain the knowledge of human search from the literature as human mind have complex set of factors running for making a decision. In view of the mechanism of ASS, it is addressed that shoppers' purchasing behavior is hard to quantify as a number of works in this case are making efforts to model an optimization solution but end in unsatisfactory (Kwon & Sadeh, 2004). Only in the project of *Alice* (Domingue, Stutt, Martins, Tan, Petursson, & Motta, 2003) was the early attempt proposed to assist shoppers by applying the methodology of knowledge engineering. *Alice* is an ASS that has a modeled knowledge base in terms of products, shopping tasks, external context, and customer profiles. The idea of the Alice project has similarities to our chapter and can be complementary. However, it is stated that it is unfeasible to find definitive knowledge source for classifying customers' according to their shopping behavior. It is agreed that there are complexities with respect to customers' behavior, but it will be beneficial to bring up a justified and experimentally proven model about shoppers' knowledge of buying. User preference modeling is investigated (Dastani, Jacobs, Jonker, & Treur, 2005), that is, the preferences of a user towards a set of items, and it can be defined in terms of information concerning either the contents or the use of the items by a society of users, and it is stated that the construction of a preference model is usually time consuming. There are two main methods to model user preferences: (1) the manual method, users can be asked by a set of questions in a consecutive manner to construct or update users' most frequent and predictable preferences. (2) The other method is referred to as the machine-learning method that composes of "like" and "dislike" elements for users, for example, collaborative-based preference modeling or content-based preference modeling methods. However, the machine-learning method is based on a justified model that provides information about shopper's preferences. In view of this point, this chapter takes the move to model shopper's preferences in a justified way. It will be done by combining studies of shoppers' behavior and presenting the acquired knowledge by the method of knowledge engineering. By this method, this study is designed to improve the effectiveness of ASS by following the customer-centric way.

Applications of ASS and the Mix of Internet Marketing Strategies

A systematic analysis about the relationships among antecedents of the determinants of online shopping based on a number of **empirical studies** is presented (Chang, Cheung, & Lai, 2005). A total of 355 articles were investigated and a number of theories were given that were continually used by researchers to investigate online customer behavior. These theories include: (1) **expectation-confirmation theory** (ECT), (2) **innovation diffusion theory** (IDT), (3) **technology acceptance model** (TAM), (4) **theory of planned behavior** (TPB), and (5) **theory of reasoned action** (TRA). It is suggested that researchers should explore new theories or frameworks and investigate online **consumer behavior** from other perspectives.

For the category of Web site features that shoppers interact with while searching online information, the information contents and Web site design have a significant relationship with the intention or usage of online shopping. However, a further study that carefully examines shoppers' types of decision-making knowledge is absent in current literature. Furthermore, there are inconsistent findings regarding the link between customer behavior and product categories in the sense that knowledge ASS should follow search information on behalf of shoppers. Some researchers state that that the type of product does not significantly influence consumers' online shopping behavior (Senecal, Kalczynski, & Nantel, 2005), while others disagree (Chang et al., 2005). This chapter intends to add more remarks on this issue. According to studies (Baker, 1999; Lin, 2004), there are generally four groups of decision styles in physical shops that are classified regarding different types of products. They are complex buying, dissonant buying, habitual buying, and variety-seeking products (Figure

1). Lin's empirical study adapts Baker's theory and experimentally examines online shopping behavior. It reveals that shopper's decision-making style remains the same for general buy and book-book buy; however, shoppers perceive certain levels of significant differences toward a number of Internet marketing terms between buying a specific product, for example, books in this case and general buy. Therefore, more detailed surveys should be carried out in terms of these four types of products before a generic model of customer's decision-making model can be finalized.

On the other hand, there are significant differences observed in human users' decision-making processes in consulting a shopping support, for example, ASS in our case (Brynjolfsson & Smith, 2000; Senecal et al., 2005). In the earlier time of adopting ASS for decision making in 1999, the branding issue was influential in human users' decision making; that is, shoppers prefer to buy products made by particular brands. After years of evolving, human users get more experiences in adopting ASS. And human users come to have more complex shopping behavior, for example, more Web pages viewed or the number of purchasing decisions per user makes increases. Also, time saving and decision quality improvement are the two main benefits shoppers are aware of after adopting ASS (Hostler et al., 2005). It is conducted an analysis of economical implications of ASS, the preference-based pricing agents in particular that infer buyer valuations for customized products (Aron, Sundararajan, & Viswanathan, 2006). They address that shoppers invariably choose less-than-ideal products for better prices. In view of this point, it is advantageous for studying what online shoppers look for, and shoppers will be more willing to pay more to get an ideal and customized product. Therefore, the online business can gain its popularity in a faster pace.

Figure 1. Decision styles of market initiators

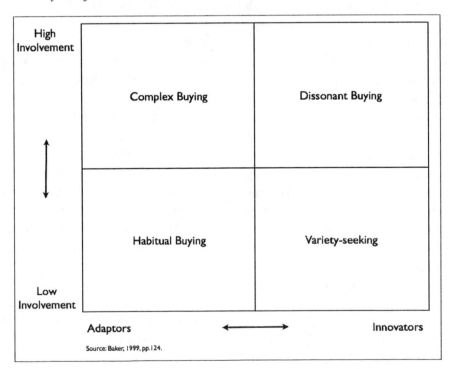

Problems of ASS and the Proposed Solution

The contents of the Web are large and grow fast. In normal occasions, human users will perform an *internal search*, for example, relying on their prior knowledge brands, and *external search*, following search results recommended by ASS. The **artificial-intelligence technology** is applied by equipping with information retrieval technology to assist human users to search for relevant information for different purposes, for example, keyword search, product search, and buyer search. It is category-related information retrieval technologies divided into two groups (Kushchu, 2005): (1) **genetic algorithms** or programming as an optimizer in the contexts of clustering, query improvement, or keyword selection; and (2) **adaptive intelligent** are employed and evolutionary methods are used as learning mechanisms. Nevertheless, problems of Web-based information retrieval stay unsolved due to the nature of Web documents. The problems are the information base, and contents of the Web comprise a number of properties; they are large size, unlabeled, dynamic, duplicated, interconnected, semistructured, distributed, heterogeneous, and language dependent; precise delivery of contents for human users becomes elusive. One major challenge ASS should try to meet is to adapt to the changing context interfaces and user requirements in the sense that ASS needs to handle incomplete, imprecise data. More often, ASS relies on contextual information in the form of human natural language as displaced in Web sites to complete job searches. In view of this, this chapter proposes that in order to make ASS more intelligent, the human-nature language, as it is used to present, Web pages should be considered and included as the knowledge base in the architecture of ASS.

The knowledge base (KB) of an ASS acts as the human brain to infer tasks that are delegated in its architecture. It is stated that generic or specific types of knowledge is required for instantiating functionalities of ASS on behalf of users, as well as the user preference model (Dastani et al., 2005). This chapter proposes the solution of knowledge engineering to model shopper's knowledge of search in the form of **ontology**, and the ASS search effectiveness should be improved by mimicking human knowledge. Regarding knowledge engineering, there are three methods of knowledge acquisition that are addressed in developing a decision support system. These methods are: (1) knowledge engineer-driven method, (2) expert-driven methods, and (3) machine-driven method (Marakas, 1999; Turban & Aronson, 2001). The first two methods are applied to identify problem characteristics, conceptualize concepts, and represent and formalize the knowledge. And these should be done by investigations, communications, and organizing all the knowledge about the problem domain. The third method, the machine-learning oriented way, can only be feasible while the formalized knowledge base had been constructed. The methodology of modeling knowledge in the form of ontology is stated as an experimental-proven method (Ding, Fensel, Klein, & Omelayenko, 2002; Sugumaran & Storey, 2002; Corby, Dieng-Kuntz, Faron-Zucker, & Gandon, 2006; Formica, 2006;). The introduction of ontology and its applications in building ASS will not be included in this chapter, as it already has gained lots of attention. Readers are encouraged to read additional references in the Additional Reading section. By applying this notion to our research, the knowledge base of the proposed ASS includes essential information to act on behalf of users. A case study of outbound group-package tour products in Taiwan is taken to investigate shoppers' buying needs while searching online information. It is aimed at demonstrating the feasibility to transform human knowledge of search into a knowledge base for assisting an ASS for online information search.

INTERNET MARKETING STRATEGIES AND THEIR LINK WITH ONLINE SHOPS DESIGN AND SHOPPERS' BUYING BEHAVIOR

This section addresses the knowledge acquiring process regarding shopper's knowledge of search. An empirical study of a focus-group is conducted to verify the research model of the knowledge base of an ASS. The architecture of the ASS is also presented. It denotes the relationship of the ASS that can perform search tasks by manipulating the knowledge base. Although the real ASS is under the experimental and constructive stage, further developing processes and the examination of its real effects with shoppers will be conducted. The current research progress is presented and has been well verified.

It is denoted that the Internet represents a sufficiently different retail environment compared to physical shops that concepts of consumer information search behavior should be revisited (Senecal et al., 2005). Internet marketing literature is a means to promote and sell products online. Three forces have frequently been addressed as the unique characteristics of the Internet: individualization, information, and interactivity. Internet marketing strategies also are recognized as a study to investigate shoppers' advocacy, reactions to stimulated transactions, and emotional experiences. These three forces have also been referred to as crucial in building the understandings of customer experience in the design of appropriate marketing programs for securing customer relationships (Kwan, Fong, & Wong, 2004). Furthermore, online shoppers' requests on Web sites are categorized as: (1) awareness, (2) exploration, and (3) commitment. In effect, typical requests of a shopper are identified as log-in, home page, browse, search, register, select, add to shopping cart, and pay. Further readings about general consumer behavior can be found in Solomon (2007).

Identifying the important semantic terms for assisting shoppers in the process of shopping is another key issue (Yuan, 2002). Shoppers' purchasing behavior is associated with the contextual or semantic terms that are listed in online stores. On one hand, these semantic terms reflect online stores' marketing strategies while these terms promote selling products online and are designed by online shops; on the other hand, shoppers' behavior is determined by assessing this contextual information. Therefore, ASS should take these semantic terms into account to match online shops' profiles with shoppers'. And these semantic knowledge should be included as the knowledge based for ASS to implement with searching tasks.

An empirical study of **online bookstores** was simultaneously conducted and pointed out that potential shoppers are more likely to shop at well-designed Web sites (Liang & Lai, 2002). It is addressed that a number of factors affect shoppers' choice of stores in three categories: (1) **hygiene**, for example, having good security, protecting customers from risks, and so forth; (2) motivation, for example, having a good search engine, providing direct shopper support; and (3) media richness, for example, providing chat rooms. It is also denoted that shoppers' visit or shopping decision has a strong link with online shops' designs. In other words, online shops' design plays the same role as the layout of a conventional store in the sense that it affects shopper's decision making and purchasing intentions.

Review of the Framework ASS Applications and Its Link with Internet Marketing Strategies

Based on Simon's **human problem-solving model**, the intelligence phase of the agent intermediary in human decision-support systems, such as the ASS in our case, is defined as three phases. The first is to *search for more information* about a problem and deliver relevant information from various sources to the user or other agents (Vahidov, 2005). The other two phases are *design* and *choice*. They are concerned with generating alternative decisions and converging the final decision respectively (Kwon et al., 2005). All of these three phases are regarded as the way to accommodate the needs of decision makers. In organizing the knowledge base of the ASS, the Simon's model is adopted mainly in recognizing users' needs and providing solutions with alternatives for users to decide the best buy for their own definitions.

User models are based on how human behaviors are represented with the purposes (Frias-Martinex, Magoulas, Chen, & Macredie, 2005). With respect to personalized information delivery for human users, the user model is considered to represent: (1) the assumptions about the knowledge, goals, plans, and preferences about one or more types of user; (2) relevant common characteristics of users pertaining to specific user subgroups or stereotypes; and (3) the classification of a user. In our case, the shoppers' decision-making models are based on marketing segmentation strategies as addressed and popularly accepted by the marketers.

Consumers' decision-making processes are investigated by a number of researchers (Frias-Martinex et al., 2005; Senecal et al.; 2005) and have been referred to as the central component of ASS framework. It is stated that consumers' decision behaviors in most cases are stimulated by inner characteristics and outer stimulus. *The inner characteristics* are referred to as the shopper's personal style in terms of what kind of information he looks for; while *the outer stimulus* are about the Web site information, in our case, that can motivate a shopper for further action to buy. As the best buy decision is made by selecting the best match between profiles of shoppers' and online shops, an intelligent ASS should take account of the knowledge of shoppers and know how to identify the profiles of online shops in terms of the

revealed Internet marketing strategies. From the Internet marketing perspective, sellers apply the market-segmentation strategies to target certain types of shoppers, and the Web sites are presented by reflecting the outer stimulus that can attract these target customers. Therefore, a review of the ASS framework and its link between Internet marketing strategies should be bridged.

Assisting Shoppers Through Ontology

Web pages are presented in two kinds: syntactic and semantic structures. The syntactic structure is hierarchical, as reflected in the use of HTML/XML, while the semantic structure generates a more complex structure; that is, it is not necessarily expressible as a simple hierarchy. Pontelli and Son (2003) propose the **semantic navigation** as a mechanism of ASS to support universal accessibility of e-commerce services. Their proposition is consistent to our work in the sense that by constructing a knowledge base that consists of the semantic terms and shoppers' navigational rules, the ASS can be really intelligent in its functionality and effectiveness. With respect to the dynamic information of Web sites, the reasoning mechanism of ASS while searching on behalf of shoppers plays an important role in bringing relevant search results. It is stated that agents carry out search tasks online by manipulating functions of the concept and weight. *Concepts* represent all the factors an agent should take into account, while *weight* denotes the casual relationship between the concepts (Miao, Goh, Miao, & Yang, 2002). In our case, concepts correspond to the notion of Internet marketing attributes that Web semantic terms represent, while shoppers' different decision styles can be translated to different sets of weights that denote Internet marketing terms. In other words, different styles of shoppers are influenced by certain set of Internet marketing terms before make buying decision. Therefore, shoppers' knowledge can be translated into concepts and weights included in the knowledge base of the ASS for conducting search tasks.

An Architecture of ASS

A matrix of Internet marketing strategies is proposed and experimentally examined in cases of general online buying and online book buying (Lin, 2004). The matrix is illustrated in Table 1. This matrix links shopper's buying behavior and can be used to identify influential factors of information that shoppers perceive important while searching information with the intention to buy. There are six categories in this matrix: (1) Web store, (2) computer-mediated electronic commerce (CMEC), (3) product, (4) promotion, (5) price, and (6) process. These six categories are transformed from conventional marketing strategies that are applicable in physical shops.

As addressed earlier, it is important to take account of Internet marketing strategies while building up the knowledge base of ASS. It has expanded the whole idea in a clearer manner such as the way online shops are designed revealing the Internet marketing matrix of an online shop. That matrix aims at promoting products and selling products. Shoppers perceive the *concept*

of Internet marketing terms in different levels of *weights* as people have their own buying styles. In general cases, shoppers have mixed styles. Nevertheless, Lin has experimentally validated the proposal of applying the market segmentation strategies such as how sellers segment different styles of shoppers on online shopping (Lin, 2004). Her results are promising, but it is still risky to only take those validated styles of shopper into consideration while the business environment changes quickly. It is proposed that leading shoppers to provide more detail of their intention or criteria to buy are beneficial. For this stage of building up a knowledge base of ASS, taking all possible factors that affect shoppers into account is the first and feasible approach. As long as the knowledge base is complete enough, the reasoning mechanism can be improved by combining other methods, such as machine learning. This is beyond the scope of this chapter. Acquiring the essential knowledge is the primary aim of this chapter.

The proposed architecture of the ASS is presented (Figure 2). The case of purchasing online group-package tour (GPT) products in the Taiwanese outbound travel market is selected as the business domain to explore. Two reasons explain

Table 1. The Internet marketing mix

Strategies / Framework	Place	People	Product	Promotion	Price	Process
Conventional marketing strategy	Physical shop	Shopkeepers	Product	Promotion	Price	Process
E-marketing strategy	Online shops	Computer-mediated electronic commerce	Product	Promotion	Price	Process

why this business domain is selected: one is due to the fact that the online travel market's popularity gains lots of attention, and the outbound group-package tour product is dominant in Taiwanese travel industry; the other reason is that travel products are referred to the type of variety-seeking product in Baker's theory (Figure 1) among the other three. Therefore, expanding this case study for building up the knowledge base of the ASS can demonstrate the feasibility of applying the proposed method in the real world.

There are six components of the ASS:

1. **Group package tour ontology (GPT ontology):** This GPT ontology consists of three elements including classes, properties, and the relationship between the two. It aims at providing the abstract concept for further inferring and filtering search tasks the ASS should complete on behalf of shoppers.
2. **Concept mapping module:** This module consists of the semantic terms that are synonyms or relevant terms with respect to the GPT ontology, and their relationships in semantic meanings are stated.
3. **Inference engine:** This engine mainly conducts the inferring tasks, including receiving the search conditions or search terms typed

by shoppers and the reasoning behind the recommended work. It is also referred to as the concept of *weights* of shoppers' decision-making processes. In this chapter, this part of knowledge is not presented but is definitely been taken for ongoing research objectives.

4. **Group package tour database (GPT database):** This GPT database contains information of online travel Web sites in Taiwan. In order to proceed with experiments in the lab for testing the effectiveness of the built ASS in this research, online data are retrieved but stored off-line for further inferring and reasoning works, as there is limited time and money. This database is maintained and updated followed by a controlled schedule by the research team.
5. **Integrated information module:** This module is used to combine results from the inference engine and GPT database before shown to shoppers.
6. **Interface design of the ASS:** The interface between shoppers and the ASS is designed in a pleasant and straightforward manner. It is shown in a Web-based format.

Figure 2. System architecture for developing the ASS for outbound group-package tour products

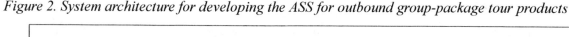
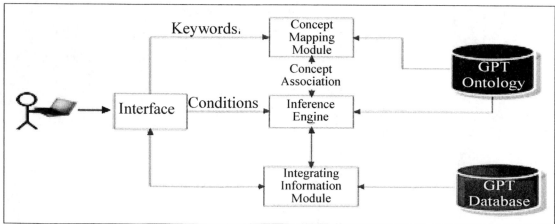

Knowledge Acquirement Process and the Experiment

In order to acquire the essential knowledge of the selected business, a knowledge acquiring process with domain experts was conducted. This **knowledge acquiring process** follows the focus-group research methodology 18 subjects. It is noted that only the product characteristics from the Internet marketing matrix are taken for validating the shoppers' knowledge of buying. This is due to the reason that investigating the business domain is the fundamental step for modeling knowledge. More concepts of knowledge can be included in the future in an incremental manner.

The experiment conducted to acquire the knowledge of ASS is composed of two parts: the first is to assess the preference level of eight categories regarding characteristics of outbound travel-package products on a five-point **Likert scale**. These eight categories are proposed by involving three experts in the alpha test. By interviewing experts, investigations of relevant literature and Web sites, these eight categories are modeled and proposed to have a link with online shoppers' buying behavior. These eight categories are: (1) destination information, (2) activity by theme information, (3) accommodation information, (4) transportation information, (5) price information, (6) shopping information, (7) information of optional tour, and (8) other information. The second part of the experiment is to analyze the consecutive level of priority of shoppers when making buying decisions. It is noted that certain levels of precedence occur in people's decision-making process, and these schemes can assist us to build up the inference engine of the ASS.

In the experiment, 18 students (8 males and 10 females) from the postgraduate school of travel and tourism management at National Chia-Yi University were recruited and each received NT$100 (equivalent to US$3.5) salary for participation. All subjects have a professional level of knowledge about the online travel market. They had done extensive research of online travel companies,

Table 2. Ontological structure of searching outbound travel-package products

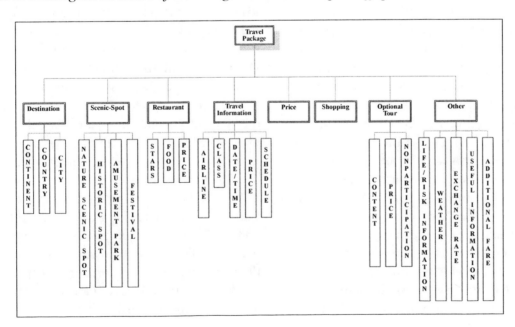

particularly the biggest Taiwanese online travel company, Lion Travel Co. (www.liontravel.com). They spent five months investigating and studying the case on the basis of the school-and-industry cooperative link. During the focus-group interview, all subjects filled in a questionnaire and shared their ideas about purchasing outbound travel-package products online. The experiment of the focus-group interview took 30 minutes to complete.

Based on the experimental results, the ontology structure of the knowledge base is built up (Table 2). This ontological structure of searching outbound travel-package products is composed of two levels. The first level consists of eight categories. All these categories represent the product characteristics of outbound travel-package products that have influences on shoppers' decision making. The second level refers to the detailed semantic attributes that expanded from each product characteristic accordingly. In total, 23 attributes are revealed and acquired.

Knowledge Editing and the Tool Used

There are a number of knowledge editing tools available worldwide but Protégé is selected in this study for editing. Protégé was initially developed by Stanford medical informatics at Stanford University School of Medicine. It is popularly utilized by knowledge engineers, researchers, and is supported by a strong community for updating this tool and inventing new plug-ins for supporting different purposes. In our case, this tool is selected in light of its extensive ability for implementing online ASS. In this stage of research, experiments are conducted off-line based on a GPT database and the knowledge base built by Protégé that can assist managing and accumulating semantic terms in ontological form. Moreover, there are a number of plug-ins built by other researchers that are associated with this tool. By installing these plug-ins, the function of search as an ASS should

be easily implemented. Protégé also provides two methods to build up ontology: (1) Protégé-Frame editor: it is mainly for building up a set of ontology in the frame-based method, and it is suitable to be applied in the architecture of open knowledge base connectivity protocol. (2) Protégé-Owl editor: it is mainly for building up a set of ontology and applying it in supporting Semantic Web. It appears that the second method is applicable to this chapter. By manipulating the knowledge base built up by Protégé with the inference mechanism provided by the ASS, it is feasible for demonstrating effective product-recommending service for shoppers.

RESULTS AND DISCUSSIONS

Based on the experimental results of a focus-group interview, the structure of a knowledge base for assisting shoppers in searching outbound travel-package products is built up. Results (Appendix 1) show that all participants agree with the importance level of all concepts as proposed in affecting intention to buy a travel product. It is noted that the concept of "optional tours" is composed of a subclass of three attributes. But there is no question for asking its importance level of the concept of "optional tours". The reason is that this information is optional in nature, and it has not happened in every case of outbound travel-package product. Results show the importance of these three attributes regarding the concept of "optional tour, and this claim is widely agreed upon. Therefore, the eight categories are proven important in this interview for affecting shoppers' decision to buy. However, results of the subclass of 23 attributes regarding these 8 categories reveal in a different way. Few negative results are collected, but the major groups of participants view all attributes in a positive way regarding their importance. The aim of this experiment was to gather the full set of knowledge that may affect a shopper's decision in selecting a group travel package; all concepts and attributes are included for the next stage of ontology editing because the knowledge base of the ASS should be complete in order to fulfill further inferring tasks in the sense of recommending products for shoppers' different needs. Furthermore, the interface design of the ASS will provide shoppers with selecting or typing optional search terms and criteria. Shoppers are free to select search criteria or terms out of their own needs. Therefore, it will be unreasonable to exclude any attribute that may affect their decisions.

For the first part of the questionnaire, the eight categories with their referred attribute, if applicable, are listed and explained:

1. **Destination information:** It includes the information of continent, country, or city where shoppers plan to go.
2. **Information of activities by theme:** This category refers to the activities or themes the shopper may want to choose. It includes selections of a nature scene, historical spot, amusement park, and festival.
3. **Accommodation information:** This concept refers to the needed information for shoppers to select a place to stay during the journey. It includes the rating information of this accommodation or hotel, price for stay, and the food provided in the place. In a normal case, the food information provided will be the serving breakfast. Some accommodations do no provide this service, but it is important for travelers to decide to stay there or not.
4. **Transportation information:** The concepts include information of the airline company that operates the journey for sending travelers from their home to the destination, class of seat on the plane travelers, original ticket price (although most package tours includes everything, still some travelers think it is important to know), and the date and time of departing and returning flight.

5. Shopping information is about the locations, the business hour, or even what is worth buying that travelers will be interested in such as shopping near the place where they stay and going on a journey.
6. Price information is about the total price every traveler should pay for the package buy.
7. **Information of optional tours:** Most travel packages offer additional tours as options for travelers to select for extra charges, for example, horse riding, spa service, water diving, or those activities for interested people but may not be demanded by everyone.
8. **Other information about the travel package:** This can be the information of life or risk, weather, the exchange rate for local currency, additional fares as tip, or other information such as visa information or travel suggestions issued by governmental tourism departments.

In the second part of questionnaire, participants rank the priority for each category, and the statistical results are shown in Table 3. In this table, the priority is ranked in an accumulative manner. Each participant ranked the priority in the order up to eight as there are eight categories for selection, and the ranking of importance for each category is counted by the accumulated votes. It is revealed that the information of destination is ranked as the most important. Therefore, the interface design of the ASS provides a selection of destination information for shoppers to input as well the time they intend to travel and the budget plan before conducting further product searches. The snapshot of the ASS interface is shown in Figure 3.

GPT Ontology

In this section, the GPT ontology is demonstrated. By following the steps addressed as essential by

Table 3. Results of the priority order of each category

Class	1	2	3	4	5	6	7	8	Order
Destination	11	3	1	3					1
Activity by theme	3	4	6	3		1	1		1, 2, 3, 4
Accommodation		1	3	6	6	1	1		3, 4, 5
Transportation		1	2	3	6	5	1		4, 5, 6
Price	4	7	4	1	3				1, 2, 3
Shopping				1	3	6	5	3	5, 6, 7, 8
Optional tour		2	2	1		5	8		6, 7
Other					1		2	15	8

Figure 3. The snapshot of the ASS interface design

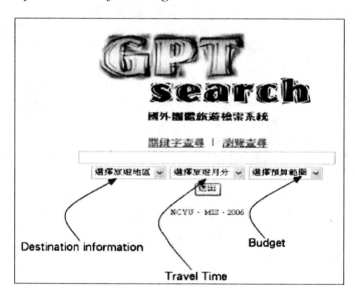

Figure 4. The first layer of the main eight classes

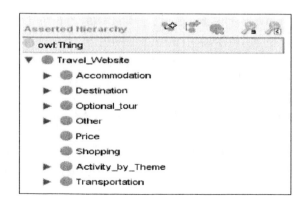

the Protégé development team, this GPT ontology can be defined by:

1. The scope of the domain refers outbound group package tour. The set of ontology are those important semantic terms of the product characteristics that have influence in shoppers/travelers decision to buy online.
2. The existence of substitute ontology. There is no existence of any substitute ontology

in Chinese regarding this domain, and it is reusable. Therefore, it strengthens the research aim of this work. It is the first attempt to investigate and contribute in this phenomenon.

3. The concepts of the domain are proposed and validated by the method of focus-group interview with experts. All possible attributes are included for presenting in the form of ontology in the sense that they affect shoppers' decision to buy.

Figure 5. The second layer of 23 attributes

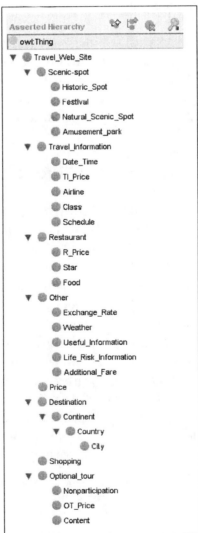

4. Define the concepts and their relationship with their subclass terms in a hierarchical order. In the current stage of research, there is only one layer of class denoted to each attribute in the GPT ontology. However, it can be reused or broadened for the future in an incremental manner. The snapshots of the first and second class of the GPT ontology are shown in Figure 4 and Figure 5 accordingly.

CONCLUSION

This chapter investigates the pros and cons of relevant literature regarding intelligent shopping support. A customer-centric solution is proposed to improve the search effectiveness of ASS by developing a knowledge base. This knowledge base consists of the representation of the product characteristics in the form of ontology as those concepts that affect shoppers' decision-making behavior. Equipped with the knowledge base, the ASS should take into account the factors shoppers consider while shopping online. By taking the case of the outbound group-package tour in Taiwan, the business domain is investigated and well defined by the experiment. A focus-group interview is applied to validate the ontological framework of the knowledge base. An architecture of the ASS is proposed, and a GPT ontology is developed. In the future, the ASS for the GPT products will be developed and tested in an off-line fashion experimentally. Further experiments with the ASS users to demonstrate its effectiveness will be conducted. Furthermore, examining the types of products according to Baker's theory will be beneficial to finalize a universal knowledge base of shopping supports. Also, the machine learning methodology is complementary with this research for providing more advanced and powerful shopping assistants.

By reading this chapter, researchers and practitioners can gain insights of ASS in electronic business in the direction of bringing technology and humanity together, taking the ground of bridging the gap between Internet marketing strategies and agent's developments. The ASS development requires more researchers to join in as it is a cross-disciplinary domain. More contributions made to improve the effectiveness of ASS in assisting shoppers in the domain of electronic commerce can be beneficial and nurturing to business as well as supporting shoppers.

ACKNOWLEDGMENT

The author acknowledges the valuable comments that were received from anonymous reviewers. This work was greatly supported by the National Science Council, Taiwan (grant code: 95-2416-H-415-008-) and the research assistant: Mr. Ming-Fong Chen and Miss Yen-Yen Chen.

FUTURE RESEARCH DIRECTIONS

A real ASS applied by the knowledge base is under the experimental and constructive stage. Further developing process and the examination of its real effectiveness with shoppers will be conducted. For the future, the evolution of the knowledge base should consider to involve experts in Internet marketing and consumer behavior. These fields can contribute more ideas about consumer decision-making behavior. As well, it can bridge the gap between consumer expectations and realities. The machine learning and text mining can be beneficial to expand the search effectiveness of ASS while interacting with shoppers for online purchasing.

REFERENCES

Amin, M., & Ballard, D. (2000, July-August). Defining new markets for intelligent agents. *IT Pro*, pp. 29-35.

Aron, R., Sundararajan, A., & Viswanathan, S. (2006). Intelligent agents in electronic markets for information goods: Customization, preference revelation and pricing. *Decision Support Systems, 41*, 764-786.

Baker, M.J. (1999). *The marketing book*. Woburm, MA: Butterworth-Heinemann.

Brynjolfsson, E., & Smith, M. (2000, July). *The great equalizer? Consumer choice behavior at Internet shopbots* (MIT Working Paper). Retrieved August 16, 2007, from http://ebusiness.mit.edu/erik/TGE%202000_08_12.html

Chang, M.K., Cheung, W., & Lai, V. (2005). Literature derived reference models for the adoption of online shopping. *Information & Management, 42*, 543-559.

Cheung, C., Chan, G., & Limayem, M. (2005). A critical review of online consumer behavior: Empirical research. *Journal of Electronic Commerce in Organizations, 3*(4), 1-19.

Corby, O., Dieng-Kuntz, R., Faron-Zucker, C., & Gandon, F. (2006). Searching the Semantic Web: Approximate query processing based on ontologies. *IEEE Intelligent Systems, 21*(1), 20-27.

Dastani, M., Jacobs, N., Jonker, C.M., & Treur, J. (2005). Modelling user preferences and mediating agents in electronic commerce. *Knowledge-Based Systems, 18*, 335-352.

Ding, Y., Fensel, D., Klein, M., & Omelayenko, B. (2002). The Semantic Web: Yet another hip? *Data & Knowledge Engineering, 42*(2-3), 205-227.

Domingue, J., Stutt, A., Martins, M., Tan, J., Petursson, H., & Motta, E. (2003). Supporting online shopping through a combination of ontologies and interface metaphors. *International Journal of Human-Computer Studies, 59*, 699-723.

Formica, A. (2006). Ontology-based concept similarity in formal concept analysis. *Information Sciences, 176*(18), 2624-2641.

Frias-Martinez, E., Magoulas, G., Chen, S., & Macredie, R. (2005). Modeling human behavior in user-adaptive systems: recent advances using soft computing techniques. *Expert Systems with Applications, 29*, 320-329.

Hawking, E. (2006, June). Web search engines: Part 1. Computer, pp. 86-88. IEEE Computer Society.

Hostler, R.E., Yoon, V.Y., & Guimaraes, T. (2005). Assessing the impact of Internet agent on end users' performance. *Decision Support Systems, 41*, 313-323.

Kushchu, I. (2005). Web-based evolutionary and adaptive information retrieval. *IEEE Transactions on Evolutionary Computation, 9*(2), 117-125.

Kwan, I.S., Fong, J., & Wong, H.K. (2005). An e-customer behavior model with online analytical mining for internet marketing planning. *Decision Support Systems, 41*, 189-204.

Kwon, O. (2006). The potential roles of context-aware computing technology in optimization-based intelligent decision-making. *Expert Systems with Applications, 31*, 629-642.

Kwon, O.B., & Sadeh, N. (2004). Applying case-based reasoning and multi-agent intelligent system to context-aware comparative shopping. *Decision Support Systems, 37*, 199-213.

Liang, T.P., & Lai, H.J. (2002). Effect of store design on consumer purchases: An empirical study of on-line bookstores. *Information & Management, 39*, 431-444.

Lin, W.S. (2004). *Framework for intelligent shopping support*. Unpublished doctoral thesis, University of Manchester, UK.

Liu, J., & You, J. (2003, April). Smart shopper: An agent-based Web-mining approach to Internet shopping. *IEEE Transactions on Fuzzy Systems, 11*(2), 226-237.

Marakas, G.M. (1999). *Decision support systems in the twenty-first century.* Upper Saddle River, NJ: Prentice Hall.

Miao, C.Y., Goh, A., Miao, Y., & Yang, Z.H. (2002). Agent that models, reasons and makes decisions. *Knowledge-Based Systems, 15*, 203-211.

Pontelli, E., & Son, T.C. (2003). Designing intelligent agents to support universal accessibility of e-commerce services. *Electronic Commerce Research and Applications, 2*, 147-161.

Senecal, S., Kalczynski, P.J., & Nantel, J. (2005). "Consumers" decision-making process and their online shopping behavior: A clickstream analysis. *Journal of Business Research, 58*, 1599-1608.

Solomon, M.R. (2007). *Consumer behavior: Buying, having, being* (7th ed.). Pearson/Prentice Hall.

Sugumaran, V., & Storey, V.C. (2002). Ontologies for conceptual modeling: Their creation, use, and management. *Data and Knowledge Engineering, 42*(3), 251-271.

Turban, E., & Aronson, J.E. (2001). Decision support systems and intelligent systems (6th international ed.). Upper Saddle River, NJ: Prentice Hall.

Vahidov, R. (2005). Intermediating user-DSS interaction with autonomous agents. *IEEE Transactions on Systems, Man, and Cybernetics-Part A: Systems and Humans, 35*(6), 964-970.

Yuan, S.T. (2002). A personalized and integrative comparison-shopping engine and its applications. *Decision Support Systems, 34*, 139-156.

ADDITIONAL READINGS

Chia-hao, C., & Yubao, C. (1996). Autonomous intelligent agent and its potential applications. *Computers & Industrial Engineering, 31*(1-2), 409-412.

Bui, T., & Lee, J. (1999). An agent-based framework for building decision support systems. *Decision Support Systems, 25*(3), 225-237.

Borst, P., Akkermans, H., & Top, J. (1997). Engineering ontologies. *International Journal of Human-Computer Studies, 46*(2-3), 365-406.

Deadman, P.J. (1999). Modeling individual behavior and group performance in an intelligent agent-based simulation of the tragedy of the commons. *Journal of Environmental Management, 56*(3), 159-172.

Fernandez-Breis, J.T., & Martinez-Bejar, R. (2000). A cooperative tool for facilitating knowledge management. *Expert Systems with Applications, 18*(4), 315-330.

Good, D.J., & Stone, R.W. (1995). Computer technology and the marketing organization: An empirical investigation. *Journal of Business Research, 34*(3), 197-209.

Gruber, T.R. (1995). Toward principles for the design of ontologies used for knowledge sharing? *International Journal of Human-Computer Studies, 43*(5-6), 907-928.

Guarino, N. (1997). Understanding, building and using ontologies. *International Journal of Human-Computer Studies, 46*(2-3), 293-310.

Hayes-Roth, B. (1995). An architecture for adaptive intelligent systems. *Artificial Intelligence, 72*, 329-365.

Keeble, R.J., & Macredie, R.D. (2000). Assistant agents for the World Wide Web intelligent interface design challenges. *Interacting with Computers, 12*(4), 357-381.

Lai, H., & Yang, T.-C. (2000). A system architecture for intelligent browsing on the Web. *Decision Support Systems, 28*(3), 219-239.

Lee, S., Seo, W., Kang, D., Kim, K., & Lee, J.Y. (2007). A framework for supporting bottom-up ontology evolution for discovery and description of Grid services. *Expert Systems With Applications, 32*(2), 376-385.

Li, S. (2000). The development of a hybrid intelligent system for developing marketing strategy. *Decision Support Systems, 27*(4), 395-409.

Mihoubi, H., Simonet, A., & Simonet, M. (1998). Towards a declarative approach for reusing domain ontologies. *Information Systems, 23*(6), 365-381.

O'Leary, D.E. (1997). Impediments in the use of explicit ontologies for KBS development. *International Journal of Human-Computer Studies, 46*, 327-337.

Pirlein, T., & Studer, R. (1995). An environment for reusing ontologies within a knowledge engineering approach. *International Journal of Human-Computer Studies, 43*, 945-965.

Raventos, P. (2006). The Internet strategy of the Costa Rican Tourism Board. *Journal of Business Research, 59*(3), 375-386.

Sheth, J.N., Sisodia, R.S., & Sharma, A. (2000). The antecedents and consequences of customer-centric marketing. *Journal of the Academy of Marketing Science, 28*(1), 55-66.

Teo, T.S.H., & Liu, J. (2007). Consumer trust in e-commerce in the United States, Singapore and China. *Omega, 35*(1), 22-38.

Umit Kucuk, S., & Krishnamurthy, S. (2007). An analysis of consumer power on the Internet. *Technovation, 27*(1-2), 47-56.

Van der Merwe, J., & von Solms, S.H. (1998). Electronic commerce with secure intelligent trade agents. *Computers & Security, 17*(5), 435-446.

Wang, K.C., Hsieh, A.T., Chou, S.H., & Lin, Y.S. (2007). GPTCCC: An instrument for measuring group package tour service. *Tourism Management, 28*(2), 361-376.

Wangermann, J.P., & Stengel, R.F. (1998). Principled negotiation between intelligent agents: A model for air traffic management. *Artificial Intelligence in Engineering, 12*(3), 177-187.

Weber, K., Story, M., & Harnack, L. (2006). Internet food marketing strategies aimed at children and adolescents: A content analysis of food and beverage brand Web sites. *Journal of the American Dietetic Association, 106*(9), 1463-1466.

Yang, T.-C., & Lai, H. (2006). Comparison of product bundling strategies on different online shopping behaviors. *Electronic Commerce Research and Applications, 5*(4), 295-304.

Zhang, Y., & Jiao, J.R. (2007). An associative classification-based recommendation system for personalization in B2C e-commerce applications. *Expert Systems with Applications, 33*(2), 357-367.

APPENDIX 1. THE QUESTIONNAIRE AND RESULTS
(1 = Strongly disagree, 2 = Disagree, 3 = Medium, 4 = Agree, 5 = Strongly agree)

1. Destination information

	Questions	1	2	3	4	5
1.	I will not look for the foreign journey information on travel Websites with a **specific continent**.	0	1	5	12	0
2.	I will look for the foreign journey information on travel Websites with a **specific country**.	0	1	1	12	4
3.	I will look for the foreign journey information on travel Websites with a **specific city**.	0	1	4	8	5
4.	I want to know relative foreign journey information in a specific destination.	0	0	2	7	9

2. Activity by theme information

	Questions	1	2	3	4	5
5.	I will search the specific **natural scene**.	0	0	6	7	5
6.	I will search the specific **historic spot**.	0	2	3	7	6
7.	I will search the specific **amusement park**.	0	2	6	7	3
8.	I will search the **specific festival**.	0	2	4	9	3
9.	The activity by theme information is my important reference information of participating in this travel journey.	0	0	2	6	10

3. Accommodation information

	Questions	1	2	3	4	5
10.	I will consult the **rating** of the hotel.	0	0	3	10	5
11.	I will consult the **food** of the hotel.	0	0	7	6	5
12.	I want to know the **price** of the restaurant.	0	0	2	6	10
13.	The relevant information of accommodation is worth consulting.	0	0	2	8	8

4. Transportation information

	Questions	1	2	3	4	5
14.	I will choose the **airline** that I want to take.	0	1	9	5	3
15.	I will choose the **cabin of the plane**.	0	1	9	6	2
16.	I want to know the **date and time** of the airplane.	0	0	2	6	10
17.	I want to know the **price**.	0	1	5	5	7
18.	I want to know **schedule** in order to understand the going on of the whole journey.	0	0	1	5	12
19.	Travel information is necessary.	0	0	2	10	6

5. Price information

Questions	1 2 3 4 5
20. The **price of the tour fee** is very important information to me.	0 0 3 5 10

6. Shopping information

Questions	1 2 3 4 5
21. I want to know the relative **shopping** information.	0 3 3 9 3

7. Information of optional tour

Questions	1 2 3 4 5
22. I want to know the optional tour **activity**.	0 0 1 5 12
23. I want to know the optional tour **price**.	0 0 3 4 11
24. I want to know the information of **nonparticipation**.	0 0 2 8 8

8. Other information

Questions	1 2 3 4 5
25. I want to know the **life and risk information**.	0 0 0 13 5
26. I want to know the **weather** at the destination.	0 0 1 5 12
27. I want to know the **exchange rate**.	0 0 3 11 4
28. I want to know the **useful information**.	0 0 1 11 6
29. I wonder whether there is **additional fare** besides traveling tour fee.	0 0 1 7 10
30. Other information is necessary.	0 0 4 13 1

Chapter IV
Agent–Based Matching of Demands and Supplies in Business Transactions Formation

Raymund J. Lin
Institute for Information Industry, Taiwan

Seng-Cho T. Chou
National Taiwan University, Taiwan

ABSTRACT

The theme of this chapter includes topics of matching, auction, and negotiation. We have shown that the knowledge of game theory is very important when designing an agent-based matching or negotiation system. The problem of bounded rationality in multi-agent systems is also discussed; we put forward the mechanism design and heuristic methods as solutions. A real negotiation scenario is presented to demonstrate our proposed solutions. In addition, we discuss the future trends of the agent technology in the e-commerce system.

INTRODUCTION

Since the commercialization of the Internet and the introduction of the World Wide Web in the early 1990s, e-commerce (EC) has expanded quickly. However, since 2000, there was a major shakeout in EC activities when hundreds of dot-com companies went out of business. By 2003, EC has continued its steady progress. Today, the Internet has changed the nature of the supply chain and created significant need for electronic markets. Therefore, most medium and large organizations and many small ones are now closely involved in e-commerce. The field of e-commerce is broad. Various business-to-consumer (B2C) and business-to-business (B2B) activities, such as e-auctions, e-marketplaces, and e-services, are mushrooming. The introduction of agent technology into the trading mechanism has improved efficiency and autonomy of the trading systems. Taking e-auctions as an example, bidding proxies have become a standard trading mechanism for auction participants.

The topics of matching, auction, and negotiation are the theme of this chapter. We assume predefined product/service ontology to exclude the need of Google-styled searching. The matching of descriptions of demands and supplies is integrated in a generic agent negotiation process, with non-negotiable terms treated as specific types of demands/supplies descriptions. The matching of negotiable demands/supplies descriptions can be quite difficult, especially when the matching involves multiple negotiating parties and multiple negotiable issues (not only prices). Agents on behalf of humans not only have to interact strategically to gain advantages in a negotiation but also have to be able to compute their chance of winning within a reasonable time. More efficient and strategy-proof mechanisms can be introduced to facilitate the negotiations; however, game-theoretic issues must be considered in the design of the agent-based trading mechanism.

The sections are arranged as follows. The second section gives background information regarding agents in e-commerce. The issues of designing negotiation agents for e-commerce are discussed in the third section. The fourth section summarizes generic solutions to the proposed issues. A real negotiation problem and its agent system designing based on the proposed solutions are presented in the fifth section. The sixth section is future trends, and conclusions are given in the last section.

BACKGROUND

Agents in E-Commerce

In e-commerce, there is a broad range of issues including security, trust, reputation, law, payment, advertising, electronic product catalogs, intermediaries, shopping experiences, and back-office management to be addressed (Guttman, Moukas, & Maes, 1999). Agent technologies can be applied to these areas where a personalized, continuously running semi-autonomous behavior is desirable. Certain characteristics determine to what extent agent technologies are appropriate, for instance, the money and time saved from partially automated processes, the risks of making a sub-optimal transaction decision, the consequences for missed opportunities, and so forth. It is suggested that exploring the roles of agent as mediators in e-commerce in the context of a common model could be useful for the research of agents in e-commerce (Guttman et al., 1999).

Some descriptive research, such as the Nicosia model (Nicosia, 1996), the HowardSheth model (Howard & Sheth, 1969), the EngelBlackwell model (Engel & Blackwell, 1982), the Bettman information processing model (Bettman, 1979), and the Andreasen model (Andreasen, 1965), attempts to capture consumer buying behavior using a model. Guttman et al. (1999) identify six common fundamental stages among the models. The six stages are potential areas to which agent technologies apply, and are briefly described in the following.

1. **Need identification:** In this stage, consumers are becoming aware of some unmet need, probably stimulated by product information.

It is called Problem Recognition in the EngelBlackwell model (Engel & Blackwell, 1982). The best example of this stage is the AdSense technology used by Google to provide ads to users based on the contents of their searching. Consumers may not be aware of the existence of a specific product that could satisfy their needs; therefore, an ad in time helps identify the needs.

2. **Product brokering:** After some unmet need is identified, consumers retrieve more information from a public database to help them determine what to buy. The information retrieval process includes evaluations of product alternatives based on consumer criteria. The result of this stage is called a consideration set of products. Amazon.com is a good example of product brokering, and it recommends products to users based on other customers' purchasing history.

3. **Merchant brokering:** Consumers combine the consideration set from the previous stage with merchant-specific information to determine who to buy from. In this stage, consumers evaluate merchant alternatives based on their criteria. Froogle is a Web site that provides both product brokering and merchant brokering. For instance, when searching "notebook," Froogle gives a list of brands and a list of stores. Users can evaluate products and merchants (stores) based on the features of products and the seller rating information.

4. **Negotiation:** In this stage, consumers and merchants together determine the terms of the transaction. Although there are conflicts of interests between the two parties, they both agree that the transaction (exchange) should be beneficial to them. A negotiation process is therefore conducted to resolve the conflicts. In traditional retail markets, terms of transactions are often fixed, leaving no room for negotiation. In markets of automobile and other highly priced products, terms

are usually negotiable. In other markets, like stocks and fine art, the negotiation is integral to product and merchant brokering. eBay is well known for its consumer-to-consumer (C2C) online auctions where buyers use bidding agents to compete for prices.

5. **Purchase and delivery:** After the negotiation, the product purchased will be delivered to the consumer immediately or sometime afterwards. The available payment or delivery options may influence the brokering and negotiation of the transaction. Agents play little or no role in this stage. However, many online stores, like Amazon, Apple, and Nintendo, provide simple delivery status reporting services. Users could use some third-party delivery tracker agent to track the deliveries.

6. **Product service and evaluation:** Product service and customer service are important in this postpurchase stage. Merchants who wish to create products or services that best fit customers' demands in the future often evaluate the customer satisfaction of the overall buying experience. Agents used in this stage are mainly for providing interactive customer services. For instance, Microsoft plans to use interactive agents for improving customer service.

Stages 1-4 make up the matching of demands and supplies in business transaction formation. Stages 1-3 are required when there is no predefined product/service ontology (common knowledge) such that buyers can describe their demands and evaluation criteria in detail. In this chapter, the assumption of predefined product/service ontology is made so that an integrated matching process including four stages at once could be explored. In this case, the need can be clearly defined, therefore no fuzzy matching is required in the first stage. The only problem is whether it is economic for such demands to be satisfied.

Buyers must make a trade-off between satisfaction and cost. Since the combination of product and merchant determines the satisfaction and cost of a transaction, the brokering of product and merchant is integrated with the negotiation for satisfaction and cost. For instance, buyers may request offers from merchants; merchants respond with proposals containing products/services with cost; buyers may accept the best offer or counter-offer; the negotiation process ends when both parties reach an agreement. In the following sections, the mechanisms for such a negotiation are discussed.

Auction

Auctions are classic one-to-many negotiation mechanisms, particularly well fitted for automation. A variety of auctions, models, and techniques exist because of the differences in the trading products, in the rules for users as well as in the pricing policies. Wurman, Wellman, and Walsh (2001) present a general parameterization of auction rules capturing the similarities and differences of several auction types. This parameterization is especially useful for the classification of research efforts in the area of auction analysis and for revealing novel auction mechanisms. More specifically, a broad categorization of auctions basing on pricing policies will identify the increasing price (or English), the sealed-bid and the decreasing price (or Dutch), the first/second-price sealed bid, the call markets and the continuous double auctions (CDA). At present, three major types of auctions are used most widely: open cry auction, first-price sealed-bid auction, and Vickrey auction (Vickrey, 1961). They are described in the following.

1. **Open cry auction:** This is a traditional type of auction. Because all bids are broadcasted to all participants in public, it is easy to satisfy the property of public verifiability.

Furthermore, open cry auctions are frequently used to sell fine arts. Two typical open cry auctions are listed.

a. **English auction:** This is the most well-known type of auction. It is used on the real world and Internet, such as eBay and Yahoo auctions. In an English auction, each bidder bids the higher price and wins at the highest price.

b. **Dutch auction:** In a Dutch auction, one auctioneer offers the goods for high price and queries each bidder whether she accepts the offer or not. If no bidder accepts the offer, the auctioneer reduces the price and then asks each bidder again. The bidder who offers to accept the price for the first time gains the product.

2. **First-price sealed-bid auction:** A first-price sealed-bid auction comprises two states, the bidding state and the opening state. In the former state, each bidder submits his bid sealed to the auctioneer. In the latter state, the auctioneer performs the procedure of opening the sealed bids, and then the bidder who bid the highest price gains the goods. Generally speaking, because the first-price sealed-bid auction requires only one round to determine the winner and the winning price, it is more efficient than the open cry auction. Governments in awarding procurement contracts commonly use first-price sealedbid auctions.

3. **Vickrey auction:** W. Vickrey proposed this type of auction in 1961 (Vickrey, 1961). It is also called second-price sealed-bid auction, which is a special case of (M+1)-price sealed auction. In a Vickrey auction, the bidder who bids the highest price gets the goods with the second highest price. In other words, the winner pays less than his bid (i.e., the highest bid). Vickrey auction

satisfies a useful property called "incentive compatibility" (Wurman, Walsh, & Wellman, 1998), so that from economic points of view, it is more excellent than first-price sealed-bid auctions (Rothkopf & Harstad, 1995; Rothkopf, Teisberg, & Kahn, 1990).

Contract Negotiation

Unlike auctions, which involve only price negotiations, the problem of contract negotiation (Klein, Faratin, & Sayama, 2003) focuses on searching for an agreement between two or more parties involving multiple issues. Therefore, issues like quality, delivery time, and warranty are discussed during a purchase negotiation. Auctions are basically distributive, which means that the interests of each party directly conflict with one another, and is hence a win-lose game. Contract negotiations, or simply multi-issue negotiations, are considered integrative (Raiffa, 1982), where all parties may find mutually beneficial outcomes, that is, win-win solutions. However, the complexity of a multi-issue negotiation increases rapidly as the number of issues increases, which means that people need more time and rationale in handling the negotiation problem. The development of negotiation support systems (NSSs) and negotiating software agents (NSAs) have proven to be able to reduce significantly the negotiation time and alleviate the negative effects of human cognitive biases and limitations (Foroughi, 1998; Lomuscio & Jennings, 2001; Maes & Moukas, 1999; Sandholm, 2000).

The construction of these automatic or semi-automatic systems is based on the concept of negotiation analysis (Sebenius, 1992), which integrates decision analysis and game theory in order to provide a meaningful support of decisions. As noted by Kersten (2001), negotiation analysis is used to generate prescriptive advice to the supported party given a descriptive assessment of the opposing parties. Problems can arise if assessment of the opposing parties is not available or

vague. This is very likely to happen in a dynamic environment, where new agents appear and leave frequently. While people are capable of intuitively acting on situations that are of great uncertainty, software agents rely on rational strategies for making decisions. Deadlocks can occur among agents if there is too much uncertainty to determine a rational strategy, especially when the uncertainty is two-sided. For one-sided uncertainty, there is a protocol through which the uninformed agent makes all the offers and the informed agent either accepts or rejects them (Vincent, 1989). Alternatively the uninformed agent can try to model the opponent using a Bayesian network or an influence diagram (Vassileva, 2002). The problem of two-sided uncertainty, on the other hand, can be addressed using recursive modeling (Gmytrasiewicz & Durfee, 1995). Nevertheless, for complex multi-issue negotiations, it is computationally intractable.

Since the tasks of finding highly effective negotiation strategies and protocols are not trivial, some research (Axelrod, 1987; Fudenberg & Levine, 1998; Oliver, 1997; Peyman, Jennings, Lomuscio, Parsons, Sierra, & Wooldridge, 2001) adopted evolutionary approaches to find effective negotiation strategies through learning. However, this approach is criticized because the complete payoff information is usually unavailable during the negotiation in most real-life cases. Besides, these evolutionary approaches assume repeated games and a static environment. The strategy learned, therefore, suffers a fixed strategy limitation, which means that they are not able to adapt to a dynamic environment. In fact, in a dynamic environment, most negotiations are one-off negotiations. This limitation has been observed in the experiments conducted in research (Goh, Teo, Wu, & Wei, 2000). Their research also indicates the existence of a fixed-pie bias, which is a tendency for agents to assume that their own interests directly conflict with those of the other party (Bazerman & Carroll, 1987; Thompson & Hastie, 1990). The fixed-pie bias will affect the efficiency of a negotiation strategy because agents assuming directly conflicted interests neglect mutually beneficial outcomes. It is therefore suggested in Goh et al. (2000) that the lack of communication between agents may have impeded information sharing and joint problem-solving possibilities.

ISSUES, CONTROVERSIES, AND PROBLEMS

Negotiations are conducted by at least two interest-conflicting parties. Information for solving the negotiation problem is not retrieved from a public database, but instead extracted from the interactions among negotiation parties. Valued information are often guarded by the parties to protect their own interests, and therefore the interactions which signal one's hidden information are often strategic. It means that parties can conceal their true intentions by avoiding certain interactions or mislead the opponents by conducting false interactions. For instance, a buyer might try to mislead a seller to believe that the deal is unacceptable by pretending to quit thereby forcing the seller to concede. These strategic behaviors make gaining complete information a difficult task. Although an agent can try to deduce the true intentions of other agents using some artificial intelligence (AI) techniques, the reasoning process is often so complicated that it cannot be completed in a reasonable time. In order to automate the negotiation process via agent technology, two issues need to be addressed. One is (for agents) how to deal with incomplete information, and the other is how to overcome the problem of bounded computing resources.

If complete information about agents' utility functions and the solution space of outcome is available to all negotiating parties, any negotiated agreement must be efficient. This is the case of *full rationality* and there is no reason for the parties to knowingly accept an inefficient agreement. Unfortunately, full rationality is rare. Agents often

make a decision in the situation of incomplete information, which leads to bounded rationality (Pietrula & Weingart, 1994). Besides rationality bounded by incomplete information, agents' power to compute responses are constrained by available resources and other external time constraints like deadlines, which make the situation even worse. These problems are discussed in the following.

Rationality Bounded by Incomplete Information

To understand the problem of incomplete information, the bilateral negotiation problem addressed in the literature is discussed. In order to model the situation of incomplete information, Rubinstein (1982) uses the method of applying a continuous distribution or discrete probabilities over the other agent's *type* and Bayesian rules to learn the type during the negotiation. The negotiation protocol used is the sequential alternating protocol (SAP). For a single-issue negotiation of price, the type of the other agent is represented by her reservation price. These distributions over the types are common knowledge. As a result, no subgame-perfect equilibrium exists, which requires that the predicted solution to a game be a Nash equilibrium (Nash, 1950) in every subgame.

Kreps and Wilson (1982) analyzed this problem using the stronger equilibrium concept of sequential equilibria. Their method requires that each uncertain player's belief be specified given every possible history, and that Bayes rules are used to make beliefs consistent. However, if the other agent's behavior deviates from the equilibrium path, an update problem may arise since non-equilibrium paths are assigned zero probability. This situation may lead to the generation of incentives for agents to deviate from the equilibrium so as to increase the number of possible outcomes (Faratin, 2000, p. 85). Deviation from equilibrium behavior cannot be ruled out in games solved via SAP with two-sided uncertainty (both parties have incomplete information). The situation can be worse in multi-issue negotiation, since the number of possible types of agents increases dramatically with the number of issues. Besides, in a multi-issue negotiation, it is not necessarily true that an agreement that is worse for one agent is best for another, or vice versa. This fact substantially increases the difficulty of the analysis of the intentions behind each offer proposed by the opponent.

Since determining a rational strategy in a multi-issue negotiation with two-sided uncertainty is not easy, most research (Chen, Chao, Godwin, Reeves, & Smith, 2002; Haigh, 1991; Matwin, Szapiro, & Sycara, 1990; Oliver, 1997) tries to reduce the uncertainty by learning from past negotiations or simulations. Unfortunately, applying this learning is inefficient in a dynamic environment, where new agents appear and leave. Learning from opponent behavior during a negotiation by modeling the behavior of that opponent is also infeasible because all known modeling techniques (Carmel & Markovitch, 1996; Gmytrasiewicz & Durfee, 1995; Zheng & Sycara, 1997) are computationally intractable when applied to a multi-issue negotiation with two-sided uncertainty.

Rationality Bounded by Computing Resources

Perfect rationality has been widely assumed in the research of self-interested agents (Ephrati & Rosenschein, 1992; Kraus, Wilkenfeld, & Zlotkin, 1995; Rosenschein & Zlotkin, 1994; Zlokin & Rosenschein, 1989), and it implies that agents always compute responses exactly and immediately. However, this is untrue in most practical situations, since agents' computing resources are often bounded or costly. It means that the running time and memory requirements of a negotiation algorithm is critical if the resource-bounded algorithm must produce an acceptable outcome

within a short time. The theory of computation complexity describes the inherent difficulty in providing scalable algorithms for such computational problems. Two types of computation complexity, time complexity and space complexity, need to be addressed. The time complexity of a problem is the number of steps that the algorithm takes to solve an instance of the problem. Given an input size n, if it requires n^2 steps to solve it, the time complexity is n^2. A notation, Big O, is used to generalize the details of actual steps required in different computers. Therefore, $O(n^2)$ describes the time complexity regardless of specific types of computer used. The space complexity, on the other hand, measures the amount of space (or memory) used by an algorithm. It is also measured with Big O notation.

The problem of bounded computing resources gets worse when negotiating agents have additional real-time pressures, such as deadlines, discounting factors, or bargaining costs. Internal (negotiating) parties or external forces may set up deadlines. For instance, a mediator sets up deadlines so as to force negotiating parties to make progress or reach an agreement by a specified date or time. It is a good weapon against time-consuming delaying agents (Admati & Perry, 1987) because agents strategically delay their counter-offers or stick to their best offers, using the Boulware strategy (Raiffa, 1982) to signal their strength of bargaining power. If there is no mediator, a party who has the power to threaten to quit the negotiation may set up internal deadlines. External deadlines, on the other hand, are usually established because of important events. For instance, an upcoming court date could encourage settlement between parties, so that they do not need to go to court. The discounting factor and bargaining cost of a negotiation will affect the value of the final deal. Since they both highly depend on time, negotiating parties are encouraged or constrained to respond to an offer within a limited time.

SOLUTIONS AND RECOMMENDATIONS

As discussed in the previous section, a negotiation agent system designer must deal with the problem of incomplete information and bounded computing resources. Information is incomplete because agents hide sensitive utility function from one another. This problem can be solved in two ways: first, enhancing the artificial intelligence of agents so that they can predict the opponents' behaviors; second, designing a system or mechanism that encourages truthful sharing of information among agents. The knowledge of game theory is therefore required if one wants to take agents' strategic interactions and incentives into consideration. Take auctions for instance, where each bidder's reservation price is hidden from the others since they all wish to negotiate for a lower price. Competing agents must formulate rational bidding strategies based on incomplete information. An English auction is a type of negotiation mechanism that leverages the competition among agents to induce true information about agents' valuation of a product.

To solve the problem of bounded computing resources, one should design a computational tractable algorithm for negotiation. The technical definition of computationally tractability will depend on the particular problem. However, it is argued that an algorithm becomes computational intractable when it takes nonpolynomial time to produce a good result. While many theoretic works may prove to solve a negotiation problem, agents cannot automate the negotiation process in polynomial time, especially for those seeking an optimal solution. For instance, given strategic behaviors among agents, it is almost impossible for an agent to gather complete information (about other agents' utility functions) during a negotiation and make an optimal decision. Doing so might take too much time or computation resources, making it uneconomic. Therefore, a good design of negotiation agents is expected to be game-theoretically sound while computationally tractable.

In Feigenbaum and Shenker's work, two separable concerns regarding the design of multi-agent system, computational tractability and incentive compatibility, are discussed in the context of traditional theoretical computer science and game theory, and are jointly addressed via the method of *mechanism design* (Feigenbaum & Shenker, 2002). The method is described in the following section.

Mechanism Design

Game theory is widely used among many subdisciplines of computer science, such as networking, DAI, and market-based computation. However, works of algorithmic mechanism design (AMD) and distributed algorithmic mechanism design (DAMD) are the first works in traditional computer science to address incentives and computation complexity simultaneously (Feigenbaum, Papadimitriou, & Shenker, 2001; Nisan & Ronen, 2001). Consider a distributed multi-agent system in which there is a set of possible outcomes O. Each of the n agents has a utility function, $u_i: O \to \Re$ where $u_i \in U$, expressing its preferences over the outcomes. A social choice function (SCF), $F: U^n \to O$, is used to map each particular instantiation of agents, who are completely described by their utility functions into a particular outcome. Since only the agents know the utilities, a system designer cannot just implement the desired outcome. A strategy-proof mechanism is required to make interacting agents produce the desired outcome.

An SCF is strategy-proof if $u_i(F(u)) \geq u_i(F(u|^i v))$, for all i and all $v \in U$, where the notation $(u|^i v)_i = v$ and $(u|^i v)_j = u_j$, for all $j \neq i$. Since each agent's utility function is the best-known utility function that could meet the desired social goal, agents have no incentive to lie. A direct mechanism, which asks agents to reveal their utility functions, is used to produce the desired outcome. However, in many cases, the desired social choice function is not strategy-proof. System designers need to design an indirect mechanism $\langle M, S \rangle$, where S is a strategy space, and $M : S^n \to O$ maps vectors of strategies into outcomes. For a given mechanism M and a given utility vector u, let the set $C_M(u) \subseteq S^n$ represent all possible strategy vectors reasonably resulted from selfish behavior. If an indirect mechanism M that implements the SCF (i.e., $M(C_M(u)) = F(u)$ for all $u \in U^n$), agents' selfish behavior results in the desired system-wide outcome. The system is then called *incentive compatible*.

The set C_M is called the solution concept. There are many solution concepts including *Nash equilibrium* concept, *rational strategies, evolutionarily stable strategies*, and *dominant strategies* (Feigenbaum & Shenker, 2002). Most AMD and DAMD literature uses the dominant strategy solution concept, which means that agents choose strategies resulting highest payoffs regardless of how other agents play. Although a multiagent system can be designed to be incentive-compatible, it could still take a long time for interactive agents to produce the desired outcome. Heuristic methods must be introduced to the system if the negotiation is time-constrained.

Heuristic Methods

Computer science brings an important consideration, computational complexity, to the game theoretic study of negotiation and mechanism design. The major means of overcoming the limitations of computational complexity in mechanism design is to use heuristic methods. Given the costs of exploring each "node" in the solution space, the methods seek to search the space in a cost-effective and non-exhaustive fashion. It means that the heuristic methods may not always produce optimal solutions, but it runs reasonably quickly. Trade-offs are presented between risks and costs. This feature is very important for negotiations that have time constraints or deadlines. A heuristic algorithm for a mechanism may be a computational realization of more informal negotiation models (e.g., Pruitt, 1981; Raiffa, 1982). Examples of such models exist in the literature (Barbuceanu & Lo, 2000; Faratin, Sierra, & Jennings, 1998; Kraus & Lehmann, 1995; Sathi & Fox, 1989; Sycara, 1989).

There are many heuristic methods that can facilitate the design of a time-constrained negotiation. They generally belong to a category of algorithms called *metaheuristics* (Blum & Roli, 2003), which are methods for solving a very general class of computational problems using user given black-box procedures as part of its implementation. Metaheuristics are commonly used for combinatorial optimization problems. Since the outcome of a negotiation is often a combinatorial mix of opinions from agents, metaheuristics suit negotiation problems. Some well-known metaheuristics include greedy algorithm, hill climbing, best-first search, simulated annealing, genetic algorithms, stochastic diffusion search, and so on. For negotiation agents, since the best strategies for a designed mechanism are known, the computational complexity lies at the complexity to compute the response based on the strategies and the complexity to compute the outcome based on the responses of all agents. Take Klein et al.'s (2003), for instance, where agents compute the response by calculating the utility value of the proposals generated by a mediator. The mediator, on the other hand, computes a new proposal based on the responses from agents using a heuristic method (hill climber or simulated annealing).

In the next section, a real negotiation problem is presented, and the solution to it based on the ideas of mechanism design and heuristic methods is proposed.

A REAL NEGOTIATION PROBLEM AND ITS SOLUTION

The Problem of Group Matching Negotiation

All the negotiation problems discussed so far are based on the concept of negotiable terms, which means that the goal of the problem is to conclude a deal; negotiating parties are invariant. In this section, a concept of negotiable *groups* is discussed. Negotiable groups are a set of negotiating groups with members moving between groups during the negotiation. Considering a scenario where three women are negotiating with a trader for buying multiple units of a product, one woman eventually quits the negotiation because all the prices proposed by the trader are unacceptable to her. This woman later joins another group to negotiate the price of a similar but cheaper product.

The scenario of negotiable groups exists because consumers are grouped according to their interests while utilizing the power of group order to negotiate for a better price. Eventually, each group will find a product standard that best fits their needs and is priced reasonably. If there are a lot of product standards with corresponding buying groups, the products are highly differentiated. However, market information is often so incomplete that it is difficult for suppliers to define a standard that can satisfy a group of the customers with an acceptable price. Inefficiency in market will result in unmatched deals and therefore a loss to both suppliers and customers. Given a group of customer g, the best product standard is defined as the following:

Definition 1. A standard $t_0 \in T$ best satisfies a group of customer g if and only if for every customer $c_i \in g$, his utility function u_i and reservation utility r_i satisfies $\forall t_j \in T, j \neq 0, u_i(t_0) \geq u_i(t_j), u_i(t_0) \geq r_i$.

Multiple standards can coexist to satisfy multiple groups of different interests. For instance, there are high-end products for the wealthier customers and normal products for ordinary customers. However, products of a standard may not exist if the group requesting that standard is too small to form a mass market. Assuming that customers can only afford a mass-market product, those who cannot form a mass market must purchase a less preferred product. In the end, only *practical standards S* that have markets may exist in the world.

Definition 2. A practical standard $s_0 \in S$ best satisfies a group of customer g if and only if for every customer $c_i \in g$ with utility function u_i and reservation utility r_i, g satisfies $\forall s_j \in S, j \neq 0, X_g(s_0) \geq X_g(s_j)$, where the notation χ_g denotes the number of the cases that $u_i(s_0) > u_i(s_j), u_i(s_0) \geq r_i$.

The meaning of $x(s)$ is the overall satisfaction rate of a practical standard s. The problem is how agent technologies can match supplies and demands by helping them identify practical standards automatically. The problem of group matching negotiation (GMN) is therefore defined as the following.

Definition 3. (GMN problem): Given a set of product standards T and a set of customers C, determine a set of customer groups G (i.e., partitions of C) and a set of practical standards S, $S \subset T$ such that for every group $g_i \in G$ there is a practical standard $s_i \in S$ that best satisfies the group. Each standard $t \in T$ has a required customer base $b(t)$, and each practical standard s_i must satisfy the required customer base $\#(g_i) \geq b(s_i)$. The notation $\#(g_i)$ denotes the number of members in g_i.

The following assumptions are made:

- **Expected utility maximizer:** Each agent wants to maximize her expected utility.

- **Mass market:** Agents can only afford a mass-market product. The constraint of the required customer base of each practical standard must be satisfied.
- **One-off negotiation:** The agents consider neither the past nor the future; each negotiation stands alone.
- **N-sided uncertainty:** Each agent has his private information (a utility function) that is unknown to the others. This N-sided uncertainty is common knowledge.
- **One purchase:** Every agent purchases only one unit of product.

A Mechanism for Multiparty Multi-Issue Mediation

The mechanism for multiparty multi-issue mediation is based on the concept of single negotiation text (Fisher, 1978). Some prior theoretical work (Lin, 2004; Lin & Chou, 2004) (mainly for bilateral negotiations) has been done to lay the foundation for this mechanism. In this section, it is shown that a multiparty extension of our previous work is also strategy-proof.

Definitions

Definition 1. A contract (i.e., agreement) D is represented by a string: $\langle b_1, b_2, ..., b_p \rangle$ where b_i is a number indicating an option that may influence the utility of an agent.

Since the facts represented by the binumbers in a contract may have different influence on the utility from different agent's perspective, they are treated as "issues" that require settling.

Definition 2. (Base utility): Let $v_A(b_i)$ be the utility generated by b_i from agent A's perspective. The base utility of a contract D to A is $\Sigma_{1 \leq i \leq p} v_A(b_i)$.

Definition 3. (Synergy): Let synergy $S = \{\langle x, f(x) \rangle: x \in \{1, 2, ..., p\}, f(x) \in \Re\}$ and let the util-

ity it generates be $e(S)$. If agent A successfully contends for a set of numbers $\{b_x: \forall \langle x, f(x) \rangle \in S, b_x = f(x)\}$, extra utility $e(S)$ is accumulated to the base utility.

Definition 4. (Total utility): Let ϕ denote the sets of synergies successfully acquired by agent A; the total utility of a contract D to A becomes: $\Sigma_{1 \leq i \leq p} v_A(b_i) + \Sigma_{S \in \phi} e(S)$ (synergies are assumed to be additive).

Multilateral Multi-Issue Negotiation

Based on the GVMP protocol proposed in Lin (2004), two mediation protocols, SVMP+ and GVMP+ for multilateral multi-issue negotiation, are proposed. Let the number of participants in the mediation process be z; the mediation processes are presented as follows.

The SVMP+ Process When utilizing the SVMP+, the mediator lets all agents simultaneously select their preferred contracts by giving them a vote. If there is no contract receiving $f \cdot z$ votes in the first round, the process continues and all agents simultaneously vote other contracts. f is a coefficient determining the threshold for accepting an agreement. Since there is no guarantee for an agreement in a multilateral multiissue negotiation, an agreement receiving $f \cdot z$ votes is considering as an acceptable negotiation outcome. The value of f is determined by the characteristics of a negotiation. Some negotiations require that all participants be accepting the negotiated outcome while some requires only a majority. Each agent can vote each contract once only. If there is one or more contacts receiving $f \cdot z$ votes, the mediation ends and the mediator randomly select one of them as the final agreement.

Theorem 1. When $f = 1$, the SVMP+ process ends within $[\frac{(z-1)}{z} \cdot \#(X)] + 1$ rounds.

Proof. The votes produced in each round is z, so at round $t \in \{1, 2, ...\#(X)\}$, there are totally $z \cdot t$ votes. If at round t there is no contract receiving z votes, then for each contract voted there are at maximum $(z-1)$ votes given. According to the pigeon hole principle, there is at least one contract receiving z votes in round t+1 if $t = [\frac{(z-1)}{z} \cdot \#(X)]$.

Definition 5. (Nash): A negotiation strategy s will be in equilibrium if the following condition holds: based on the assumption that other participants use strategy s, this participant prefers s to any other strategy.

Theorem 2. The best-contract strategy is in equilibrium at every round of the SVMP+ process.

Proof. Assume that other participants use the best-contract strategy. If this participant does not use the best-contract strategy and her best contract is D_i at round t, then she votes for a not-voted contract D_j such that $j \neq i$, $U_B(D_j) < U_B(D_i)$. If D_i already received $(z-1)$ votes, then B loses at least utility of $U_B(D_j) - U_B(D_i)$. If D_i received $k < (z-1)$ votes, then the probability that D_i receives $(z-1)$ votes in this round is $(\frac{1}{\#(X)-t})^{z-1-k}$ (N-sided uncertainty). This participant's expected utility compared with the utility if she uses the best-contract strategy: $(\frac{1}{\#(X)-t})^{z-1-k} \cdot (U_B(D_j) - U_B(D_i)) < 0$. Therefore, this participant prefers the best-contract strategy to any other one. This decision of using the best-contract strategy does not depend on the negotiation history, so the strategy is in equilibrium at every round of the SVMP+. A subgame-perfect equilibrium is found.

Theorem 3. The SVMP+ mediated process produces a Pareto-optimal outcome if the information on indifferences is disclosed.

Proof. If a participant is willing to disclose the information on indifferences, he reveals other contracts that are indifferent in utility from his perspective. Therefore, when the mediated outcome is D_m, if a participant reveals other contracts that are utility-indifferent from him for others to consider. There may be k other contracts receiving z votes after the information is disclosed. The mediator randomly selects one of them as the final agreement $D_{m'}$. Since the first participant disclosing this information will not benefit from this disclosure (while others may benefit from it), the mediator must randomly determine a sequential order for all participants to disclose their information on indifferences. That is to say, according to the sequential order determined by the mediator, each participant discloses his information on indifferences until a better agreement is found or that all participants disclose their information on indifferences.

Let participants' utility functions be U_1, U_2, ... If $D_{m'}$, which is not a Pareto-optimal outcome; there exists a D_x such that $(U_1(D_x), U_2(D_x),..., U_z(D_x)) \gg (U_1(D_{m'}), U_2(D_{m'}), ..., U_z(D_{m'}))$. It means that $(U_1(D_x) > (U_1(D_{m'})$ and $U_2(D_x)) \geq (U_2(D_{m'})$ and ... and $U_z(D_x)) \geq (U_z(D_{m'}))$ or $(U_1(D_x) \geq U_1(D_{m'})$ and $U_2(D_x)) > (U_2(D_{m'})$ and ... and $U_z(D_x)) \geq (U_z(D_{m'}))$ or ...

According to the SVMP+ process, if the information on indifferences is disclosed, $U_1(D_x)) < (U_1(D_{m'})$ or $U_2(D_x)) < (U_2(D_{m'})$ or a conflict occurs. Therefore, $D_{m'}$ is a Pareto-optimal outcome.

Since there are no incentives for agents to disclose the information on indifferences, the SVMP+ process cannot produce a Pareto-optimal outcome. However, the mediated outcome is still good because the set of utility-indifferent contracts is small compared to the set of all contracts. Just like the SVMP process, the SVMP+ process is time-consuming and reveals too much information. The number of units of information revealed in each round in the SVMP process is $\#(X)-t$. In the worst case, a mediation process ends until turn $[\frac{(z-1)}{z}\bullet\#(X)]+1$ and the amount of information is up to $\frac{1}{2}([\frac{(z-1)}{z}\bullet\#(X)]+1)$ $((\#(X)-1+[\frac{(z-1)}{z}\bullet\#(X)]+1))$.

The GVMP+ Process. In the GVMP+ process, the mediator first devises randomly a set of contracts Y_1 for all protagonists to consider. Suppose that there are z participants in the mediation process, they must select $[\frac{(z-1)}{z}\bullet\#(X)]+1$ contracts from the set (the selection subprocess). The mediator will generate a new set $Y_{1'}$ of the returned $z\bullet([\frac{(z-1)}{z}\bullet\#(X)]+1)$ contracts. This new set is then recombined and mutated (the heuristic method of genetic algorithm). A new set Y_2 is created by randomly selecting z samples from $Y_{1'}$. For simplicity, the number of contracts in Y is assumed to be a multiple of z. It is expected that, at round i, Y_i will contain "good-enough" potential Pareto-optimal contracts when i is sufficiently large. The mediator then finishes the negotiation by applying SVMP+ to the last Y_i.

Theorem 4. Assume that all participants' preference relations are strict orders and agents vote according to them. Let $S(Y_1)$ be the mediated outcome generated by applying the SVMP+ process (using f = 1) to the contract set Y_1. If $S(Y_1)$ is the j_1th preferred contract in Y_1 according to participant 1's preferences and is the j_2th preferred contract in Y_1 according to participant 2's preferences and so on, then the number of the contract $S(Y_1)$ will be accumulated exponentially in the following generations $(Y_2,Y_3,...)$ up to $\frac{1}{z}(\#(Y_1)-(j_1+j_2+...+j_z)+z)$ given that the recombination rate and mutation rate are both zero.

Proof. By Theorem 1, $j_i < ([\frac{(z-1)}{z}\bullet\#(Y)]+1)$. Therefore, the number of $S(Y_1)$ in $Y_{1'}$ (denoted as $\#(S(Y_1),Y_{1'})$) equals or exceeds z. Since the

recombination rate and mutation rate are both zero $\#(S(Y_1),Y_2)$ equals or exceeds $\frac{z}{z-1}$. If $((j_i-1)+\#(S(Y_1),Y_2))\le\frac{(z-1)}{z}\bullet\#(Y)$ for all $i\in$ 1, 2, ..., z, $\#(S(Y_1),Y_{2'})=z\bullet\#(S(Y_1),Y_2)$ and $\#(S(Y_1),Y_3)$ $=\frac{z}{z-1}\bullet\#(S(Y_1),Y_2)$. In other words, for $k\in N$, the number of contract $S(Y_1)$ in Y_k accumulated exponentially up to $(\frac{z}{z-1})^k$. However, $\#(S(Y_1),Y_k)$ has a upper bound of $\frac{1}{z}(\#(Y_1)-(j_1+j_2+...+j_z)+z)$, because there are (j_i-1) contracts more preferred than $S(Y_1)$ from participant i's perspective.

Theorem 4 implies that potential Pareto-optimal contracts receive exponentially increasing "trials" in subsequent generations of a GVMP+ process. In addition to that, the recombination process and mutation process provide a possibility for locating new potential Pareto-optimal contracts. For $k\in N$ and $k>1$, if a new contract becomes the $S(Y_k)$, its number increases exponentially (if not disrupted) as described in Theorem 4. Restated, GVMP+ process locates potential Pareto-optimal contracts using a GA-like searching process.

Definition 5. When participating in a GVMP+ process, an agent k ($k\in\{1,2,...,z\}$) is said to implement a better-set strategy if and only if according to her preferences, the set $BS_k(Y)$ is selected in the selection subprocess; $BS_k(Y)$ is defined as follows: for each contract $D_i\in BS_k(Y)$ and each contract $D_j\in(Y-BS_k(Y))$, $U_k(D_i)\ge U_k(D_j)$.

Hypothesis. The better-set strategy is in equilibrium at every round of the GVMP+ process.

By Theorem 4, the GVMP+ process locates potential Pareto-optimal contracts using a GA-like searching process. Assume that other participants use the better-set strategy. At round t, if this participant k does not use the better-set strategy and the set of contracts proposed by him is $AS(Yt)$, there are $D_i\in ASk(Yt)$ and $D_j\in(Yt\ ASk(Yt))$ such that $U_k(D_j)>U_k(D_i)$. Since there is a possibility that $S(Y_i) = D_i$ which makes participant k lose a chance to increase the number of the good contract D_j in Y_{i+1}, it is participant k's best interests to stick to the better-set strategy.

Modeling Bargaining Power

To introduce the concept of bargaining power to the proposed multilateral multi-issue mediation process, weighted votes are given to all the participants. Let the proportion of bargaining power of the z participants be $\{w_1 : w_2 : ... : w_z\}$. In the proposed SVMP+ process, votes from participant k are weighted w_k; it means that the contract voted by participant k receives w_k votes. If a contract received $f(w_1 + w_2 + ... + w_z)$ votes, it is accepted.

The System for Group Matching Negotiation

The system for matching group products is based on a real negotiation scenario of travel packages. In this scenario, a travel agent offers customizable vacation packages to our institute. Choosing different hotels, activities, and restaurants can customize the packages. The agent offers a good discount for group orders (minimum group size is 20). Since all members in our institute would like to have discounts on their preferred customized travel packages, they have to negotiate for a set of customized packages in which each package gets at least 20 participants. No doubt that some members have to sacrifice a little satisfaction for discounts. Based on the proposed mediation theory, a travel package negotiation system is constructed to solve the problem.

The negotiation system enables mass customization. Our legacy travel package voting system allows users to vote for a package among several others. The travel packages for selection are few and only one travel package gets implemented. The new system, on the other hand, allows users to customize packages in advance, and lets

them negotiate for a group order. For instance, if a package customized by A did not get 20 votes, A could sacrifice his satisfaction by accepting a less preferred (but similar) package proposed by another user. By introducing multiple issues (hotels/activities/restaurants) in this negotiation, users are allowed to trade between issues when it is necessary. For example, if user A prefers a good hotel to a good activity while user B prefers a good activity to a good hotel, A and B could compromise with each other by accepting a proposal with A's preferred hotel and B's preferred activity. The complex negotiation process is fully automated and the only action done by users is to enter their preferences in the beginning of the negotiation. There are eventually eight negotiable issues in our first experiment: day 1 activity, day 1 lunch, day 1 hotel, day 2 activity, day 2 lunch, day 2 hotel, day 3 activity, and day 3 lunch.

The overall negotiation process has five stages (Figure 1).

1. **Collecting customer preferences:** Users must enter their preferences over several options regarding different issues. A Web page is constructed to facilitate the input. The information collected includes preference over options in several issues and the weights of issues (Figure 2).

2. **Agent negotiation:** Using the GVMP+ mediation process, agents continuously revise a set of proposals until the time is up. A set of proposals (product standards), which are candidate travel packages, are produced in this stage.

3. **Agent voting:** Agents start voting on the candidate proposals (the SVMP+ process). Once a proposal receives enough votes (minimum group size), it gets implemented.

4. **All customer groups found:** The system checks whether all customers are in group. If there are still customers not grouped, the system resets the vote counter and goes back to Stage 3.

Figure 1. The negotiation process

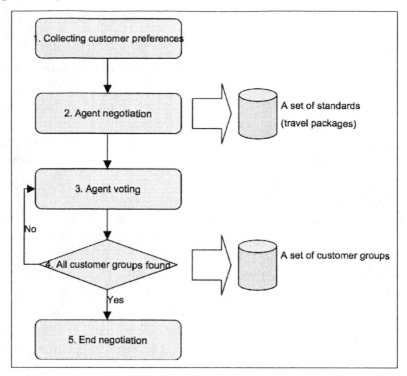

5. **End negotiation:** The negotiation ends. All proposals (travel packages) get implemented.

In order to get user preferences, questionnaires are used to survey the data. Since users may not wish to express their preferences via complex utility functions. We provide four ways for them to quickly express their preferences. Besides the preference information, users have to give weights over issues, so agents can decide how to trade between issues. The questionnaire is shown in Figure 2. Since the original system uses Chinese interfaces, the English translations to them are tagged besides the correspondents.

1. Simply choose "no preference." Agents will definitely trade the vote of this issue for others.
2. Choose your most preferred option. Agents will try to negotiate for an agreement with this option. If impossible due to small minority, agents then trade this vote for others.
3. Rank all available options. Agents try to negotiate an agreement according to the ranks of options until all options run out, then trade this vote for others.
4. Give scores to all available options. This precise utility information allows agents to trade between issues more efficiently. For instance, an option with score 98 will convince agents to trade the vote of this option for two 49-scored options of another two issues.

FUTURE TRENDS

Three trends regarding the use of agent technologies in e-commerce are identified. They are described in the following.

Mechanism Design as a Mean to Trust

Agent technologies have been widely used in e-commerce systems. However, there is still a trust issue when using a highly autonomous agent system to support business transaction formation. The trustworthiness of a new agent technology will affect the diffusion of the technology. The introduction of mechanism design enables an agent designer to build an incentive-compatible agent system such that agents' strategies are predictable. Since agents' strategies are predictable, the algorithm for agent automation tends to be simple and easily understood by humans. A human trusts bidding proxies in eauctions, because their method to bid increasingly is easily understood. Consequently, people delegate the bidding job to proxies without fearing that the agents might be doing something wrong. It is argued that the "side effect" of trust created by mechanism design could be the key to a successful multi-agent negotiation system. The travel package negotiation system proposed in this chapter is another example. The agents are only voting according to users' preferences, and this behavior is easily understood and trusted by humans.

Figure 2. Preferences and issue weights

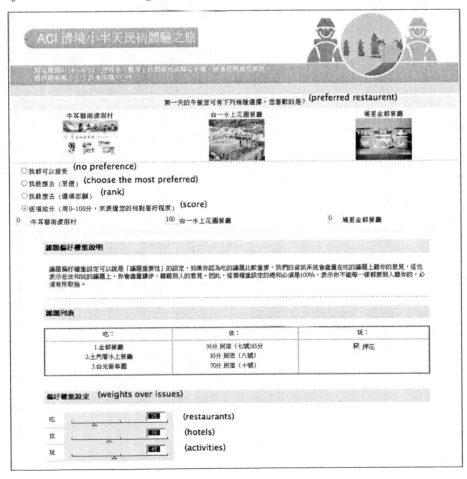

Automated Matching of Supplies and Demands

According to the CBB model, there are four stages (need identification, product brokering, merchant brokering and negotiation) related to the matching of supplies and demands. If there is no common ontology among the four stages, consumers would have to switch between different ontologies when using different automation systems in different stages. Since the recent trend of demand driven supply chain has been widely acknowledged, the concepts of customer-made products and architecture of participation become popular. Under such architecture, enterprises encourage customers to describe their needs via a common ontology so that the supply chain could adjust easily according to the needs. The standardization of product components makes integrated matching across the four stages possible. This highly automated agent-based matching of supplies and demands could improve the effectiveness and efficiency of trading.

Intelligent Interface for Describing Preferences

The development of a highly integrated agent-based matching system raises a new problem of capturing users' preferences. The preferences need to be adequately described before agents can fully automate the matching. Requesting a full description of a user's preferences or utility function is either impossible or costs too much. Besides, users should be allowed to trade between costs and risks. A "lightweight" or intelligent input interface for describing preferences could greatly save users time. In our travel package negotiation system, users are allowed to express their preferences by choosing the most preferred option, instead of giving scores to each of the options. It is believed that the research of using an intelligent interface for describing preferences could be an important area requiring our attention.

CONCLUSION

In this chapter, agent-based matching systems involving product brokering, merchant brokering, or negotiation are discussed. The general problems and issues arising in the matching are presented, including (agent's) rationality bounded by incomplete information and rationality bounded by computing resources. It is suggested that the approaches of mechanism design and heuristic methods are solutions to the problems. Consequently, a real negotiation problem is presented, with the design of an agent-based matching system demonstrated using the proposed approaches. Our travel package negotiation system demonstrates the possibility of building a negotiation system supporting mass customization.

REFERENCES

Admati, A., & Perry, M. (1987). Strategic delay in bargaining. *Review of Economical Studies, 54*, 345364.

Andreasen, A. (1965). Attitudes and customer behavior: A decision model. In L. Preston (Ed.), *New research in marketing.* California Institute of Business and Economics Research.

Axelrod, R. (1987). The evolution of strategies in the iterated prisoner's dilemma. In L. Davis (Ed.), *Genetic algorithms and simulated annealing.* London: Pittman.

Barbuceanu, M., & Lo, W. (2000). A multiattribute utility theoretic negotiation architecture for eletronic commerce. In *Proceedings of the 4th International Conference on Autonomous Agents,* Barcelona, Spain (pp. 239-247).

Bazerman, M., & Carroll, J. (1987). Negotiator cognition. In B. Staw & L.C. (Eds.), *Research in organizational behavior* (9). Greenwich, CT: JAI Press.

Bettman, J. (1979). *An information processing theory to consumer choice.* AddisonWesley.

Blum, C., & Roli, A. (2003). Metaheuristics in combinatorial optimization: Overview and conceptual comparison. *ACM Computing Surveys, 35*(3), 268-308.

Carmel, D., & Markovitch, S. (1996). Learning models of intelligent agents. In *Proceedings of the 13th National Conference on Artificial Intelligence and the 8th Innovative Applications of Artificial Intelligence Conference.*

Chen, J.H., Chao, K.M., Godwin, N., Reeves, C., & Smith, P. (2002). An automated negotiation mechanism based on co-evolution and game theory. In *Proceedings of the ACM Symposium on Applied Computing*, Madrid, Spain (pp. 63-67).

Engel, J., & Blackwell, R. (1982). *Consumer behavior* (4th ed.). CBS College Publishing.

Ephrati, E., & Rosenschein, J.S. (1992). *Reaching agreement through partial revelation of preferences.*

Faratin, P. (2000). *Automated service negotiation between autonomous computational agents.* Unpublished doctoral dissertation, University of London.

Faratin, P., Sierra, C., & Jennings, N. (1998). Negotiation decision functions for autonomous agents. *International Journal of Robotics and Autonomous Systems, 24*(34), 159-182.

Feigenbaum, J., Papadimitriou, C., & Shenker, S. (2001). Sharing the cost of multicast transmissions. *Journal of Computer and System Sciences, 63*, 2141.

Feigenbaum, J., & Shenker, S. (2002). Distributed algorithmic mechanism design: Recent results and future directions. In *Proceedings of the 6th International Workshop on Discrete Algorithms and Methods for Mobile Computing and Communications.*

Fisher, R. (1978). *International mediation: A working guide.* New York: International Peace Academy.

Foroughi, A. (1998). Minimizing negotiation process losses with computerized negotiation support systems. *The Journal of Applied Business Research, 14*(4), 1526.

Fudenberg, D., & Levine, D.K. (1998). *The theory of learning in games.* Cambridge: M.I.T. Press.

Gmytrasiewicz, P., & Durfee, E. (1995). A rigorous operational formalization of recursive modelling. In *Proceedings of the 1st International Conference on Multiagent Systems* (pp. 125-132).

Goh, K., Teo, H., Wu, H., & Wei, K. (2000). Computer supported negotiations: An experimental study of bargaining in electronic commerce. In *Proceedings of the 21st Annual International Conference on Information Systems*, Brisbane, Australia.

Guttman, R., Moukas, A., & Maes, P. (1999). Agents as mediators in electronic commerce. In M. Klusch (Ed.), *Intelligent information agents* (chap. 6). Berlin: Springer.

Howard, J., & Sheth, J. (1969). *The theory of buyer behavior.* John Wiley & Sons.

Kersten, G.E. (2001). Modeling distributive and integrative negotiations. *Group Decision and Negotiation, 10*(6), 493-514.

Klein, M., Faratin, P., & Sayama, H. (2003). Negotiating complex contracts. *Group Decision and Negotiation Journal: Special Issue on eNegotiations.*

Kraus, S., & Lehmann, D. (1995). Designing and building an automated negotiation agent. *Computational Intelligence, 11*(1), 132-171.

Kraus, S., Wilkenfeld, J., & Zlotkin, G. (1995). Multiagent negotiation under time constraints. *Artificial Intelligence Journal, 75*(2), 297-345.

Kreps, D., & Wilson, R. (1982). Sequential equilibria. *Econometrica, 50,* 863-894.

Lin, R.J. (2004). Bilateral multiissue contract negotiation for task redistribution using a mediation service. In *Proceedings of the Agent Mediated Electronic Commerce Workshop in AAMAS.*

Lin, R.J., & Chou, S.C.T. (2004). Mediating a bilateral multiissue negotiation. *Electronic Commerce Research and Applications, 3*(2).

Lomuscio, M., & Jennings, N. (2001). A classification scheme for negotiation in electronic commerce. In F. Dignum & E.C. Sierra (Eds.), *Agent Mediated Electronic Commerce: A European Agentlink Perspective.* Springer-Verlag.

Maes, R., & Moukas, A. (1999). Agents that buy and sell. *Communications of the ACM, 42*(3), 81-91.

Matwin, S., Szapiro, T., & Haigh, K. (1991). Genetic algorithms approach to a negotiation support system. *IEEE Transactions on Systems, Man, and Cybernetics, 21*(1), 102-114.

Nash, J.F. (1950). The bargaining problem. *Econometrica, 18,* 155-162.

Nicosia, F. (1996). *Consumer decision processes: Marketing and advertising implications.* Prentice Hall.

Nisan, N., & Ronen, A. (2001). Algorithmic mechanism design. *Games and Economic Behavior, 35,* 166-196.

Oliver, J. (1997). A machine learning approach to automated negotiation and prospects for electronic commerce. *Journal of Management Information Systems, 13*(3), 83-112.

Peyman, F., Jennings, N.R., Lomuscio, A.R., Parsons, S., Sierra, C., & Wooldridge, M. (2001). Automated negotiation: Prospects, methods and challenges. *International Journal of Group Decision and Negotiation, 10*(2), 199-215.

Pietrula, M.J., & Weingart, L.R. (1994). Negotiation as problem solving. In *Advances in Managerial Cognition and Organizational Information Processing.* JAI Press.

Pruitt, D.G. (1981). *Negotiation behavior.* Academic Press.

Raiffa, H. (1982). *The art and science of negotiation.* Cambridge: Harvard University Press.

Rosenschein, J.S., & Zlotkin, G. (1994). *Rules of encounter.* The MIT Press.

Rothkopf, M., & Harstad, R. (1995). Two model of bidtaker cheating in vickrey auctions. *Journal of Business, 68,* 257-267.

Rothkopf, M.H., Teisberg, T.J., & Kahn, E.P. (1990). Why are vickrey auctions rare? *Journal of Political Economy, 98*(1), 94-109.

Rubinstein, A. (1982). Perfect equilibrium in a bargaining model. *Econometrica, 50* (1), 155-162.

Sandholm, T. (2000). Agents in electronic commerce: Component technologies for automated negotiation and coalition formation. *Autonomous Agents and MultiAgent Systems, 3*(1), 73-96.

Sathi, A., & Fox, M.S. (1989). Constraint directed negotiation of resource allocation. In L. Gasser & M. Huhns (Eds.), *Distributed artificial intelligence II* (pp. 163-195). Morgan Kaufmann.

Sebenius, J.K. (1992). Negotiation analysis: A characterization and review. *Management Science, 38*(1), 18-38.

Sycara, K. (1989). Multiagent compromise via negotiation. In L. Gasser & M. Huhns (Eds.), *Distributed artificial intelligence II* (pp. 119-139). Morgan Kaufmann.

Sycara, K. (1990). Negotiation planning: An AI Approach. *European Journal of Operational Research, 46,* 216-234.

Thompson, L., & Hastie, R. (1990). Social perception in negotiation. *Organizational Behavior and Human Decision Processes, 47,* 98-123.

Vassileva, J. (2002). Bilateral negotiation with incomplete and uncertain information. In P.G.S. Parsons & M. Wooldridge (Eds.), *Game theory and decision theory in agentbased systems.* Kluwer Academic Publishers.

Vickrey, W. (1961). Counterspeculation, auctions, and competitive sealed tenders. *Journal of Finance, 16,* 8-37.

Vincent, D. (1989). Bargaining with common values. *Journal of Economic Theory, 48,* 47-62.

Wurman, P., Walsh, W., & Wellman, M. (1998). Flexible double auction for electronic commerce: Theory and implementation. *Decision Support System, 24,* 17-27.

Wurman, P., Wellman, M., & Walsh, W. (2001). A parameterization of the auction design space. *Games and Economic Behavior, 35,* 304-338.

Zheng, D., & Sycara, K. (1997). Benefits of learning in negotiation. In *Proceedings of the 14th National Conference on Artificial Intelligence* (pp. 36-41).

Zlokin, G., & Rosenschein, J.S. (1989, August). Negotiation and task sharing among autonomous agents in cooperative domains. In *Proceedings of the 11th International Joint Conference on Artificial Intelligence,* Detroit, MI (pp. 912-917).

Chapter V
Evolutionary Auction Design for Agent–Based Marketplaces

Zengchang Qin
University of California, USA

ABSTRACT

Market mechanism or auction design research is playing an important role in computational economics for resolving multi-agent allocation problems. In this chapter, we review relevant background of trading agents, and market designs by evolutionary computing methods. In particular, a genetic algorithm (GA) can be used to design auction mechanisms in order to automatically generate a desired market mechanism for electronic markets populated with trading agents. In previous research, an auction space model was studied, in which the probability that buyers and sellers are able to quote on a given time step is optimized by a simple GA in order to maximize the market efficiency in terms of Smith's coefficient of convergence. In this chapter, we also show some new results based on experiments with homogeneous and heterogeneous agents in a more realistic auction space model. This research provides a way of designing efficient auctions by evolutionary computing approaches.

INTRODUCTION

In the first generation of e-commerce, bidders are generally humans who typically browse through well-defined commodities with fixed prices via the Internet (e.g., Amazon.com). Just like the traditional marketplace, purchases are done with the prices made by sellers; buyers and sellers still have little freedom in transactions.

For customer-to-customer (C2C) e-commerce (e.g., eBay.com), sellers and buyers actually do the traditional trades, but through a new and more efficient medium—the Internet. Their freedom is also limited because both sellers and buyers still use the traditional methods of browsing to look for the goods they want. With the advent of agent technology, software agents can act as real-world traders. In comparison to human traders, software agents have the advantages of being very fast, cheap, and offer a tightly controlled environment in which a diverse range of experiments can be performed. A **trading agent** may represent a company or a customer hunting for maximized utility which means profit for the sellers or savings for the buyers. In this scenario, freedom can be greatly increased by allowing negotiation between opposite traders (i.e., sellers vs. buyers) in a large predefined cyberspace for transactions. This space can either be embedded in or independent of the current Internet. As a result, commerce may become much more dynamic and the market less frictional. This kind of commerce is referred to as agent-mediated e-commerce or the second generation of e-commerce (He, Jennings, & Leung, 2003a). Since the traders search in a very large space of commodities for matching their preferences, how to efficiently search this space and what protocols the traders have to follow in order to have a trustworthy and efficient market are all key problems for this new research area. In this chapter, a method of using **genetic algorithms** (GAs) for trading protocol designs, or market mechanism designs, is discussed.

Trading agent design is an important research branch in both multi-agent systems and computational economics. Many different types of **trading agents** have been proposed. There is even a trading agent competition (TAC) held annually since 2000.[1] In this chapter, we explore the possibility of using evolutionary computing for auction designs but not the **trading agents** themselves, so that only two simple types of agents are used: **zero intelligence** (ZI) and **zero intelligence plus** (ZIP) agents. ZI agent was proposed by Gode and Sunder (1993) who presented results that appear to indicate that a random guessing strategy can exhibit human-like behavior in **continuous double auction** (CDA) markets. However, Cliff (1997) indicated that the price convergence of ZI traders is predictable from *a priori* analysis of the statistics

of the system, so that a more complex bargaining mechanism or some "intelligence" may be necessary. Consequently, a type of agent with simple machine learning techniques was developed and referred to as **zero intelligence plus** (ZIP) agents. Further experiments by Das, Hanson, Kephart, and Tesauro (2001) showed that ZIP agents outperform their human counterparts.

Market mechanism design is an important topic in classical economics, computational economics, and finance (LeBaron, 2000; Wurman, 2001). Using **genetic algorithms** for market mechanism designs was first proposed by Cliff (2001). A continuous **auction space** model was proposed and explored by a GA to maximize the market efficiency. Some hybrid auctions with more desirable market dynamics, which had never been found in the real world, were discovered (Cliff, 2003). However, the **auction space** model proposed by Cliff is not an exact analogue to single-sided real world auctions such as the **English auction** (EA) and **Dutch flower auction** (DFA). Qin and Kovacs (2004) proposed a new **auction space** model which is an exact analogue to real world auctions. In this chapter, we give a review of the research of using GAs for automatic auction designs and present some new results by using heterogeneous **trading agents** (i.e., a mixture of ZI and ZIP agents), which shows that the optimal auction for heterogeneous agents could be a human-designed one while the optimal auction for homogeneous agents is a hybrid one, given the same **supply-demand** schedule.

In the second section, a brief introduction on experimental economics, auctions, and the measure for evaluating the market performances are given. In the third section, we give a detailed description of the **trading agents** used in this chapter. The fourth section gives the idea of using GAs for auction designs, and the fifth section gives the results based on the experiments with homogeneous and heterogeneous trading agents.

BACKGROUND ECONOMICS

Supply and Demand

In every classical economic model, demand and supply always play prominent roles. There is a joke about teaching a parrot to be an economist simply by teaching it to say "supply and demand." Supply is used to describe the quantity of a good or service that a household or firm would like to sell at a particular price. Demand is used to describe the quantity of a good or service that a household or firm chooses to buy at a given price. For a buyer, with increasing of quantity of the commodity, he will be inclined to bid a lower price to make a purchase, but with the less quantity of commodity, he has to increase his bid price. Because buyers want to make purchases at lower prices so that the demand curve slopes downward. For sellers, if the commodity is at a higher price, they will be inclined to sell as many as they can; this keeps the supply curve slope upward (Staglitz & Driffill, 2000).

The intersection of the supply curve and demand curves is called the equilibrium, and the corresponding price and quantity are called, respectively, the *equilibrium price* and the *equilibrium quantity* (see Figure 1).

Figure 1 depicts a qualitative relation of supply and demand: Equilibrium occurs at the intersection of the demand and supply curves, at point *E*. At any price above E, the quantity supplied will exceed the quantity demanded, and the market will be in a state of excess supply. For example, at price P_1, from the demand curve, we learn that, for a particular good, consumers only need this good with quantity of Q_3; however, the producers provided the quantity of Q_2; Apparently $Q_3 < Q_2$; the difference of Q_2 and Q_3 is the excess supply of the market. At any point below E, the quantity demanded will exceed the quantity supplied, and the market is in excess demand. For example, at the price of P_2, consumers need Q_4 in quantity while the producers only provide Q_1 of goods. $Q_4 > Q_1$, the difference is the excess demand of the market.

In the case of prices beyond the equilibrium, the market will self-correct them to the equilibrium by an "invisible hand." At an equilibrium price, consumers get precisely the quantity of the good they are willing to buy at that price, and sellers sell out the quantity they are willing to sell at that price. Neither of them has any incentive to change. In a competitive market, the price actually paid and received in the market will tend to the equilibrium price. This is called the law of supply and demand (Staglitz & Driffill, 2000).

Figure 1. Supply and demand curves cross at the equilibrium

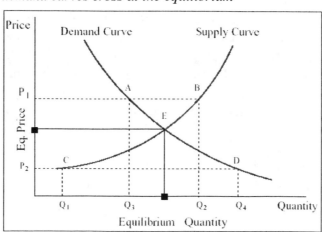

Experimental Economics

Experimental economics started out with Chamberlin (1948) who attempted to examine the characteristics of a market institution under controlled conditions of his classroom. Smith (1962) contributed to the foundation of experimental economics[2] by demonstrating that markets consisting of small numbers of traders could still exhibit equilibration to the values predictable from classical microeconomic theory. In Smith's experiments, only one kind of good is traded in a market populated with a number of traders, and each trader (seller or buyer) has a private limit price for a unit, that a seller could not sell less than and a buyer could not pay more than. Typically, different traders had different limit prices and the distribution of limit price is determined by a **supply-demand** schedule. In each experiment, the trading lasted for a few "days." In each day, a predefined number of transactions were given and the rights to quote were distributed to traders. Trades operated under a specific market mechanism, for example, CDA. In each trade, the transaction prices were logged to study the convergence property of the market. In a given **supply-demand** schedule with n transactions, the *coefficient of convergence* is introduced to measure the deviation of transaction prices from the theoretical market equilibrium price p_0 given a series of transaction prices p_i for $i = 1,..., n$. The **coefficient of convergence** α is calculated as follows:

$$\alpha = \frac{100 \cdot \delta_0}{p_0} \tag{1}$$

where

$$p_0 = \frac{1}{n} \sum_{i=1}^{n} p_i \tag{2}$$

and

$$\delta_0 = \sqrt{\frac{1}{n} \sum_{i=1}^{n} (p_i - p_0)^2} \tag{3}$$

The e-market discussed in this chapter is based on **Smith's experiment** and the **coefficient of convergence** (α measure, see equation (1)) is used to evaluate the efficiency of a market. In our experiments, we will follow Smith's setting for a market in which the **trading agents** negotiate with each other. The behaviors of agents in trading will be introduced with details in the subsequent chapters.

Market Mechanisms

In economics and game theory, interactions of traders consist of two components: a protocol and a strategy. Protocol defines the valid behavior of traders during the interaction. It is set by the marketplace owner and should be known publicly for all the participants. Strategy is privately designed by each agent to achieve their negotiation objectives within a protocol (He, Leung, & Jennings 2003b). Moreover, the effectiveness of the strategy is very much dependent on the protocol: an optimal strategy for one protocol may perform badly for other protocols. In a marketplace, an auction is the protocol by which buyers and sellers interact in this marketplace. Strategy is the adaptive behavior or "intelligence" of traders such as the ZIP agents' updating rules that will be discussed later.

There are many types of auctions in economics. Three types of auctions will be used in this chapter:

1. **English auction** (EA), in which sellers keep silent and buyers quote increasing bid prices, and the buyer with the highest bid is allowed to buy.

2. **Dutch flower auction** (DFA), in which buyers keep silent and sellers quote decreasing offer prices and the seller with lowest offer is allowed to sell. EA and DFA are also called single-sided auctions because either buyers or sellers are active but not both.

3. The continuous double auction (CDA), one of the most popular of all auctions, allows buyers and sellers to continuously update their bids/offers at any time in the trading period. The bids/offers are quoted simultaneously and asynchronously by buyers/sellers. At any time, the sellers/buyers are free to accept the quoted bids/offers.

Trading Agents

Market mechanism design addresses the problem of designing an auction in which the agents' interaction generates a desirable macroscale outcome, by assuming the trading agents are self-interested. How to design self-interested trading agents is a key issue in both computer science and economics. Many types of trading agents have been proposed (e.g., Das et al., 2001; Cliff, 1997; He et al., 2003b). In this section, we will use ZI and ZIP agents with the following reasons. First, they are simple and easy to be implemented. The second reason is that they show human-like behavior and even outperform humans (Das et al., 2001). The third is that we are studying market design but not the agent design. Therefore, we only consider these two types of simple **trading agents** in our experiments of auction designs.

ZI and ZIP Agents

Zero-intelligence (ZI) agents were initially proposed by Gode and Sunder (1993) to study the relationship between limited rationality and competitive equilibrium. Their fundamental discovery is that within the classical double auction market institution, only the weakest elements of rationality need to be present for markets to exhibit high allocative efficiency and price convergence (Brewer, Leung, & Jennings, 1999). ZI agents quote prices randomly in a predefined range of prices determined by the given **supply-demand** schedule. They are also called random traders in some other literatures (McCauley, 2005).

Zero-intelligence-plus (ZIP) agents, an augmented version of ZI agents with simple machine learning technique, are fully described in Cliff (1997). Like ZI agents, each ZIP agent maintains a profit margin within which to quote a price. Different from ZI agents, ZIP agents update their price margins based on previous transaction prices and the qualitative decisions of when to

Algorithm 1. Pseudocode for updating rules of ZIP traders

For sellers:

IF (the last shout was accepted at price q)
THEN (any seller s_i for which $p_i \leq q$ should raise its profit margin)
· if (the last shout was a bid)
· then (any active seller s_i for which $p_i \geq q$ should lower its margin)
· or else
   o   if (the last shout was an offer)
   o   then (any active seller s_i for which $p_i \geq q$ should lower its margin)
For buyers:

IF (the last shout was accepted at price q)
THEN (any buyer b_i for which $p_i \geq q$ should raise its profit margin)
· if (the last shout was an offer)
· then (any active buyer b_i for which $p_i \leq q$ should lower its margin)
· or else
   o   if (the last shout was a bid)
   o   then (any active buyer b_i for which $p_i \leq q$ should lower its margin)

---

alter their profit margin based on the following four factors:

1. Whether the current agent is active or not.
2. Whether the last quote in the market is an offer or a bid.
3. Whether the last quote was accepted or rejected.
4. Whether the last quote is greater or less than the price the agent currently quotes.

The pseudocodes for updating rules are described in Algorithm 1.

A high-level description of the parameters for ZIP traders is described as follows. Each ZIP trader i is given a private secret limit price, i, which for a seller is the price below which it must not sell and for a buyer is the price above which it must not buy (based on **Smith's experiment**). Each ZIP trader i maintains a time-varying profit margin $\mu_i(t)$ and generates quote-prices $p_i(t)$ at time t according to $p_i(t)=\lambda_i(1 + \mu_i(t))$ for sellers and $p_i(t)=\lambda_i(1 - \mu_i(t))$ for buyers. Trader i is given an initial value $\mu_i(0)$ (when $t = 0$) which is subsequently adapted over time using a simple machine learning technique known as the *Widrow-Hoff* (W-H) rule (Hertz, Krogh, & Palmer, 1991) which is well used in back-propagation neural networks. The W-H rule has a "learning rate" β_i that governs the speed of convergence between trader i's quote price $p_i(t)$ and the trader's idealized target price $\tau_i(t)$

which is determined by a stochastic function of last quote price with two small random absolute perturbations: $A_i(t)$ and $R_i(t)$. $A_i(t)$ is generated uniformly from the interval $[0,C_a]$ denoted by $U[0,C_a]$ for sellers and $U[-C_a, 0]$ for buyers. $R_i(t)$ is generated from $U[1, 1+C_r]$ for sellers and $U[1-C_r, 1]$ for buyers. Here C_a and C_r are called system constants. To smooth over noise in the learning, there is an additional "momentum" γ_i for each trader. For each ZIP agent i, its adaptation is governed by three real-valued parameters: learning rate β_i, momentum γ_i and initial profit margin $\mu_i(0)$. Because of the randomness and the uncertainty involved in trading, a trader's values for these parameters are assigned at initialization, using uniform distributions: for all traders, β_i is assigned a value at random from $U[\beta_{min}, \beta_{min} + \beta_\Delta]$; and β_i is from $U[\beta_{min}, \beta_{min} + \beta_\Delta]$, and $\mu_i(0)$ is from $U[\mu_{min}, \mu_{min} + \mu_\Delta]$. Table 1 summarizes the methods for setting these parameters.

Genetic Algorithms for Parameter Settings

According to the above discussion, in order to initialize an entire ZIP trader market, it is necessary to specify values for the six market initialization parameters: β_{min}, β_Δ, γ_{min}, γ_Δ, μ_{min}, μ_Δ plus the other two system constants C_a and C_r. Clearly, any particular choice of values for these eight parameters can be represented as a vector V:

Table 1. Equations of parameter settings for ZIP agents: Parameters of ZIP traders are generated differently according to the type of traders (i.e., buyers or sellers)

	Seller	Buyer
β	$\beta_i=U[\beta_{min}, \beta_{min}+\beta_\Delta]$	$\beta_i=U[\beta_{min}, \beta_{min}+\beta_\Delta]$
γ	$\gamma_i=U[\gamma_{min}, \gamma_{min}+\gamma_\Delta]$	$\gamma_i=U[\gamma_{min}, \gamma_{min}+\gamma_\Delta]$
μ	$\mu_i(t)=U[\mu_{min}, \mu_{min}+\mu_\Delta]$	$\mu_i(t)=U[-\mu_{min}-\mu_\Delta, -\mu_{min}]$
$R_i(t)$	$R_i(t)=U[1.0, 1.0+C_r]$	$R_i(t)=U[1.0-C_r, 1.0]$
$A_i(t)$	$A_i(t)=U[0.0, C_a]$	$A_i(t)=U[-C_a, 0.0]$

$$V = [\,\beta_{min},\ \beta_\Delta,\ \gamma_{min},\ \gamma_\Delta,\ \mu_{min},\ \mu_\Delta,\ C_a,\ C_r\,] \in R^8$$

which corresponds to a single point in the 8-dimensional space of possible parameter values. A **genetic algorithm** can be used to explore this parameter space. Following **Smith's experiment**, the fitness for each individual was calculated by monitoring price convergence in a series of n CDA market experiments, measured by weighting Smith's measurement of convergence on a given **supply-demand** schedule. Each experiment lasted k "days" and the score of experiment number e is:

$$S(V_i, e) = \frac{1}{k}\sum_{d=1}^{k} w_d \alpha(d) \qquad (4)$$

where $\alpha(d)$ is the value of α and w_d is the weight on the day d. According to the experiments in Cliff (1997), all experiments last for six days and we place a greater emphasis on the early days of trading. The weights are set as follows: $w_1 = 1.75$, $w_2 = 1.50$, $w_3 = 1.25$, and w_4, w_5, and w_6 are all equal to 1.00. The fitness of the genotype V_i is evaluated by the mean score of n experiments:

$$F(V_i) = \frac{1}{n}\sum_{e=1}^{n} S(V_i, e) \qquad (5)$$

where $n = 50$ based on the empirical research in Qin (2002), which reported that the average of *50* independent runs of the trading experiments are fairly stable. The lower fitness market has: the sooner the market approaches to the **equilibrium,** the smaller price variances the market has. More experimental results will be given in the fifth section.

EVOLUTIONARY DESIGN OF AUCTIONS

The most desirable market is the one with least transaction price variance to the theoretical **equilibrium** price determined by the market's **supply-demand** schedule. **Smith's experiment** (1962) qualitatively indicated that the relationship of the **supply-demand** schedule has an impact on the way in which transaction prices approached the **equilibrium**, even with a small number of participants; such a market would converge to **equilibrium** after only a small number of trading periods if the supply and demand remained constant. This experiment was conducted by using ZI (Gode & Sunder, 1993) and ZIP agents (Cliff, 1997), respectively. We have discussed the approach of using GAs to optimize parameters for ZIP agents. In this section, we will introduce a continuous **auction space** model (Cliff, 2001) characterized by a parameter that needs to be optimized by GAs. Because this parameter determines the auction type in the market, so this research is referred to as evolutionary auction design.

Hybrid Auctions

CDA is generally considered to be a more efficient auction than single-sided auctions such as the **English auction** and **Dutch flower auction**. Consider the case when we implement CDA. At time t, either a seller or a buyer will be selected to quote. This means that sellers and buyers have a 50/50 chance to quote. We use Q_s to denote the probability of the event that a seller offers. Then in CDA, $Q_s = 0.5$. For the **English auction**, $Q_s = 0$, and for the **Dutch flower auction**, $Q_s = 1$, which means sellers cannot quote and sellers are always able to quote, respectively. The inventive step introduced in Cliff (2001) was to consider the Q_s with values of *0.0, 0.5,* and *1.0,* not as three distinct market mechanisms, but rather as the two endpoints and the midpoint on a *continuous auction space* (see Figure 2).

Q_s, with values other than *0, 0.5,* and *1,* for example, $Q_s = 0.2$, can be interpreted as follows: on the average, for every 10 quotes, there will be

Figure 2. A schematic illustration of the continuous auction space model

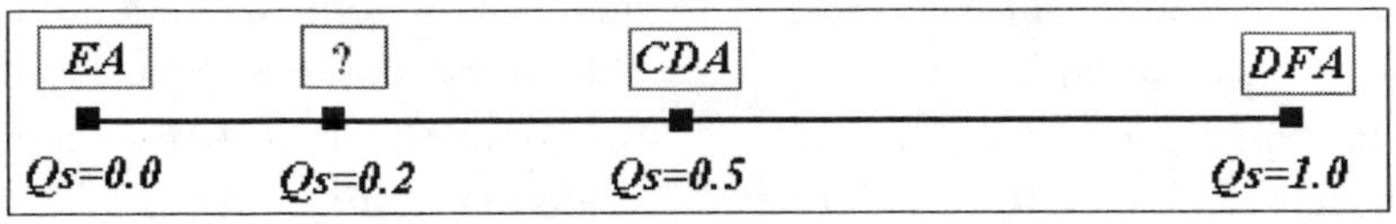

only 2 from sellers while 8 are from buyers. This also means that for a particular significant time *t*, the probability of a seller quoting the trader is *0.2*. This kind of auction is never found in human-designed markets. However, no one knows whether this kind of hybrid mechanism in which $Q_s \neq 0, 0.5$ or *1.0* is preferable to human-designed ones. This motivates us to use a GA to explore Q_s ranging from *0* to *1*. This gives us the following genotype based on the old one by adding a new dimension Q_s (which is now 9-dimensional):

$$V = [\beta_{min}, \ \beta_\Delta, \ \gamma_{min}, \ \gamma_\Delta, \ \mu_{min}, \ \mu_\Delta, \ C_a, \ C_r \ C_a, \ C_r, \ Q_s]$$
$$\in \ \boldsymbol{R^9}$$

In the experiments by Cliff (2003), four **supply-demand** schedules are tested and two **hybrid auctions** are found. However, the continuous space model is not an exact analogue of the EA and DFA. In CDA, by *50%* chance a trader, say a seller, will be chosen to quote at the price *q(t)*. Then any buyers *j* with $p_j(t)$ across *q(t)* can accept this quote. All buyers with prices above *q(t)* will form a candidate price list denoted by *P(t)*. From this list, a buyer is selected randomly to accept the *q(t)*. This is indeed a good analogue to CDA. Now, consider an extreme case of $Q_s = 0$. If a seller quotes *q(t)*, any buyers with price above *q(t)* have equal chances of being selected to make this transaction according to the mechanism of the old **auction space**. However, this is not the case in the real-world **English auction** where the buyer with the highest bid is the only one allowed to make the deal and others with the nonhighest price have no chance to make this deal. Similar to DFA, when a buyer makes a bid, only the seller

with lowest offer price can be allowed to make this deal. Since there is a lot of randomness and uncertainty involved in the experiments and the **auction space** used is not a faithful version of single-sided auctions. A question may arise as to whether the hybrid market mechanism is an artifact of the unfaithful **auction space** used. To answer this, a more realistic model of the **auction space** is proposed in the following section.

Realistic Auction Space Model

Suppose the last quote *q(t)* is an offer and there are *m* buyers with prices above *q(t)*. We put these prices in decreasing order into a set called the *sorted price list* denoted by *SP*, so that $SP = \{p_1, ..., p_m\}$ where $p_i \geq p_{i+1}$ for *i* = 1, ... , *m*–1. If the last quote is a bid, we put the prices of sellers in increasing order so that $p_i \leq p_{i+1}$ for *i* = 1, ... , *m*–1. We then propose a function θ that is defined as follows: when the last quote is from a seller,

$$\theta = \begin{cases} 2Q_s & if : Q_s \leq 0.5 \qquad (6) \\ (m-1)/m & otherwise \end{cases}$$

and when the last quote is from a buyer,

$$\theta = \begin{cases} 2(1-Q_s) & if : Q_s > 0.5 \qquad (7) \\ (m-1)/m & otherwise \end{cases}$$

θ is used to construct a *restricted price list* $RP = \{p_1, ..., p_z\}$ where,

$$z = R(m \bullet \theta) + 1 \qquad (8)$$

R(x) is a function for returning the nearest integer for a real number *x*. The restricted price list (*RP*) is a subset of the sorted price list (*SP*). *RP* contains a number of traders with higher bid prices (for buyers) or lower offer prices (for sellers) according to Q_s. In this **auction space**, a trader to accept the last quote is chosen randomly from *RP*. By restricting the price list of potential traders, the new **auction space** is an exact analogue to single sided auctions as well as CDA. Auctions within this **auction space** model are more realistic to real-world auctions than Cliff's model. For example, when $Q_s = 0.5$, we can obtain $z = m$ so that *RP* = *SP* according to Equations 6, 7, and 8. This means, for CDA, the new **auction space** is the same as the old **auction space**. When $Q_s = 0$, we obtain $RP = \{p_1\}$ where p_1 is the highest price of a set of able buyers, and only the buyer with this price can make this transaction. This is an exact analogue to **English auctions**.

Let us consider an example of a **hybrid auction**. Suppose $m = 5$ (there are five potential traders selected for the last quote) and the last quote is from a seller. When $Q_s = 0.1$, we obtain $\theta = 2 \cdot 0.1 = 0.2$, and $z = R(5 \cdot 0.2) + 1 = 2$, so that $RP = \{p_1, p_2\}$. In Cliff's model, $\{p_1, \ldots, p_5\}$ has the same possibility to be selected to make the deal. However, in the new model, because Q_s is near to **English auction**, the buyers with higher prices (i.e., p_1 and p_2) have more chances to be selected. By using this new **auction space**, the **hybrid auctions** still can be found by simple GAs. More details are in Qin and Kovacs (2004).

EXPERIMENTAL STUDIES

In this section, we conduct a series of evolutionary auction design experiments populated with homogeneous and heterogeneous agents in the realistic **auction space** model. All experiments are based on four **supply-demand** schedules taken from previous research (Cliff, 2003; Qin, 2002; Qin & Kovacs, 2004): SD_1, SD_2, SD_3, and SD_4 (see Figure 3). Each schedule of supply and demand curves is stepped. This is because the

commodity is dealt in indivisible discrete units, and there are only a small number of units available in the market. Thus, supply and demand in this simple market differs appreciably from the smoothly sloping curves of an idealized market. For a given **supply-demand** schedule, the **trading agents** will negotiate with each other following **Smith's experiment**. The transaction prices will be recorded for each trading day. The score for this experiment is calculated by equation (4). We run 50 **Smith's experiments** and the fitness value is calculated by equation (5) with predefined ZIP parameters.

Genetic Algorithms for Parameter Settings

Experimental results based on ZIP agents with GA designed parameters are shown in the following figures. Figure 4A shows the $\langle$ values for 6-day trading experiments with ZI agents given a particular **supply-demand** schedule SD_1 (see Figure 3). The overall fitness according to equation (5) is *43.884*. With "educated guessing" (Cliff, 2001), we can obtain a more efficient market with fitness value *8.904* (Figure 4B). Figure 5 shows the results with ZIP agents whose parameters are optimized by a simple GA over *200* generations (Figure 5A) and *600* generations (Figure 5B), respectively. The fitness values are *8.011* and *4.674* where the latter is much more efficient than other experiments. The parameter settings for the GA are shown in Table 2. These results show that GAs can be used for design parameters of ZIP agents in order to get desired market dynamics. Markets populated by ZIP agents with optimized parameters are much more efficient than the one with ZI agents and ZIP agents with "educated guessing" parameters. Figure 6 also shows a series of transaction prices from one of 50 6-day trading experiments for ZI agents (Figure 6A) and ZIP agents with GA designed parameters (Figure 6B), respectively, given the **supply-demand** schedule SD_1. From the figures, we can see that ZIP agents

Figure 3. Supply-demand schedule SD_1, SD_2, SD_3, and SD_4: there are 22 trading agents, 11 sellers, and 11 buyers, each of them initialized with one unit of goods, their limit prices are distributed as supply, and demand curves show. The vertical axis represents price and the equilibrium price is 2.00 for all these four schedules.

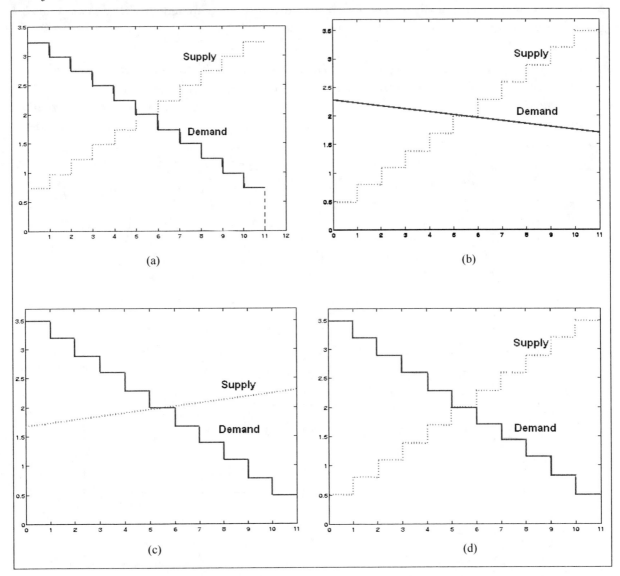

have faster and smaller price deviations from the **equilibrium** price compared to ZI agents. This is because ZIP agents employ a more intelligent method for quoting.

Figure 7 shows the transaction prices for the heterogeneous agents (a mixture of equal numbers of ZI and ZIP agents) given SD_1. We can see that the convergence and the speed of convergence

for the heterogeneous agents are between ZI and ZIP agents. Since the transactions may occur between two ZIP agents, two ZI agents or one ZI and one ZIP agent, we want to study the behaviors of each type of agent. Now we introduce another parameter $\lambda_{ZI/ZIP}$ that is defined as follows:

$$\lambda_{ZI/ZIP} = \frac{\alpha_{ZI}}{\alpha_{ZIP}} \qquad (9)$$

This parameter can be considered as a measure of the divergence of the transaction prices from the **equilibrium** price (see second section), so that α_{ZI} is the measure of divergence for ZI agents and α_{ZIP} is for ZIP agents. Therefore, the $\lambda_{ZI/ZIP}$ value measures the ratio of the divergence of ZI and ZIP agents. When $\lambda_{ZI/ZIP} > 1$, it implies that the divergence of ZI agents is bigger than ZIP agents. Otherwise, ZI agents have smaller divergence compared to ZIP agents. Figure 8

Figure 4. Experiments based on ZI agents (A) and ZIP agents with "educated guessing" parameters: the results show the average (Ave.) value and the average plus and minus standard deviation of α value in each trading day over 30 experiments for the elite individual. "+s.d." means the average plus standard deviation, and "–s.d." means minus standard deviation. Horizontal axis is trading period (i.e., a trading "day"), and vertical axis is the α measure.

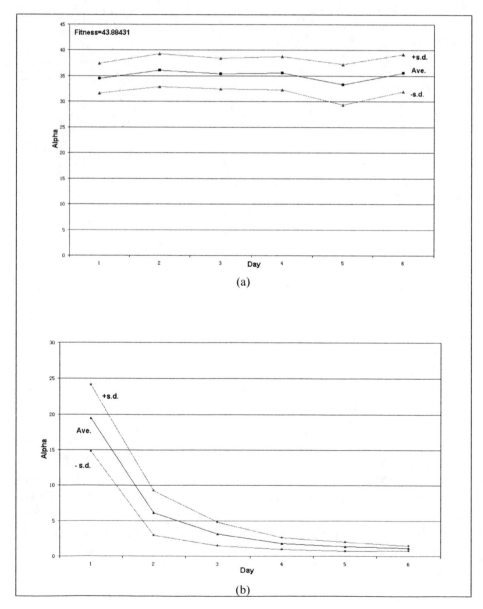

shows the average value of $\lambda_{ZI/ZIP}$ over *500* runs of 6-day trading experiments on SD_1. The average values are approximately 1 for each day, which means both ZI and ZIP agent have nearly the same divergence on each trading day on average.

Evolutionary Auction Design with Heterogeneous Agents

In the market evolution experiments, a simple GA is used to minimize the fitness value (see equation (5)) given *50* independent runs of trading experiments. The values for key parameters of GA are

Figure 5. Experiments based on ZIP agent with optimized parameters with a GA over 200 generations (A) and 600 generations (B): results are the average value of α with it plus and minus the standard deviation over 30 experiments.

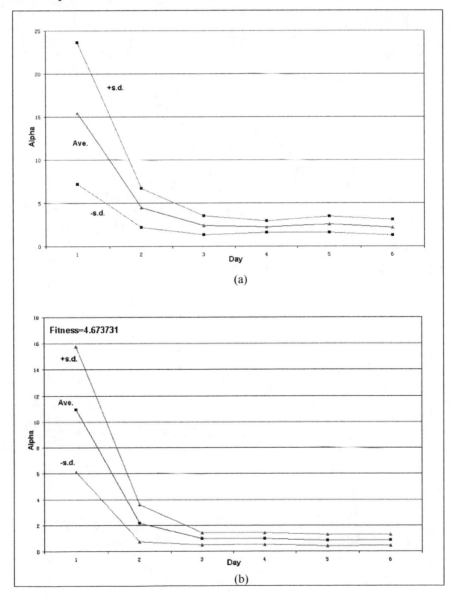

(a)

(b)

Table 2. Parameter settings for the GAs used in the evolving market mechanism experiments

GA-Parameter	Value	GA-Parameter	Value
Population	20	Crossover Rate	0.7
Num. Parameters	8	Mutation Rate	0.015
Max Generation	600	Bits per Parameter	8
Elitism	YES	Selection Method	Rank

Figure 6. A series of transaction prices for ZI agents (A) and ZIP agents (B), respectively, given a particular supply-demand schedule SD$_1$ (see Figure 4) in a 6-day trading experiments.

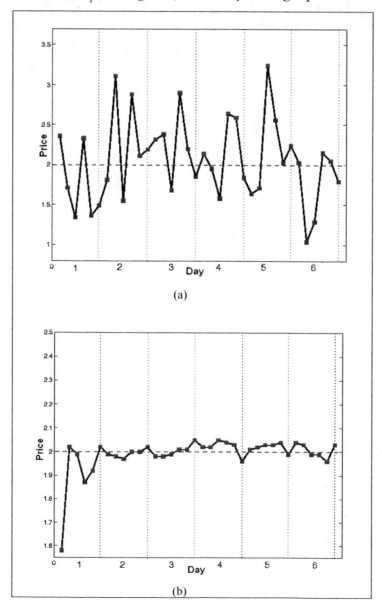

(a)

(b)

given in Table 2. However, one of the drawbacks of using a GA is that it cannot be guaranteed that the solution on which the population eventually converges is a global rather than a local optimum. Thus, we gain formal simplicity at the cost of computation. We run the entire process of evolution many times independently and reduce the effect of mutation as time goes by, to encourage convergence. The results of Q_s represented here are based on *25* independent runs of the GA on the given four **supply-demand** schedules (SD_1 to SD_4) based on the realistic **auction space** model. The average results on homogeneous and heterogeneous agents with standard deviation through generation *600* are shown in Figure 9.

As we can see from the figures, although Q_s values converges to real-world auctions in three of the four given schedules, we still found a **hybrid**

Figure 7. The series of transaction prices for heterogeneous agents (a mixture of half ZI and half ZIP agents) on 6-day trading experiment given SD$_1$

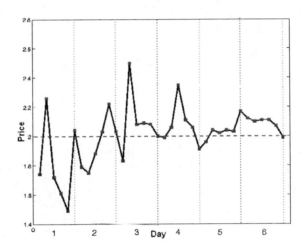

Figure 8. The average value of $\lambda_{ZI/ZIP}$ with standard deviation over 500 experiments

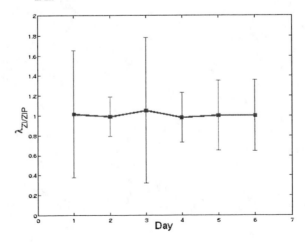

auction in SD_4. Comparing the ZIP agents in the old **auction space** and the new **auction space**, the only difference is SD_3. Both in the old auction and new **auction space** with ZIP agents, there were **hybrid auctions** found by GAs. Cliff, Walia, and Byde (2003) presented a result of using only ZI agents given SD_3 and the **hybrid auction** was found. However, the Q_s values for these **hybrid auctions** are different: $Q_s = 0.39$ for experiments with ZI agents only and $Q_s = 0.23$ for ZIP agents in the new **auction**

space. Here in the experiment with heterogeneous agents which are a mixture of ZI and ZIP agents, the optimal auction is CDA but not a hybrid one. The reasons for this outcome are unknown and we will consider it as a future work.

CONCLUSION AND DISCUSSION

In this chapter, we reviewed the method of using

Figure 9. Comparisons of evolutionary trials of Q_s for ZIP (dot lines) and heterogeneous agents given SD_1 (A), SD_2 (B), SD_3 (C), and SD_4 (D) though 600 generations

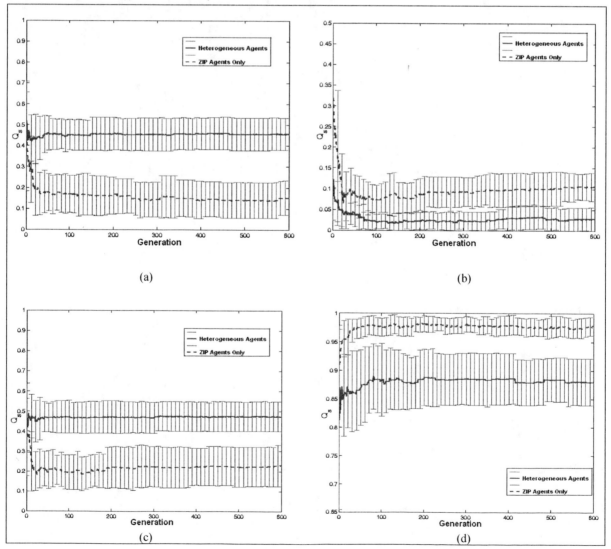

genetic algorithms for mechanism designs. Based on a new model of **auction space** proposed in Qin and Kovacs (2004), the parameter that determines the activities of traders are optimized by GAs in order to obtain the most desirable market dynamics. In our experiments, we tested with homogeneous and heterogeneous **trading agents**. We found that although the two types of agents have different "intelligence," empirical studies suggest the convergence of transaction prices to the **equilibrium** price is almost the same. Based on the evolving market mechanism experiments with heterogeneous agents, we found that in three of the four given **supply-demand** schedules, the Q_s values converge to real-world auctions such as CDA and EA. In the last schedule SD_4, the auction with the most desirable market dynamics was a **hybrid auction**. We conclude that human-designed auctions are likely to be efficient for many **supply-demand** schedules, but not for all, and evolved auctions can be the optimal one for some **supply-demand** schedules.

We would like to point out that this is not a trivial academic point: although the efficiency of automatically designed markets is only a few percentage points better than those designed by humans, the economic consequences could be highly significant. For instance, the total value of trades on the CDA-based New York Stock Exchange (NYSE) for the year 2000 was US$11060 billion[4] and if only 0.1% of the liquidity could be saved by using a market employing an efficient automatically designed mechanism, there would be the value of around US$10 billion. Although the experiments presented in this chapter are far from real-world market, we provide a new way of using evolutionary computing methods for designing market mechanisms. At the current stage, this is academic research without economic consequences though it could potentially be useful.

Based on the experiments presented in this chapter, whether an optimal auction is a **hybrid auction** depends on the predefined **supply-de-**mand schedule. The relation between **supply-demand** schedules and optimal auction still needs more investigations and we will leave it as future work.

ACKNOWLEDGMENT

The author thanks Tim Kovacs for his guidance and help on writing this chapter. Dave Cliff and Andrew Byde in HP Labs Bristol have been very supportive in the early stages of this research. Finally, my thanks go to the anonymous reviewers for their insightful suggestions on this chapter. This work was conducted when the author was with the Department of Computer Science and the Department of Engineering Mathematics of Bristol University, UK.

REFERENCES

Brewer, P. J., Huang, M., Nelson, B., & Plott, C. R. (1999). *On the behavioral foundations of the law of supply and demand: Human convergence and robot randomness* (Social Science Working Paper 1079). California Institute of Technology.

Chamberlin, E. H. (1948). An experimental imperfect market. *Journal of Political Economy, 56*(2), 95-108.

Cliff, D. (1997). *Minimal-intelligence agents for bargaining behaviors in market-based environments* (Tech. Rep. No. HPL-97-91). Hewlett-Packard Laboratories.

Cliff, D. (2001). *Evolution of market mechanism through a continuous space of aution-types* (Tech. Rep. No. HPL-2001-326). Hewlett-Packard Laboratories.

Cliff, D. (2003). Explorations in evolutionary design of online auction market mechanism. *Electronic Commerce Research and Applica-*

tions, 2, 162-175.

Cliff, D., Walia, V., & Byde, A. (2003). Evolved hybrid auction mechanisms in non-ZIP trader marketplaces. In *Proceedings of the IEEE International Conference on Computational Intelligence for Financial Engineering* (CIFEr03), Hong Kong, China.

Das, R., Hanson, J. E., Kephart, J. O., & Tesauro, G. (2001). *Agent-human interactions in the continuous double auction.* Proceedings of the IJCAI-2001, Seattle, USA, 2001.

Gode, D. K., & Sunder S. (1993). Allocative efficiency of markets with zero-intelligence traders: market as a partial substitute for individual rationality. *Journal of Political Economy, 101*(1), 119-137, 1993.

He, M., Jennings, N., & Leung, H. F. (2003a). On agent-mediated electronic commerce. *IEEE Transactions on Knowledge and Data Engineering, 15*(4).

He, M., Leung H. F., & Jennings, N. (2003b). A fuzzy logic based bidding strategy for autonomous agents in continuous double auctions. *IEEE Transactions on Knowledge and Data Engineering, 15*(6), 1345-1363.

Hertz, J., Krogh, A., & Palmer, R. G. (1991). *Introduction to the theory of neural computation.* Addison-Wesley.

LeBaron, B. (2000). Agent-based computational finance: suggested readings and early research. *Journal of Economic Dynamics and Control, 24*, 679-702.

McCauley, J. L. (2005). Making dynamics modeling effective in economics. *Physica A.*

Qin, Z. (2002). *Evolving marketplace designs by artificial agents.* Unpublished master's dissertation, University of Bristol, Computer Science.

Qin, Z., & Kovacs, T. (2004). Evolution of realistic auctions. In M. Withall & C. Hinde (Eds.), *Proceedings of the 2004 UK Workshop on Computational Intelligence*, Loughborough, UK (pp. 43-50).

Smith, V. L. (1962). An experimental study of competitive market behavior. *Journal of Political Economy, 70*, 111-137.

Stiglitz, J. E., & Driffill, J. (2000). *Economics* (chap. 4: demand, supply and price). W. W. Norton & Company, Inc.

Wurman, P. R. (2001). A parameterization of the auction design space. *Games and Economic Behavior, 35*, 304-338.0

ENDNOTES

[1] Web site of TAC: http://tac.eecs.umich.edu/

[2] Research in statistical physics (McCauley, 2005) shows that the convergence of transaction prices is due to the stochastic dynamics and put doubt on the existence of "invisible hand."

[3] Smith won the 2002 Nobel Prize in Economics for his contribution in experimental economics and the study of alternative market mechanisms.

[4] New York Stock Exchange, *Stock Market Activity Report*, 2002.

Section III
Agent Technologies in Electronic Business Processes

Chapter VI
An Inter–Organizational Business Process Study from Agents Interaction Perspective

Joseph Barjis
University of Wisconsin–Stevens Point, USA

Samuel Chong
Accenture, UK

ABSTRACT

It is observed that agent (or software agent) based systems largely imitate organizations of human actors. Thus, the nature of agent-based systems can be better understood by first studying the ordinary human actors or organizations that own the agent-based systems. In this chapter, we first study agent systems and discuss characteristics of software agents. Then we introduce a generic pattern of agents interaction derived from the communication patterns of human actors. Agent-based systems are studied in the context of inter-organizational business process using diagrams and notations adapted by the authors. The methods and concepts used in this chapter are based on the semiotics approach and the language action perspective. For the illustration of our concept of agent-based systems, we discuss a case study conducted based on a real life business.

INTRODUCTION

Organizations of the 21st century are a dynamic, distributed, and complex network of interactions. Daily business operations of these organizations are governed, controlled, and accomplished not only by human actors but to a great extent, if not fully, by software agent systems (e.g., placing or receiving orders, processing payments, sending reminders, alerting of opportunities or new offers, gathering business data or analyzing customer data, using a search engine to find a book at

Amazon.com, using a loan approval tool to obtain a pre-approval in a bank, and so forth).[1]

The main objective of this chapter is to provide some insight into agent systems in an inter-organizational business setting. In this chapter, we define a software agent as a semi or fully autonomous IT system that has been delegated authority to cooperate with other agents to accomplish the common goals of the human actors. Thus, components of modern organizations consist of not only human actors but, to a certain extent, also consist of simplified agent-based systems (e.g., search engines) or more advanced agent-based systems (e.g., automated travel booking agent).

Agent-based systems have always being loosely classed under the artificial intelligence (AI) branch, and, over the years, the research in the area adopted a research paradigm and approach that is tailored to AI. Many social aspects that such systems can bring to society have been largely left ignored or seen as unimportant. For example, a human can merely delegate the authority to carry out a transaction but not the responsibility. The mainstream approaches focus heavily on capturing the technical obligations of software agents during the design. Many of these approaches are based on the object-oriented paradigm and extend an object to include additional properties such as *preferences, moods, and technical obligations.* An instance of these technical obligations is the obligation for a seller agent to send an auto-reply to the buyer agent. What is considered less important in these approaches is the capturing and formalization of exceptions to obligations and social penalties for the human actors when obligations or deadlines are missed. After all, one cannot bring an agent to court to settle any dispute when things go horribly wrong. In this scenario, fallback mechanisms that require the interventions of the human agents need to be put in place. Among others, this chapter argues that the "soft" factors, such as the simple example given, need to be fully considered during the design of such systems. The rest of the chapter is devoted to a discussion and application of two proven theories, namely organizational semiotics (OS) and the language action perspective (LAP), and how they can be or have been adapted over the years by the authors to enable the social aspects to be included in the design of agent-based systems, while not ignoring the important contributions made by the mainstream design approaches. In fact, this chapter furthers the research findings and results of Barjis, Chong, Dietz, and Liu (2002). The adoption of these two sets of theories to agent-based systems is novel and is the first treatment of its kind to understand the design issues of agent-based systems. The novelty of our approach is that an analyst can check and execute her agent system model via a simulation tool before designing or implementing an actual system. This will prevent system developers from the expensive trial and error design approach.

By tackling a few aspects of human actors' interaction, inter-organizational business processes, and agent systems, we put forward the following main hypotheses that this chapter aims to prove:

- To fully understand the social aspects of agent systems, we need to start from studying the communication pattern between human agents and software agents.

- A clear distinction between interactions and actions between customer and provider, and among the members of organizations will enable an agent-based system to better coordinate and support inter-organizational business functions.

- To further understand the current issues that agent based systems face, taxonomy and classification of agent-based systems and their purposes in our society need to be established.

- A clear distinction between the authority and responsibility of agents will enable a socially responsible agent-based system to be design and developed.

- To provide a systematic approach to capture and formalize the social obligations to enable a seamless transition from design to development.

SOFTWARE AGENTS IN E-COMMERCE SYSTEMS

A key characteristic of the e-commerce world is that companies will have to be more customer-oriented to gain competitiveness. However, customer behavior cannot be accurately predicted using traditional analytic methods like forecasting or budgeting (Papazoglou, 2001). In the past few years, software agent technology has caught the attention of system designers and developers of e-commerce systems. Software agents are software entities to which one can delegate some routine or time consuming tasks. Technically, a software agent can be defined as a software entity that has been programmed to possess some form of autonomy to perform tasks on behalf of the human users. Software agents can work in a distributed environment and are helpful in helping the human users to keep track of dynamic changes continuously (e.g., stock prices). In complex software agents, notions such as intentions and beliefs are captured from the human users and programmed into the software agent so that automatic manipulation and reasoning can be performed on a rational basis.

Software agents were first used to automate routine tasks, for example, filtering e-mail (Maes, 1994), scheduling meetings (Mitchell, Caruana, Freitag, McDermott, & Zabowski 1994), collecting newsgroup articles (Lang, 1995), and so on. With the rapid growth of the Internet, the World Wide Web (WWW) and e-commerce software agents now enable the organizations to make their goods or services available to customers by performing some of the tasks in the buying and selling process. Software agents in e-commerce can help customers to automate many of the routine tasks, such as information finding, information filtering, price comparison, and so on. In this respect, software agents are sometimes referred to as mediators in e-commerce systems (Guttman, Moukas, & Maes, 1998; Nwana, Rosenschein, Sandholm, Sierra, Maes, & Guttman, 1998). Agent-based systems are used to refer to any e-commerce systems that incorporate software agents. Agent-based systems are concerned with combining human and software agents to solve a variety of problems in e-commerce. It is on this notion of agent-based e-commerce systems that the rest of the discussions in this chapter will be based: collaboration and interaction of human and software agents.

As the richness and complexity of the information available to the customers has grown, the value in deploying software agents in e-commerce or e-business systems has also grown.

In the following sections, we first introduce and discuss organizational semiotics and LAP Net as a modeling technique to visualize inter-agents communication and interaction and inter-organizational business processes. The resulting models are executable models to study agents' behavior in e-business applications. Finally, a case study with inter-organizational character will be studied with application of semantic analysis, norm analysis, and LAP Net methods.

ORGANIZATIONAL SEMIOTICS

In order to study the complex relationship between a software agent and a human actor, semiotics, the discipline of the study of signs, is adopted. Semiotics was founded by Peirce (1960). The sign is an important notion in the tradition of semiotics. In a semiotic sense, signs can be words, images, sounds, gestures, objects, or anything that carries meaning with it. For example, the sign of a red man on a traffic light signifies that it is not safe to cross the road. Such signs are not studied in isolation, but in consideration of the social context in which the signs find their uses. The semiosis process is an important sense-making process in semiotics and reflects that the meaning of a sign is not contained within it, but arises in its interpretation. This is best represented in a triangle, as shown in Figure 1.

The bottom lefthand corner of the triangle shows the notion of sign. This sign could be anything that refers to another concept other than itself. The object in the righthand corner of the triangle is the concept that the sign refers to. The dotted line linking the sign and object indicates that the relationship between them is always subjectively established through an interpretant, which is the most important notion in the triangle. The interpretant is the sense made of the sign in the mind of the interpreter (agent). It reflects the importance of the process of sense-making and demonstrates that the meaning of a sign is not contained within it, but arises in its interpretation. It is important to note that an agent's interpretation of signs is dependent on the agent's social context, which includes cultural and social norms. Norms include explicit and implicit conventions and rules that govern people's behavior.

The theory of semiotics has given rise to a new branch of study known as *organizational semiotics*. Through the observation of signs and business norms that govern their meanings and uses, practitioners are able to study the requirements of an organization before plunging into the unknown world of software development. In 2001, Chong successfully adapted organizational semiotics into the study of agent-based systems. This piece of research work has shed new light and highlighted the areas that was largely ignored by researchers and practitioners alike and was the first systematic treatment of its kind to progress the advance of agent-based system design and development.

Generic Agent Communication Patterns

To make software agents use languages or signs which are consistent with the way the human agents would, we first need to understand how signs are created, used, manipulated, and understood by the human agents themselves. This will ensure that software agents will not use signs in a way that is contrary to the human agents' natural way of using them. Therefore, we are interested in one aspect of analysis and design, which is studying how signs (and meanings through them) are conveyed by natural language and what aspects of language need to be captured so that through the representation of data, an agent based system can function as an effective substitute for natural human communication. The language to be studied

Figure 1. The process of semiosis

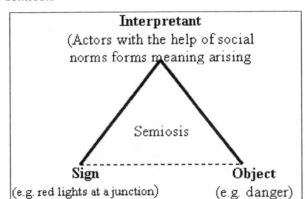

covers not only languages of different countries, but also the technical vocabularies of different professions and organizations.

A key characteristic of software agents is their ability to exchange messages among themselves. This characteristic is to be differentiated from message passing in the object-oriented sense, because objects respond to messages by invoking certain functions within them. Software agents, on the other hand, communicate intentions through messages and need to share a common agent communication language (e.g., KQML and ACL) that they all understand.

To be able to communicate effectively, software agents must have the ability to understand the semantic meanings of each other's message. This is normally facilitated by employing a common vocabulary of terms to eliminate the problem of inconsistencies and arbitrary variations of terms. Without this shared vocabulary that systematically defines the meaning of terms, achieving true interoperability between agent systems has little

prospect. The shared definition of terms is commonly known as the "ontology."

The second key characteristics of software agents are their ability to perform tasks on behalf of their human agents autonomously or semi-autonomously (somewhat dangerous if left to its own devices). The obligations, which may imply the obligations that must be fulfilled by some responsible human agents or software agents, are the resulting conditions of the communication acts between software agents. Understanding the communication pattern among software agents is therefore important because the means of communication are employed to create, modify, and discharge social obligations (Liu, 2000).

The communication process needs to be conceptualized to understand the agent communication pattern. From an organizational semiotics perspective, we are interested in finding out how signs are manipulated in agent-to-agent communication. Figure 2 exhibits the usage of signs in agent communication. When observing

Figure 2. Usage of signs in agent communication

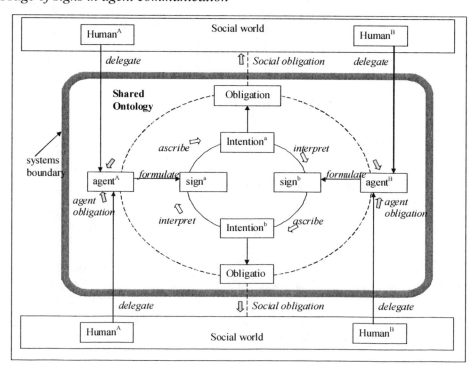

communication between two software agents, one can notice that each software agent actually formulates meaningful electronic signs to communicate intentions to the recipient software agent. The intentions, in the form of speech acts (Searle, 1969) such as promise, may have the effect of creating obligations (shown by the dotted line) for some responsible parties to perform some tasks. These obligations can be further broken down into two types: *agent obligations* and *social obligations*. An agent obligation usually compels some software agent to perform some task. A social obligation, on the other hand, has the effect of constructing or changing the social world, for example, by creating a commitment or changing a legal relationship in business affairs.

Based on such an understanding, we can further expound our understanding of the communication pattern. It appears that one is intrinsically aware that when we promise to wash the clothes, there is an obligation for us to fulfill the promise (assuming it is given in good faith), even though it is not written in black and white. This suggests that there exists a field of force that makes us conscious of our obligations. The principle behind this awareness that in turn leads to our behavior (i.e., fulfilling the obligation) lies in the behavioral norms of our society. By and large, behavioral norms are like an unseen force that compels us to discharge our obligations, though there may be exceptions from time to time.

The foundation for effective agent-to-agent communication on behalf of the human agent lies in the same principle. It involves a highly complex norm-governed form of behavior. Software agents are only assistants to the human agents.

Organizational Semiotics Methods

Using the philosophical understanding described in the third section, several methods have evolved underneath the umbrella of the approach, most notably of which is the semantic analysis and norm analysis. Semantic analysis maps out a well-defined organizational structure, which clearly identifies the role of software agents and their reporting human actors. This enables designers to identify the types of software agents to support the business functions of an organization at the outset.

While semantic analysis focuses on the "what" of an organization and the type of agent-based system needed to support it, norm analysis focuses on the "how." Documents that have been identified during the earlier process are studied further to identify multiple sets of norms that govern the understanding and use of them. Oftentimes, even within the same organization, the same document may have conflicting interpretations to different departments, and norm analysis enables these conflicts to be identified and resolved at the outset. Other types of norms identified during this stage include behavior, perceptual, and cognitive norms. These different sets of norms are the essence that enables different human actors with different backgrounds, job duties, and knowledge to cooperate together in a business setting to achieve a common business goal. The final product of norm analysis is a formalized expression of norms which will enable a seamless transition to the software development process.

Philosophical Assumptions of Semantic Analysis

A key concept in semiotics, first coined by Gibson (1968, 1979) as *affordance*, an affordance refers to conditions made possible by a preceding condition (e.g., swim is an affordance made possible by the existence of water). They can be graphically depicted as shown in Figure 3.

The entity on the lefthand side of the diagram is called the *antecedent* while the one on the right is referred to as the *dependent*. The existence of the dependent is reliant on its antecedent. This is a very simple concept but offers a powerful way of charting out the dependencies of human actors, their actions, and resulting terms which lends

Figure 3. Antecedent and dependent

itself very readily to agent-based systems design. If one were to remove any of the antecedents, the dependent(s) vanishes as well.

If we examine carefully how the knower knows things, we note two routes. First, the knowing agent can get involved directly in some discernible experience or behavior to obtain knowledge of the world (e.g., drive, eat, sit, swim). Alternatively, the knowing agent can know something by referring to signs and information that they or others supply about their direct involvement in the former case (e.g., a driving license, a restaurant receipt, a cinema ticket, a swimming pool entrance ticket). We refer to the first mode of obtaining knowledge as *substantive behavior* and the generation of signs as *semiological behavior*. The semiological behavior involves the creation of signs and assigns meanings to them, which can be used to help us establish and understand coordinated substantive behavior among members of a society. We shall illustrate this using a more familiar example.

Consider ourselves in a position trying to type a resume using our computer. We can go on to experience substantive behavior, such as switching on the computer, activating the word processor program, typing, checking, printing, and so on. These substantive behaviors, of course, are possible on the condition that we can recognize a computer and know what it is capable of achieving. In the process of experiencing these substantive behaviors, we also experience semiological behavior by generating information and signs, such as the printout of the resume. These signs can be used by us, or others, to gain knowledge by referring to them. The printout of the resume, for example, reveals our educational background and work experiences.

Note from the above account that substantive behavior usually does not exist alone. This is because as far as our substantive behavior is concerned, the only things with which we have direct contact are our experiences and direct involvement such as kicking and typing. However, these experiences and involvement are confined to the "here-and-now." Other types of information, constituted from semiological behavior, are needed in order to bind prior and new knowledge to the present knowledge (e.g., a resume tells people the historical facts of the schools that one attended). When we get involved in semiological behavior, we generate signs which carry meanings.

This type of philosophical assumption is useful in understanding agent-based systems because one could never detach a semiological behavior from a responsible human agent. In other words, from a resulting sign, it is possible to trace the human agent that generates the sign through his semiological behavior.

Philosophical Assumptions of Norm Analysis

Stamper (1997) and Chong (2001) identified five types of norms, each of which governs human behavior from different aspects. *Perceptual norms* are concerned with how signals from the environment are perceived by people through their senses such as sight, sound, and taste. *Evaluative norms* deal with why people have certain opinions, values, and objectives. *Cognitive norms* explain how one incorporates existing beliefs and knowledge to interpret what is perceived in order to gain an understanding. *Intentional norms* are concerned with how intentions are formed before deciding on a course of action. *Behavioral norms* govern human behavior within regular patterns. Out of these five norms, behavioral norms are more observable and are the ones that affect and regulate humans' behavior in an organization.

Each of these norms manifests either a mental state or a constraint on the behavior that we will adopt. Perceptual norms concern the way in which we divide up the world into the phenomena to which we attach names such as marriages, poverty, and copyright. Evaluative norms allow us to make judgments about what we have felt and recognized. They function after we have sensed a phenomenon or event. Based on our existing or newfound beliefs, intentional norms steer our intentions. Cognitive norms and behavioral norms can be recognized because their consequent parts affect our beliefs and behavior respectively. For example, in a society that has a high crime rate, it is not uncommon to find that people have false beliefs about the reliability of making credit card payments through the Internet. This general belief may affect the behavior of organizations. As a result, an organization may permit customers to check their market position before making any payment.

Table 1 lists these five categories of norms that will be useful in our analysis of agent-based e-commerce systems.

Conducting Semantic Analysis

The task set for semantic analysis is to understand the business domain by having the meaning of every term of the business domain clearly defined and to have the conceptual dependencies of the terms clearly established. The result of semantic analysis can be graphically depicted as an ontology chart. For brevity, we only show a simplified version of the ontology chart of ABC Supermarket in Figure 4. Semantic analysis can be carried out in four phases as seen from the figure. The ontology chart is a result of respecting the viewpoints of different social groups and deciding which term is important in a social group.

Problem Statement

The semantic analysis begins with the phase of problem statement, which is concerned with developing an understanding of the business requirements. The problem statement is usually given to the analyst by the stakeholders of a company such as the management and employees, in the form of a written document where the requirements are defined. However, in many cases, the written document is not sufficient to provide a complete picture of the business. Therefore, the analyst may be required to acquire additional documents or to carry out interviews with the potential users in order to gather auxiliary information. To demonstrate the semantic analysis method, we make up a simplified version of the problem statement of a fictitious supermarket named ABC:

ABC Supermarket owns and sells a wide range of products. Each product has a unique product description and price per quantity. The accounting department of ABC Supermarket is responsible

Table 1. Categories of norms (Adapted from Stamper et al., 2000)

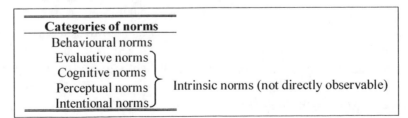

Figure 4. Four phases of semantic analysis

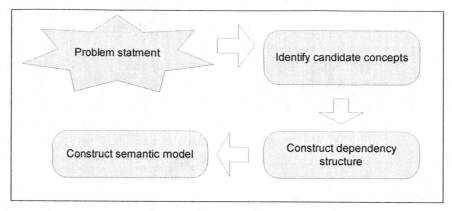

for collecting the payment from the credit card company or the bank. Customers who want to buy some products will search for them from the aisles, put them into their shopping trolley, and will then proceed to the checkout counter for payment. The customer (or the payer, at this point) may choose to pay by several options, including credit cards, debit cards, or cash. ABC Supermarket will check the validity of the credit or debit card number and expiry date.

Identification of Candidate Semantic Entities

When confronted with the business situation in the problem statement, the first step in carrying out a useful analysis of the statement is to identify the different semantic entities (Liu, 2000). This phase is concerned with studying the problem statement to identify potential semantic entities that will be helpful in the analysis later on. Examples of useful semantic entities include verbs, nouns, prepositional verbs, pronouns, or articles, which may indicate possible human agents, software agents, and their potential behavior. Note that behavior here includes the software agents' act of performing speech acts on behalf of the human agents. Every word is useful in analysis,

so one has to be cautious when choosing to ignore a word.

The five semantic entities to be identified are *human agents, software agents, potential behavior* or *speech acts, terms,* and *determiners.* Human agents are the humans or organizations responsible for the behavior of their software agents. Software agents depict the software entities needed to execute the tasks on behalf of their human agents. For example, the software agent on behalf of the human agent may play both the role of a buyer and a holder of a valid credit card. Potential behaviors or speech acts are the possible actions that the software agents can perform. Terms are involved in a communication and are necessary components of any successful communication. Determiners are the set of attributes that determine the characteristics and identity of its parent node. The identification of the terms and determiners is the first step to identifying the required ontology terms. A further analysis of the ontology terms will however be conducted during the norm analysis phase. All the potential behaviors of the software agents, along with their responsible human agents, are specified in Table 2.

When identifying the potential semantic entities from the problem statement, the following rules can be used. Verbs that have the effect

Table 2. Semantic entities table

	Potential behaviour/speech acts (verbs)	Software agent (executor)	Responsible human agent	Initiating human agent
B1	searches	buyer	customer	customer
B2	sells	seller	ABC	ABC
B3	buys ← @ searches	buyer	customer	customer
B4	pays ← @ buys	payer	customer	customer
B5	owns	credit card owner	customer	customer
B6	owns	products owner	ABC	ABC
B7	validates	seller	ABC	ABC

of changing the social world can be classified as the potential behavior of the software agent, for example, "pays" changes the ownership of a product, or "searches" returns some information of value to the users. Nouns that describe the human users or the organization can be classified as the responsible human agent. Sometimes, the responsible human agent is also the initiating human agent of an agent's behavior but this is not always the case. Other types of nouns are classified as terms. Nouns can identify software agents that can have the abilities to accomplish the potential behavior. For example, "payer" has the ability to perform "pays." However, some software agents may have to be introduced and linked to the potential behavior if they cannot be found in the problem statement, for example, "buyer" and "seller." The two "owner" software agents are named "credit card owner" and "products owner" in order to distinguish them. In Table

2, the behavior "pays" is depicted with an arrow sign and a "@" sign, followed by "buys," thus showing that the behavior "buys" triggers the behavior "pays."

It may be impractical for the analysts to know at this stage all the terms and the determiners that are related to them, since there may be a lack of useful information at this point in time. Sometimes, the problem statement may be too vague to gain a full understanding of the term. At this phase, the analysts may make an attempt to list all the terms and their determiners they could possible know (e.g., those listed in Table 3). It is likely that the analysts may leave out some essential details, since this level of analysis is still at a very abstract level. In the norm analysis stage, additional information has to be gathered from a more thorough study of the norms that will serve to specify the meanings of each term.

Table 3. Term table

	Terms (nouns)	Determiners/part of
T1	credit/debit card	#credit/debit card number
T2	products	#product description
		#price per quantity
		#telephone numbers
T3	aisle	#shelf

Semantic Entities Grouping

The specific semantic entities will be grouped according to their generic-specific relationship. The grouping is interpreted from the left to the right. Semantic entities that are on the right are dependent on the existence of the semantic entities on their left to exist. Any semantic entity whose existence is dependent on the existence of other semantic entities is known as the dependent, while the parent concept of the dependent is known as the antecedent. In this phase, the dependents will be identified and connected to their antecedents. As a result, sketches of fragmentary dependency structures are constructed in Figure 5.

These fragmented pieces of dependency structures help one to develop a better picture of the dependency and generic-specific relationship between the semantic entities. The structures identify the five semantic entities in circles (human agents), semicircles (software agents), nodes (potential behavior), nodes (terms), and nodes preceded by the "#" sign (determiners). Once the piecemeal dependency structures are conceived, the next step is to group them together into an integrated chart known as an ontology chart in order to form a unified view of the entire problem domain.

Construction of Ontology Chart

The key to connecting the fragmented dependency structure lies in the dependency of the fragments. When connecting them, some additional semantic entities may be added to present a sound and integrated chart. For instance, "Society" is introduced into the model as the root which functions as the ultimate antecedent because all knowledge in the problem domain is dependent on the existence of a shared understanding of the fundamental cultures or norms in a society. Without these fundamentals, there would not be any agreed meanings of "person" and "buyer."

From the ontology chart in Figure 6, the roles, potential behavior of software agents, and the responsible human agents can be identified. The ontology chart has to be read from the left to the right. The left to right concept shows the generic-specific relationship or the dependency of each semantic entity. Semantic entities that are on the right are dependent on the existence of the semantic entities on their left to exist. The semantic entities in circles are usually the organizations or human agents that are responsible for the behavior of their software agents. The roles of the software agents that are required in the system can be identified from the semicircles. Nodes

Figure 5. Fragmentary dependency structure

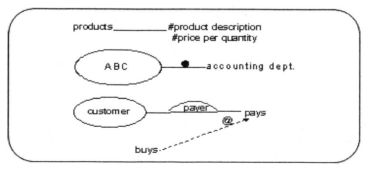

that represent verbs reflect the potential behavior and actions that the software agents can perform on behalf of their human agent. Notice that the existence of the behavior "sells" is dependent on the existence of the antecedents "owner" and "owns." The behavior "buys," on the other hand, is dependent on the existence of the antecedents "customer" and "owns." This implies that selling is only possible for the owner who owns the products, while buying is for any customer. The behavior "buys" and "sells" are referred to the behavior "owns." This means that when people are trading, it is the ownership rather than the physical product itself that is dealt with. The behavior "buys" leads to the existence of the behavior "pays" if the behavior "buys" is satisfactorily completed. Therefore, a dotted line with an "@" sign is used to indicate that the completion of the behavior "buys" activates the existence of the behavior "pays." Nodes indicated with an "#" sign before it are the determiners. Determiners are the set of attributes that determine the characteristics and identity of its parent node.

The ontology chart serves as a unifying framework for the next phase of capturing *intrinsic norms* to specify the meaning of the identified terms and capturing *behavioral norms* for specifying the constraints associated with the potential behavior, which will be discussed in the next section.

Conducting Norm Analysis

Norm analysis is carried out on the basis of semantic analysis. The ontology chart delineates the area of concern in the business environment. However, certain issues are not dealt with by semantic analysis. Norm analysis serves to examine the ontology chart in greater depth, the result of which will also be used in the interface requirements analysis. Norm analysis is useful for studying agent-based systems from the perspective of the human agents' behavior which is governed by norms. From this perspective, these norms can be specified in natural language and further translated into any programming language for the

Figure 6. Ontology chart of ABC

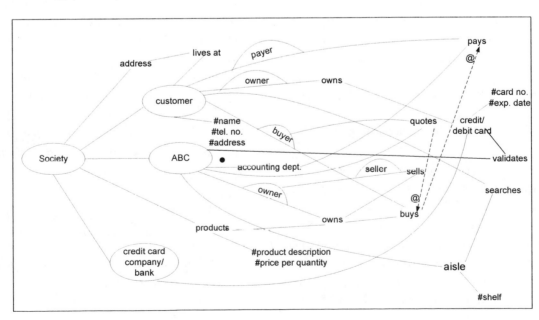

execution of the software agent. Underlying each potential behavior and term in the ontology chart, there is a set of associated behavioral norms and intrinsic norms, respectively.

Although the ontology chart offers an understanding of what potential behavior is available, the detailed rights and constraints associated with the realization of each behavior are not covered. The potential behavior specified in the ontology chart only shows the link between a behavior and its antecedent agent without imposing any further constraints. However, existence of the antecedent agent may not necessarily lead to the existence of the dependent behavior. We propose that the rights and constraints associated with the existence of each behavior could be better understood by eliciting the behavioral norms from the human agents. Behavioral norms are the ones that affect and regulate human agents' behavior in an organization. They allow the human agent to judge the situation and decide whether to be involved in particular behavior. They have a prescriptive function in governing the behavior of the human agents and are expressed in the form of "you are obliged, permitted, or forbidden to behave in a certain way." The constraints under which the human agents will behave is studied so that the behavior of the software agent will be consistent with the human agents if the human agents have to do the same tasks themselves. In general, behavioral norms have the following structure:

```
IF <certain conditions apply>
    THEN <agent>
        is <obliged/permitted/forbidden>
            To perform <action/speech act>
```

Intrinsic norms, on the other hand, can be found underlying each term in the ontology chart. They can be specified from the human agents to enhance the user interface of the software agents and their knowledge of the environment. They have a descriptive function in depicting the mental states of individual human members of a society. For example, intrinsic norms enable members to

recognize the term "customer," expecting that it is capable of buying a product and have an intention of making the actual payment. Intrinsic norms govern how the human agents perceive, evaluate things, and determine how they expect things to happen around them. They are elicited directly from the human agents and play an essential role in maintaining the stability in the environment in which the software agents work.

In particular, perceptual and evaluative norms can be elicited from the human agents to understand and determine the business context, from which the jargon and business language obtain their meaning. This issue will be dealt with in the contextual HCI semiotics analysis phase which is out of scope in the chapter. Cognitive and intentional norms, on the other hand, reveal the beliefs and intentions of the human agents and are vital to the successful performance of the tasks at hand. For example, if we know that a human agent will have an intention to pay for their goods either by cash, check, credit card, or debit card, the software agents that are developed must be capable of supporting such intentions. The beliefs and intentions, when elicited can be programmed into the software agents and form part of the knowledge of the human agents who the software agents represent. This knowledge is a mentalistic metaphor and is in no way implying that software agents are humanlike. The knowledge includes the kind of beliefs and intentions to expect from the human agents, and how it should cope with and support the expected beliefs and intentions of the human agents in order to create the predictability and the stability in the environment in which the software agents work. Intrinsic norms have the following structure:

```
IF <certain conditions apply >
    THEN <agent>
        Adopts <an attitude>
            Towards < some consequences or proposition>
```

At this stage, we are interested in eliciting as many potential behavioral and intrinsic norms from the ontology chart as we can. However, norm analysis does not stop at identifying and capturing the behavioral and intrinsic norms. A complete analysis of the norms has to be performed by taking a step further in the analysis of these norms to deal with more complex issues in the subphase of norm analysis: *behavioral analysis*.

Adopting the two structures, ABC Supermarket may state the intrinsic and behavioral norms as shown in Textbox 1 (the norms shown here are not exhaustive):

Once the intrinsic norms are specified, they are fed back into the analyst's search for the meaning of the terms, which was left incomplete in Table 3 due to a lack of information. The terms and their determiners or subcomponent (part-of relation)

are specified in Table 4. The accounting department cannot be considered a determiner of ABC Supermarket; instead, it is more appropriate to see it as a part of ABC as can be revealed from IN3 in Textbox 1.

Behavioral Analysis

When we design the behavior of software agents, we are also defining much about the responsibilities of their human agents. Performance of the behavior "pays" by a software agent, for example, may establish or invoke responsibilities between the human agents. The analysis therefore focuses on the responsible human or software agents.

Textbox 1. Norm specification

```
Perceptual Norms:
IN1 IF ABC owns a product,
          THEN seller agent
                    will recognise
              t          hat it has a product desc. and a price.
IN2 IF customer owns a credit card,
          THEN buyer agent
     w               ill recognise
     t               hat there is a credit card number and expiry date.
IN3 IF ABC is receiving payment for the product,
          THEN payer agent
     w               ill recognize
     t               hat it is paying to ABC accounting department.
Cognitive Norm:
IN4 IF customer wants to buy a product,
          THEN payer agent
                    will believe
              t          hat it is possible to pay by credit card or debit card.
Behavioural Norms:
BN1 IF customer had requested information about products
          THEN seller agent
                    is obliged
     t                         o provide information on products.
BN2 IF customer had not selected any products or delivery date,
          THEN buyer agent
                    is forbidden
     t                         o checkout.
BN3 IF customer wants to make payment,
          THEN buyer agent
                    is obliged
     t                         o provide valid credit/debit card number and expiry date.
BN4 IF customer had submitted valid credit/debit card number and expiry date,
          THEN buyer agent
                    is obliged
                         to confirm payment or delivery date.
BN6 IF ABC had made a quotation,
          THEN seller agent
                    is obliged
              t          o sell at the agreed price.
BN6 IF customer had sent credit/debit card details,
          THEN customer
                    is obliged
                         to make the actual payment.
```

Table 4. Term table (revised)

Terms (nouns)		Determiners/part of
T1	credit/debit card	#credit/debit card number
		# expiry date
T2	products	#product description
		#price per quantity
T3	customer	#name
		#address
		#telephone numbers
T4	Sainsbury's	accounting dept. (part of)
T5	aisle	#shelf

Responsibility in this respect can be viewed from two perspectives: the obligations incurred on the software agents and the social obligations incurred on the human agents.

In agent-based systems, software agents adopt the behavior found in the ontology chart in order to perform the delegated tasks. The performance of these behaviors may incur an obligation on the software agents to perform other actions or speech acts. In other cases, however, the human agents' actions are needed to fulfill the social obligations that arise from the task performed by their software agents. It is important for any company to know what kind of social obligation is incurred on them in order to gain the trust of their customers, such as keeping their obligation to sell the goods at a quoted price. The obligations can be determined by the behavioral norms associated with each behavior. Capturing the behavioral norms to identify the obligations of the software and human agents enables the analysts to get a fresh approach to looking at such issues.

In this phase, we are interested in capturing the obligations that arise due to the interaction between the software agents and the human agents. The flow of interaction is depicted as an interaction diagram in Figure 7. In contrast to the interaction diagram in object-oriented design, which shows the sequence of messages passed between objects, the interaction diagram resulting from norm analysis is totally free from imple-

mentation details. One therefore does not need to have in mind any implementation strategy. It can be drawn without implying that there should be some sequence of operations or some aggregation of links between objects or entities.

In the interaction diagram, the software agents are seen as helpers to the human agents. It shows the obligations (perlocutionary effects) on both the human and software agents that arise due to the interaction between the software agents. There are two types of obligations: the ones depicted by the grey arrow are the social obligations on the human agents, while the dotted arrows with a bold bar are the obligations on the software agents. The gray rectangle depicts the system boundary, through which the human agents communicate with their software agents. The rounded symbols depict the behavioral norms that specify the responsible software or human agents. The submission of the credit card details from the payer software agent to the seller software agent, for example, creates two obligations. These obligations are represented by the rounded symbol named BN4 and BN6. BN4 specifies that there is an obligation on the seller software agent to confirm the payment details and delivery date, while BN6 specifies that there is a social obligation on the customer to make the actual payment.

The following presents the relationship that exists between the delegation and obligation in the interaction diagram.

Figure 7. Interaction diagram

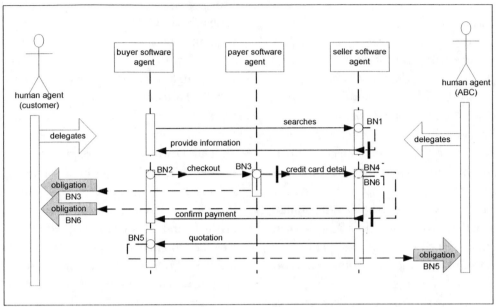

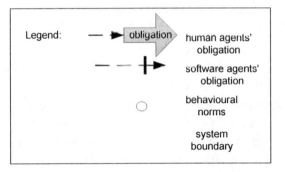

When $x = i_n$, $n \leq 2$; i_1 = human agent, i_2 = software agent

$Agent_i.Delegate_{j, k,}$

THEN

$Agent_j.Obl_{x, k}$

(means that when $agent_i$ delegates a task k to $agent_j$, $agent_j$ is obliged to perform task k on behalf of $agent_i$ which is denoted by x).

From the above, the following relationship is obtained:

$\forall Agent_1, i \neq j, i = x;$

$\forall Agent_i \{Delegate_{j, k}\} \Rightarrow Agent_j \{Obl_{x, k}\}$

For instance, two examples of the obligations captured in the interaction diagram can be expressed syntactically as:

$Customer_1.\{Delegate buyer_2, search\} \Rightarrow Seller_2.\{Obl Customer_1, provide information\}$

(when human agent customer delegates the task search to software agent buyer, this implies that software agent seller is obliged to perform the task provide information on behalf of human agent customer).

Customer$_1$.{ Delegate payer$_2$, pays } $\Rightarrow$ payer$_2$.{ Obl Customer$_1$, provide valid credit/debit card details and expiry date}

(when human agent customer delegates the task pays to software agent payer, this implies that software agent payer is obliged to perform the task provide valid credit/debit card details and expiry date on behalf of human agent customer).

The interaction diagram provides a graphical and syntactical means of specifying the obligations of both the human and software agents. It is important to make this distinction because the software agents only act on behalf of the human agents. Where there are some obligations that can be dealt with by the software agents, the human agents need to deal with the social obligations incurred on them through their software agents.

One of the limitations of the ontology chart is that it is charted as a "timeless" design artifact. Like the ontology chart, the interaction diagram introduced is also "timeless" because there was no mention of what information is needed for a behavior to occur. In order to gain an understanding of the order and constraints under which the behavior in the ontology chart would be executed, we introduce the *Language Action Petri Net* to deal with the time aspect, because in actual business practice, business behavior occurs in relation to time. In this phase, one needs to re-examine the problem statement, ontology chart, and the norms specification for the potential behavior and the enabling conditions or constraints for their occurrence and put them together. The end result is a sequence diagram which shows the sequences and the enabling conditions of the behavior adopted by the software agent.

LANGUAGE ACTION PETRI NET

In LAP Net, the core concept is of the business transaction concept introduced within the DEMO methodology (Dietz, 1994) and further developed and discussed in Dietz (2006). Often, the terms transaction, business transaction, and essential transaction are used interchangeably. We, most probably, will follow the same interchange when referring to a transaction, although we would prefer to refer to it as an *essential transaction*, because it is thought that the essence of an organization captured in essential transactions is a constant and remains independent of the realization aspects or the supporting IT.

The DEMO transaction concept is based on the "theory of communicative action" by Habermas (1984). What follows is an illustrative introduction to the transaction concept using artifacts and constructs adapted by the authors. Readers, interested in more in-depth study about the transaction concept, are referred to the above cited original works by Dietz. We have adapted the Petri net notations and extended them as modeling constructs. Assuming that readers are familiar with the basic concepts of Petri nets that are widely used in systems analysis and design, we skip their introduction. Interested in Petri nets, readers are referred to Murata (1989), Peterson (1981), and Reisig (1985), for example.

The transaction concept is based on the idea that an agent system and its underlying business processes can be better understood through the observation and analysis of interaction between the agents involved. Thus, this concept implies communication as a tool for capturing action patterns. In this context, the notion of communication is not just an exchange of information, but sharing of action-triggering meaning such as negotiation, coordination, agreement, and commitment that led to certain actions. In turn, these actions create new facts, deliver results, and accomplish the mission of an agent system, an organization.

Transactions are patterns of interactions and actions (see Figure 8). The *action* is the core of a business transaction and represents an activity that brings about a new fact changing the state of the world. An *interaction* is a communicative act involving two agents (actor roles) to coordinate or negotiate. An example of an interaction is a request made by one actor towards another actor that leads to creation of a new fact, for example, "calculating a quote for a new insurance policy." Other examples of interactions are clicking "apply" or "submit" buttons in an electronic form, searching online databases, inserting debit card into an ATM to withdraw cash, or pushing elevator's summon button.

Each business transaction is carried out in three distinct phases, the *order phase*, the *execution phase*, and the *result phase*. These phases are abbreviated as O, E, and R, correspondingly, and constitute the OER paradigm (Dietz, 1994). Note the order (O) and result (R) phases are interactions, and the execution (E) phase is an action, therefore they are illustrated using different colors and shapes (the Execution phase is represented by a rectangle with slightly rounding angles and colored in blue or grey in grayscale print). Namely, these three phases are a distinct feature of the LAP Net that claims to be a business process modeling technique vs. just a process modeling. These three phases not only allow for the boundary of an agent (or business unit) to be clearly defined, but also to depict interaction and action as a generic pattern involving (social) agents.

Now, based on Figure 8, we try to introduce the further notions of the transaction concept along with the Petri net notations we adapted. A distinct notion of the transaction concept is the role of agents involved in a transaction. Each transaction is carried out by two agents. The agent that initiates the transaction is called the *initiator* of the transaction, while the agent that executes the transaction is called the *executor* of the transaction. Since the order (O) and result (R) phases are interaction between the two agents, their corresponding transitions are positioned between the two agents. However, the execution (E) phase is an activity solely carried out by the executor, and, therefore, its corresponding transition is positioned within the confines of the executor. In case of multiple agents, they will be conveniently denoted by the letter A and numbered (A1, A2, A#).

In the Figure 8, interactions between the two agents are illustrated at a high or essential level. In effect, each of the two phases (O and R) may involve a series of back and forth interactions (request, offer, counter-offer, negotiation, decline, etc.). A complete state-transition schema for the conversation for action can be found in Winograd and Flores (1986, p. 65) and a transaction in Dietz (2006, chap. 10).

Another distinction is made between simple and composite transactions. Agents' interactions may be arbitrarily complex, nested, extensive, and multilayered (hierarchical). A complex collaboration typically consists of numerous transactions that are chained together and nested into each

Figure 8. A transaction diagram: (a) detailed; (b) compact

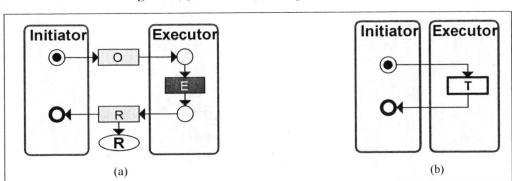

(a)   (b)

other. *Simple* transactions do not involve, that is, trigger or cause, other transactions during their execution (like in Figure 8). In *composite* transactions, on the other hand, one or more phases will trigger further nested transactions. For instance, agent A1 contacts agent A2 to reserve a hotel room (we denote this request as Transaction 1, or T1). Agent A2 receives the request, checks the room availability, but in order to complete the request, it has to request the first agent (A1) for a payment guarantee (we denote this second request as Transaction 2, or T2). For agent A2 to complete the reservation task, first the payment transaction should be completed. A close look at the reservation process reveals that, in fact, the payment transaction, T2, is rather an interaction between the hotel and a credit card company (a typical inter-organizational process). Thus, the process rather involves three agents: A1 (customer or guest), A2 (hotel receptionist), and A3 (credit card company). The interaction process between the three agents is illustrated in the form of a nested transaction in Figure 9.

In this manner, any complex process with any number of actors and outcomes can be modeled and illustrated. However, for more complex processes, one needs to use the compact notation of a transaction in order to keep the model better managed and controlled. The compact notation is useful for those transactions that are simple (not nesting further transactions). If a compact notation is used, by a convention, the whole transaction is positioned within the confines of the executing agents.

Another notion, a typical phenomenon in process modeling, is of probability of some activities: optional transactions that may take place depending on some conditions. To indicate that a transaction is an optional one, a small decision symbol (diamond shape) is attached to its initiation (connection) point as illustrated in Figure 10. In order to transform this optional transaction construct into standard Petri net semantics, a traditional XOR-split that could be modeled by one place that leads to two transitions is used. It requires addition of a dummy transition as demonstrated in the figure (notice the tiny rectangle with no labels). A dummy transition is meant that it has zero duration and consumes no resources.

Finally, there are situations that a process may halt and result in a termination. For example, if there is no room available, then the payment transaction is not initiated at all. This situation is modeled through a place identified as "decision state" graphically represented via a circle with decision symbol (diamond shape) within it (see Figure 10). As seen, for the transformation of a decision state into standard Petri net semantics, a traditional XOR-split that could be modeled by one place that leads to two dummy transitions is used. Depending on the value of the state, the process either proceeds or terminates (circle with cross).

Through these few simplified constructs and minimodels, we aimed to introduce how LAP Net can capture typical situations in inter-organizational business processes and agents interaction, provide sound concepts based on agents

Figure 9. Nested transactions with three actors

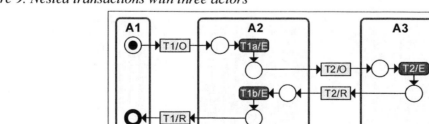

Figure 10. Standard Petri net representation of (a) an optional transaction and (b) a decision state

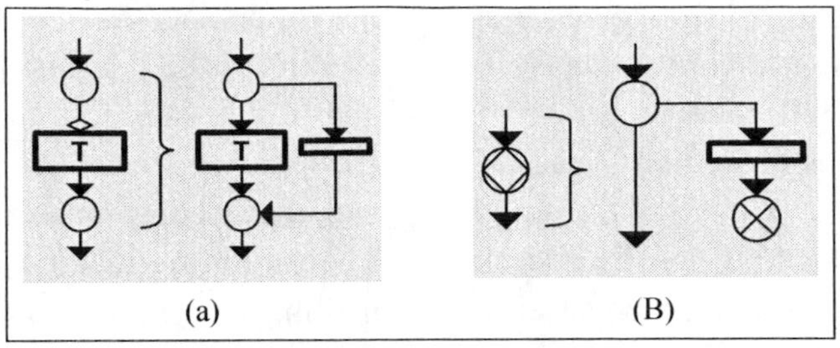

(a)   (B)

interaction, and ultimately contribute towards a more accurate agent-based e-business application design.

Due to the generic business phenomena that it studies, LAP Net lends itself very easily to the study of inter-organizational business transactions. In this chapter, the above concepts are applied to agent-based system, and a systematic approach is introduced to enable practitioners to distinguish between interactions and actions. It is argued that only when this distinction is made will practitioners be able to design and develop an agent-based system that not only facilitates controlled inter-organizational interactions, but also is responsible and knows well in advance the change in the state of affairs for its business organization which arises from business transactions.

ORGANIZATIONAL SEMIOTICS AND APPLICATION OF LAP NET

In this section, we demonstrate the application of both LAP Net and organizational semiotics, more precisely, semantic analysis and norm analysis. Throughout, using the "Family Health Care Center" case, we demonstrate how the methods and modeling techniques derived from the two frameworks can be integrated into a whole.

Case Study: Family Health Care Center

FHCC was founded in 1992 in response to a growing need for a comprehensive family-oriented health care center in the community to provide an extensive variety of health care services. Starting the practice with one physician and one nurse practitioner in 1992, currently FHCC has four physicians, two nurse-practitioners, and a number of nurses, technologists, and administrative employees. Altogether, FHCC clinicians and staff include more than 30 people. The center is managed by a professional director with a master's in business administration (MBA). Although the center's business process is fairly complex, what follows is a simplified description of business processes pertinent to the patients' examination.

Patient Examination Process (Problem Statement)

In order to be examined by a doctor, a patient needs to make an appointment beforehand. The only situation in which "walking in" is allowed, is in an emergency situation. Sometimes, the FHCC may see regular patients who walk in, but this action results in putting the staff behind schedule for those who made appointments. The daily routine at the FHCC, in regard to patients' examinations, starts with printing the appointment list, having the corresponding patient's charts (history) ready at the front desk (reception) and creating a superbill (face-sheet or multilayer bill) for each patient. A patient, upon arrival, signs in on the "Check In" sheet at the front desk and waits in the waiting room to be called. Meanwhile, the front desk forwards the patient's chart and a face-sheet to the nurse's desk so that the first available nurse may deal with the patient.

The nurse calls the patient and conducts a preliminary general checkup (blood pressure, EKG, basic lab work) and records chief complaint(s) and reason(s) for the visit. After completing this preliminary exam, the nurse escorts the patient to an available examination room and places the chart into the designated box at the door of the examination room. By the established procedures, posting the chart at a specific door indicates to the corresponding doctor which patient must be seen next. Several examination rooms are available in the center, and several physicians function at the same time. If no room is available, the patient is asked to wait in the internal waiting room (behind the front desk), and the chart is queued on the nurses' desk for a room to become available. The doctor examines the patient and updates the patients chart if any prescription is issued, diagnosis is made, referral is given, or if any other notes are taken. After completing the examination, the patient is given a copy of the face-sheet and escorted to the side desk to check out. The patient goes to the side-desk to check out, to make the (co-)payment relevant to the service delivered, and, if needed, to make a follow-up appointment. The examining doctor or the assisting nurse, after making all the updates to the chart, returns it to the storage location.

In most cases, a patient's visit to the doctor represents a routine reason (high blood pressure, diabetes, and/or infections). The FHCC is capable of providing most of the services and treatments a patient may need; however, in rare cases, patients may need further examination by external health care providers (subspecialists) or advanced diagnostic equipment such as a CAT scan, available elsewhere.

In this case, the FHCC, after providing a preliminary diagnosis, schedules an appointment with the external health care provider based on the availability of the network provider. Some procedures such as a CAT scan may require the insurance company's pre-approval, in which case the FHCC first requests pre-approval and then makes the appointment arrangement. Usually, this takes a day or two, and a nurse will make the arrangements. Finally, either the FHCC or the external health care provider itself informs the patient about the new appointment.

Table 5. Semantic entities of FHCC

	Potential behaviour/speech acts (verbs)	Software agent *(executor)*	Responsible human agent	Initiating human agent *(initiator)*
B1	Make appointment	Patient agent	Receptionist	Patient
B2	Request healthcare	Patient agent	Receptionist	Patient
B3	Conduct test ←@ Request healthcare	**Not delegated**	Physician	Physician
B4	Arrange external test ← @ Conduct test	Appointment agent	Receptionist	Physician
B5	Request pre-approval	Approval agent	Receptionist	Nurse
B6	Pays bill	Payment agent	Patient	Receptionist

Identification of Business Transactions Using Semantic Analysis

Table 5 shows the list of potential agents and behavior/speech acts that are identified from the FHCC's problem statement.

Table 6 lists some of the terms that need to be understood by all parties in the problem statement.

Figure 11 illustrates the ontology chart of the FHCC.

As mentioned before, the ontology chart serves as a unifying framework for capturing *intrinsic norms* and *behavioral norms*. For the interest of space, a detailed explanation of the ontology chart is not given. Please refer to see the third section for an explanation of how the chart is to be interpreted. It is worth noting that in Chong (2001), it has been shown the ontology chart feeds directly into the development of an ontology for an agent-based system. This is a unique treatment to ontology development, which normally involves *ah-hoc* identification of terms in the ontology.

Table 6. Term table of FHCC

Terms (nouns)		Determiners/part of
T1	Appointment list #	date/time
T2	Patient's chart #	patient's health history
T3	Super-bill #	patient's name
T4	Check in sheet	#date/time
T5	Patient's chart #	blood pressure, #EKG, #basic lab work

Figure 11. Ontology chart of FHCC

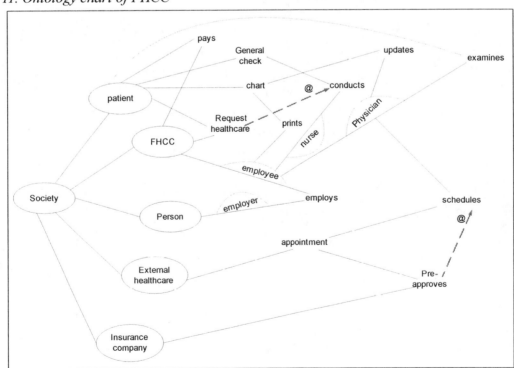

Figure 12 defines the responsibilities of the human agents and obligations of their software agents. For example, the behavior "request appointment" by the patient agent invoke responsibilities between the patient and the physician. Both have a responsibility to see each other at the agreed prescribed time, and such social responsibility needs to be captured and codified, perhaps in the form of sending a reminder to both human agents before the appointment date.

Responsibility in this diagram can be viewed from two perspectives: the obligations incurred on the software agents and the social obligations incurred on the human agents.

Two examples of the syntactic expression of the diagram are as follows:

$Patient_1.\{Delegate\ patient\ agent_2,\ search\} \Rightarrow patient\ agent_2.\{Obl\ Customer_1,\ retrieve\ patient's\ chart\}$

Textbox 2. Norm specification extract of FHCC[2]

```
Perceptual Norms:
IN1 IF a patient is registered
        THEN Patient agent
                will recognise
                        t       hat a patient has a patient's chart.
IN2 IF a patient wishes to make an appointment,
        THEN Patient agent
                will recognise
                        t       hat FHCC has an appointment list.
IN3 IF a patient is receiving payment for the product,
        THEN payment agent
                will recognize
                        that it is paying to FHCC accounting department.
Cognitive Norm:
IN4 IF a patient is paying via credit card,
        THEN payment agent
                will believe
                        t       hat the credit card number and expiry date are valid.
Behavioural Norms:
BN1 IF a patient is making an appointment,
        THEN Patient agent
                is obliged
                        to retrieve an available date from the Appointment agent.
BN2 IF a patient has made an appointment
        THEN Appointment agent
                is obliged
                        to retrieve the patient's chart and create the super-bill
BN3 IF a patient has requested for healthcare,
        THEN Appointment agent
                is obliged
                        t       o book patient in for his/her appointment.
BN4 IF a physician has requested for external test,
        THEN Appointment agent
                is obliged
                        to propose an appointment date.
BN5 IF an external test requires pre-approval
        THEN Appointment agent
                is obliged
                        to request pre-approval from the insurance company.
BN6 IF a patient wants to make payment via credit card,
        THEN buyer agent
                is obliged
                        to validate credit/debit card number and expiry date.
BN7 IF a patient has given an invalid credit card,
        THEN buyer agent
                is forbidden
                        to confirm payment.
BN8 IF a patient has been given an appointment date
        THEN patient
                is obliged to
                        to attend appointment
BN9 IF a patient has been given an appointment date
        THEN physician
                is obliged to
                        to see patient
```

Figure 12. Interaction diagram of FHCC

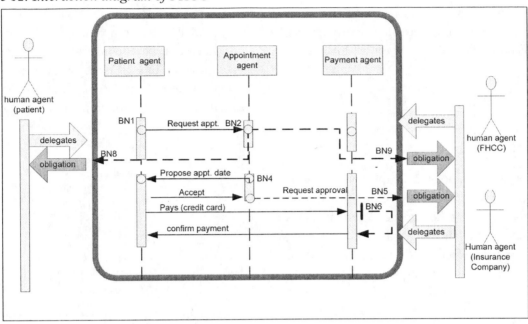

(when human agent patient delegates the task search to software agent patient agent, this implies that software agent patient agent is obliged to perform the task retrieve patient's chart on behalf of human agent patient).

Patient$_1$.{ Delegate patient agent$_2$, pays } $\Rightarrow$ payment agent$_2$.{Obl Customer$_1$, validate credit/debit card details and expiry date}

(when human agent customer delegates the task pays to software agent payer, this implies that software agent payer is obliged to perform the task provide valid credit/debit card details and expiry date on behalf of human agent customer).

One particular point to note about the information gathered from semantic and norm analysis is that they are timeless; that is, they do not tell the analyst much about the execution order. LAPNET repositions the above information into business transactions and constructs a temporal model of these transactions.

Making an appointment is the first activity in the series of processes taking place in the "patient examination." By making an appointment via an appointment that the patient will have access to via the Internet, the appointment agent has to retrieve a patient's chart and create the superbill, hence a new fact (result) is created, and this new fact is that a new appointment is made and recorded into the system. In this activity, the patient is the initiator, and the receptionist who talks to the patient (and who delegates this responsibility to an appointment agent) is the executor. So this activity comprises the first business transaction (T1) in the process of "patient examination":

T1:	making an appointment
Initiator:	patient
Executor:	FHCC (receptionist)
Fact:	a new appointment is made
Behavior Norm:	#1 and 2

Based upon a previously made appointment, a patient visits the medical practice for health care.

His arrival to the practice is marked by signing the "sign-in" sheet (which is automated by interacting with the appointment agent on the computer), which is considered a request for health care:

T2:   requesting health care
Initiator: patient
Executor: FHCC (physician)
Fact: patient is given health care
Behavior Norm   #3

Patient examination is a complex process nesting quite a few activities, therefore Transaction T2 is called a composite transaction and includes a number of other transactions. The following transactions are nested within T2, that is, initiated during the execution of transaction T2:

T3:   conducting general physical test
Initiator: FHCC (physician)
Executor: FHCC (nurse)
Fact: general physical test is conducted
Behavior Norm: N/A[3]

T4:   arranging an external appointment
Initiator: FHCC (Physician)
Executor: Specialist (external provider)
Fact: an external appointment is made
Behavior Norm: #4

Some of the external services may require a pre-approval of the insurance company; therefore, before T4 can be completed, FHCC needs to initiate a request to the insurance company for a pre-approval. Thus, T4 is also a composite transaction that nests Transaction T5. This interaction is facilitated via the appointment agent.

T5:   requesting a pre-approval
Initiator: FHCC (Nurse)
Executor: Insurance
Fact: a pre-approval is granted
Behavior Norm: #5

T6:   paying the bill
Initiator: FHCC
Executor: patient
Fact: the service is paid
Behavior Norm: #6 and 7

Summarizing the above descriptions, the entire health care delivery process consists of six essential business transactions.

A Model of the PEP Using LAP Net

Now that all the transactions are identified, and the relevant agents and their roles are defined, an analyst, using the LAP Net elements, can

Figure 13. Detailed model of PEP

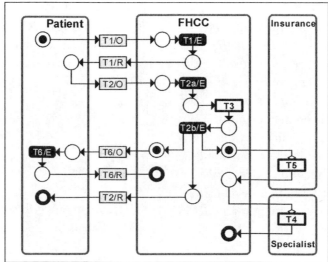

easily construct a model of the PEP or a series of models with different levels of detail. Although constructing a complete detailed model should not cause any problems, for more simplicity, let us illustrate how one can begin with a high-level model and move on toward a more detailed one. We will call this an incremental or top-down modeling approach in which the first model will not decompose any of the composite transactions. Each subsequent model will decompose one or more transactions until all the transactions are shown in the model. Due to space restriction in this chapter, we illustrate only one model, a detailed one (see Figure 13).

The above model is an example of inter-organizational business process involving human actors as well as software agents. Since most of the interactions between the organizations are facilitated by IT, it represents an example of e-business. For instance, requesting insurance pre-approval is based on a standard protocol using medical codes. It shows that when making a new appointment with a subspecialist (T4), it may be necessary for the medical center to first request the patient's insurance provider to pre-approve a visit and/or treatment (T5). If this is the case, then T5 must be completed prior to the initiation of T4. Note, both Transaction T4 and Transaction T5 are optional transactions. The above model actually has three atomic processes: the first process (delivering health care) starts with making an appointment (T1/O) and ends with delivering health care (T2/R). The second process (making a payment) starts only when the delivery health care process is completed. Finally, the last process (making an appointment with a specialist) starts when the patient is seen by a physician who defines if the patient condition demands further examination.

COMPARISON WITH OTHER METHODS

After an extensive comparison with existing methods (Chong 2001), the authors concluded that existing methods have difficulties capturing the profound social relationship and context between the software agents and the human agents. The proposed method in this chapter took a fundamentally different approach from the other methods in supporting the design of agent-based systems. This section evaluates the proposed method in this chapter from the perspective of the functions that are not offered by existing methods.

The agent-oriented-relationship (AOR) method (Taveter & Wagner, 2000) provides minimal support for the analysis of the proposed system at the conceptual level. No useful techniques or guidelines are given. It begins by modeling the proposed system as a collection of entities, objects, events, actions, and so forth. It is arguable that, as a result, one has to be preoccupied with substantial implementation details at a very early stage. For instance, the analyst has to specify the one-to-one or one-to-many relationship between the entities. The proposed method in this chapter supports the analysis right from the problem statement and gradually works out the necessary requirements for the system.

The Gaia method (Wooldridge, Jennings, & Kinny 2000) is divided into an analysis and a design stage. The method, however, does not facilitate the social obligations of the human agents to be expressed. Substantial extension is required to incorporate this function in order to provide a more realistic view of the business domain, where the human agents play a central and essential role.

The AMT method (Wooldridge et al., 2000) captures the architectural aspects of the entire agent-based system (external level) and the specific software agent design (internal level). It makes one important assumption; that is, the types and roles of software agents are known beforehand. Nevertheless, AMT is silent on the basis of which these software agents are derived.

The MaSE method (DeLoach, 2001) has one similarity with the proposed method; namely, it has a design phase that aims to capture the basic types of software agents. However, the design phase gives no guidelines to support the aim of the identification of the types of software agents. In contrast, the proposed method handles this by offering a systematic and progressive way of analyzing and identifying the types of software agents. Last, MaSE has nothing to say about the basis on which the terms for the ontology are identified.

Similar to the proposed method in this chapter, the AOM method (Elammari & Lalonde, 1999) provides a technique to identify the potential software agents and their high-level behaviour patterns. Also, it places no implementation assumptions or constraints on the analysts. However, the basis on which the ontology terms can be identified is not made explicit. Last, for the AOM method to be widely accepted for designing agent-based systems, major overhaul is needed in order to incorporate the functionality of capturing the social obligations of the human agents.

CONCLUSION

This chapter provided a comprehensive introduction to two theoretical concepts adapted by the authors for agent-based e-commerce and e-business systems study. The organizational semiotics is discussed and illustrated as a useful concept for understanding agent behavior, role, and responsibilities. However, in order to build an intuitive model in respect to time, it is demonstrated that the LAP Net models are a very useful component in this study. The LAP Net models are

diagrammatically rich and easy to communicate to nontechnical users, and, at the same time, they provide sufficient notations to model complex hierarchical agent based e-business systems.

One of the advantages of the resulting models is their capability to be executed for validation and verification, even, animation to demonstrate the interaction of agents.

However, an interesting application area for the proposed study would be to demonstrate how the studied methods can be applied to more complex real-life situations. In this respect, of special interest is the study of inter-organizational interactions (or cross-applications interactions) with multiple partners and several layers of hierarchy.

REFERENCES

Alderson, A., & Liu, K. (2000). Reverse requirements engineering: The AMBOLS approach. In P. Henderson (Ed.), *Systems Engineering for Business Process Change* (pp. 196-208). London: Springer.

Barjis, J., Chong, S., Dietz, J.L.G, & Liu, K. (2002, September). Development of agent-based e-commerce systems using semiotic approach and DEMO transaction concept. *International Journal of Information Technology and Decision Making, 1*(3), 491-510.

Chong, S. (2001). *DEON: A semiotic approach to the design of agent-mediated e-commerce systems.* Unpublished doctoral thesis, Staffordshire University.

Chong, S., & Liu, K. (2000a). A semiotic approach to the design of agent-mediated e-commerce. In E.D. Falkenberg, K. Lyytinen, & A.A. Verrijn-Stuart (Eds.), *Information Systems Concepts: An Integrated Discipline Emerging* (pp. 95-114). Boston: Kluwer Academic Publishers.

Chong, S., & Liu, K. (2000b). A semiotic approach for distinguishing responsibilities in agent-based systems. In K. Liu, R. Stamper, R. Clarke, & P. Andersen (Eds.), *Organization Semiotics* (pp. 173-186). Kluwer Academic Press.

DeLoach, S.A. (2001). Analysis and design using MaSE and agentTool. In *Proceedings of the 12th Midwest Artificial Intelligence and Cognitive Science Conference (MAICS 2001).*
Dietz, J.L.G. (1994). Business modeling for business redesign. In *Proceedings of the 27th Hawaii International Conference on System Sciences* (pp. 723-732). Los Alamitos, CA: IEEE Computer Society Press.

Dietz, J.L.G. (2006). *Enterprise ontology: Theory and methodology.* Springer.

Elammari, M., & Lalonde, W. (1999, June 14-15). An agent-oriented methodology: High-level and intermediate models. In *Proceedings of the International Workshop on Agent-Oriented Information Systems (AOIS '99)*, Heidelberg, Germany.

Gibson, J.J. (1968). *The senses considered as perceptual systems.* Allen & Unwin.

Gibson, J.J. (1979). *The ecological approach to visual perception.* Mifflin Company.

Goldkuhl, G. (1997). *The six phases of business processes: Further development of business action theory.* Linkoping University, Centre for Studies on Human, Technology and Organization (CMTO), Sweden.

Guttman, R., Moukas, A., & Maes, P. (1998). Agent-mediated electronic commerce: A survey. *Knowledge Engineering Review, 13*(2), 147-159.

Habermas, J. (1984). *The theory of communicative action: Reason and rationalization of society.* Cambridge: Polity Press.

Kinny, D., Georgeff, M., & Rao, A. (1996). A methodology and modelling technique for systems of BDI agents. In W. van der Velde & J. Perram (Eds.), *Agents Breaking Away: Proceedings of the 7th European Workshop on Modelling Autonomous*

Agents in a Multi-Agent World (MAAMAW '96) (LNAI 1038). Heidelberg, Germany: Springer-Verlag.

Lang, K. (1995). Newsweeder: Learning to filter net-news. In *Proceedings of the 12th International Conference Machine Learning* (pp. 331-339). San Fransisco: Morgan Kaufmann.

Lenat, D.B. (1994). CYC: A large-scale investment in knowledge infrastructure. *Communications of the ACM, 38*(11), 33-38.

Liu, K. (2000). *Semiotics in information systems development.* Cambridge: Cambridge University Press.

Liu, K., Alderson, A., & Qureshi, Z. (1999a). Requirements recovery of legacy systems by analysing and modeling behavior. In *Proceedings of the International Conference on Software Maintenance* (pp. 3-12). Los Alamitos: IEEE Computer Society.

Liu, K., Sharp, B., Crum, G., & Zhao, L. (1995). Applying a semiotic framework to re-engineering legacy information systems. In *Proceedings of the 1st International Conference on Organizational Semiotics*, University of Twente, The Netherlands.

Maes, P. (1994). Agents that reduce work and information overload. *Communications of the ACM, 42*(3).

Mitchell, T., Caruana, R., Freitag, D., McDermott, J., & Zabowski, D. (1994). Experience with a learning personal assistant. *Communications of the ACM, 37*(7), 80-91.

Murata, T. (1989, April). Petri Nets: Properties, analysis and applications. *Proceedings of the IEEE, 77*(4), 541-580.

Nwana, H.S., Rosenschein, J., Sandholm, T., Sierra, C., Maes, P., & Guttman, R. (1998). Agent-mediated electronic commerce: Issues, challenges and some viewpoints. In *Proceedings of the*

Second International Conference on Autonomous Agents (pp. 189-196). ACM Press.

Papazoglou, M.P. (2001, April). Agent-oriented technology in support of e-business. *Communications of the ACM, 44*(4), 35-41.

Peterson, J.L. (1981). *Petri net theory and the modeling of systems.* Englewood Cliffs, NJ: Prentice Hall.

Reisig, W. (1985). Petri Nets: An introduction (eatcs monographs on theoretical computer science). In W. Brauer, G. Rozenberg, & A. Salomaa (Eds.). Berlin: Springer Verlag.

Searle, J.R. (1969). *Speech acts: An essay in the philosophy of language.* London: Cambridge University Press.

Stamper, R.K. (1988). *MEASUR.* Enschede, The Netherlands: University of Twente.

Stamper, R.K. (1992). Language and computing in organised behavior. In R.P. van de Riet & R.A. Meersman (Eds.), *Linguistic Instruments in Knowledge Engineering* (pp. 143-163). Elsevier Science Publishers B.V.

Stamper, R.K. (1997). Organizational semiotics. In J. Mingers & F. Stowell (Eds.), *Information Systems: An Emerging Discipline.* London: McGraw-Hill.

Taveter, K., & Wagner, G. (2000). Combining AOR diagrams and Ross Rusiness rules' diagram for enterprise modelling. In *Proceedings of the International Bi-Conference workshop on Agent-Oriented Information Systems 00' (AOIS00),* Stockholm, Sweden and Texas.

Winograd, T., & Flores, F. (1986). *Understanding computers and cognition: A new foundation for design.* Norwood: Ablex.

Wooldridge, M., Jennings, N.R., & Kinny, D. (2000). The Gaia methodology for agent-oriented analysis and design. *Journal of Autonomous Agents and Multi-Agent Systems, 3*(3), 285-312.

ENDNOTES

[1] In the author's view, a software agent can be a simplistic software component such as a search engine or a complex software component that has some form of artificial intelligence incorporated into it. One common pattern from the use of software agents is that humans merely delegate their duties but not their responsibilities. They usually result in a change in the social obligation which needs to be discharged by the human agents.

[2] For brevity and illustration purposes only, an extract of the norms are shown in the textbox.

[3] A software agent is not delegated the responsibility of conducting a general physical test.

Chapter VII

Applications of Agent–Based Technology as Coordination and Cooperation in the Supply Chain Based E–Business

Golenur Begum Huq
University of Western Sydney, Australia

Robyn Lawson
University of Western Sydney, Australia

ABSTRACT

This chapter explores the utilization of a multi-agent system in the field of supply chain management for electronic business. It investigates the coordination and cooperation processes, and proposes and discusses a newly developed model for an enhanced and effective cooperation process for e-business. The contribution made by this research provides a theoretical solution and model for agents that adopt the enhanced strategy for e-business. Both large organizations and SMEs will benefit by increasing and expanding their businesses globally, and by participating and sharing with business partners to achieve common goals. As a consequence, the organizations involved will each earn more profit.

INTRODUCTION

Today's Internet-connected world has created an enormous revolution among business organizations. Nowadays, running a global business electronically is one of the most important emerging issues. Many researchers and software developers have been investigating and developing software tools and mechanisms that allow others to build distributed systems with greater ease and reliability for conducting e-business.

When a computer system acts on our behalf, it needs to interact with another computer system that represents the interests of another party, and these interests are generally not the same. In this context, Wooldridge (2002) specifies:

It becomes necessary to endow such systems with the ability to cooperate and reach agreements with the other systems, in much the same way that we cooperate and reach agreements with others in everyday life. This type of capability was not studied in computer science until very recently. (p. 3)

Traditional purchasing and selling for business-to-business (B2B) and business-to-consumer (B2C) have been conducted through different complex processes involving negotiation, as well as cooperation and coordination. It was quickly realized that e-commerce represents a natural, and potentially very lucrative, application domain for multi-agent systems. Artificial intelligence (AI) has been largely focused on the issues of intelligence in individuals, but surely a large part of what makes us unique as a species is our social ability. Not only can we communicate with one another in high-level languages, we can cooperate, coordinate, and negotiate with one another. As many species have a strong social ability (e.g., birds) like this, we also need cooperation and coordination in multi-agent systems to conduct fruitful, successful, and sustainable e-business.

It has been found that cooperation and coordination are important issues in conducting e-business. In recent years, there have been many research studies in e-business negotiation, but there is little work in e-business negotiation through cooperation and coordination. For example, large organizations mostly have enough products to sell. On the other hand, small and medium enterprises (SMEs) that are suffering from a lack of capital cannot compete with large organizations. However, some SMEs want to purchase products from large organizations and sell them to their customers. Another example is supply chain management (SCM) where at each and every stage (for instance, procurement of material, transformation of material to intermediate and finished goods, and distribution of finished products to customers) cooperation and coordination are needed. In these cases, they can cooperate with each other by exchanging products, and a deal between them can be made because both participants are able to "fine-tune" their profit. That means they can work together to achieve particular goals.

Therefore, if we can perform this type of activity electronically, it will be easier and faster, and, at the same time, very complex issues can be avoided. To perform these activities electronically using a cooperation and coordination process, models need to be investigated for performing flexible and reliable tasks. Many different disciplines including sociology, political science, computer science, management science, economics, psychology, and system theory are dealing with fundamental questions about coordination in one way or another. Furthermore, several previous writers have suggested that theories about coordination are likely to be important for designing cooperative work tools (Finnie, Berker, & Sun, 2004; Holt, 1988; Winogard & Flores, 1986). Therefore, it is possible to develop computer-supported cooperative work with the prospect of drawing on a much richer body of existing and future work in the application of multi-agents in supply chain based e-business.

The main objective of this chapter is to explore the operation of a multi-agent system in supply chain management for electronic business. It focuses on the coordination and cooperation processes, and discusses a newly developed model for an enhanced and effective cooperation process for e-business. The main contribution of this research is a theoretical solution and the model for agents that adopt this strategy for their e-business transactions. Both large organizations and SMEs will benefit, as the strategy will enhance their global business by participating and sharing with other businesses to achieve common goals. As a consequence, the organizations involved will be more profitable and competitive.

The chapter is organized as follows: first, factors in conducting e-business are discussed. Then agent-based technology is outlined as a multi-agent system that is necessary for a supply chain system. A definition/theory of coordination is introduced, and some related work on coordination and cooperation is reviewed. The next section discusses cooperative problem-solving processes. Then a theoretical model and architecture on coordination and cooperation is explained in the context of trading agent competition supply chain management (TAC/SCM). The concluding section provides an overview of the chapter.

FACTORS IN CONDUCTING E-BUSINESS

The following factors have been identified in conducting e-business:

General Problems

- **Finances:** It has been found that some SMEs do not have enough resources to conduct e-business; however, they are particularly interested in being involved. Therefore, large organizations and SMEs have a good opportunity to work together to conduct global e-business.
- **Price war:** When a buyer seeks goods through an Internet catalogue, for various reasons, the price of some products are too cheap, while others are too expensive. As a result, customers feel a level of confusion about making the right decision.
- **Postpurchase/local customer service:** It has also been found that if somebody buys goods from the Internet, the company may not have a local retailer in that city. In

this case, if any problem is found with the goods, postpurchase/local customer service becomes a complex issue to solve. As a result, some customers are not interested in buying goods from the Internet. Therefore, currently, local retailers need to stock similar goods. Moreover, to conduct e-business globally, many retailers need to participate. For that reason, cooperation is required for transactions with large organizations and with SMEs.

- **Lack of a pricing strategy:** In the real world, a pricing strategy is an important issue. To develop an effective pricing strategy, sometimes an incentive like a discount is needed. This is possible when a manager thinks its time to give a discount via a special promotion or to clear old stock. It is also possible to implement a pricing strategy in the online world.
- **Lack of customer satisfaction:** From the above points, customers can feel dissatisfied.

Problems in Supply Chain Management

- **Lack of information sharing:** Information sharing is one of the most significant issues in SCM and plays an important role. As for example, a retailer such as K-Mart may place huge orders for a particular product for their planned promotion. If suppliers had prior knowledge of this promotion, they also could plan for a production increase.
- **Lack of information access limitation and lack of transparency:** At times, users are unable to find an exact outcome due to restricted access to some information. This results in a lack of transparency. As a result, it obstructs making the right decision within the right time frame.
- **Lack of sharing the benefits of coordination equitably:** The coordination benefits

are not being shared equitably in the supply chain, which is a challenge (Chopra & Meindl, 2003, p. 503). Consequently, if agents agree to work together, the problem can be resolve accordingly.

- **Lack of agreement to work together:** Agreements are not generally found in real-world SCM. This is due to one stage of the supply chain having objectives that conflict with other stages that generally have different owners. For this reason, the main objective of each owner is to maximize its own profit. As a result, this diminishes the overall supply chain profit. Today, the supply chain is comprised of potentially hundreds or even thousands of independently owned enterprises. For instance, Ford Motor Company has thousands of suppliers from Goodyear to Motorola. To make an overall profit for the supply chain, the partners need to reach an agreement for working together. This can lead to the overall profit being maximized. Therefore, each participant in the cooperative venture will benefit accordingly.
- **Lack of communication among business organizations/supply chains (level of product availability):** Good communication can yield good results. Companies in the supply chain often do not communicate through the various stages of the supply chain and are unwilling to share information. As a result, companies become frustrated with the lack of coordination.
- **Timely manner:** Sometimes, some information is not accessible in a timely manner. Therefore, this can obstruct the right decision being made in a timely fashion.
- **Lack of use of technology to improve connectivity in the supply chain.**
- **Lack of trust:** Because of the above obstacles, trust is decreased and frustration appears at various stages of the supply chain, making coordination efforts much more difficult. On the other hand, high levels of trust involve the belief that each stage is interested in the other's welfare and would not take actions without considering the impact on the other stages.

If the organizations work together electronically towards some shared common goal, then there is a possibility that the problems defined above can be fully or partially overcome.

Benefits of Conducting E-Business

The following are the expected benefits in conducting e-business when organizations work together:

- **Reasonable and flexible price:** If different organizations work together, they will be able to sell goods at a reasonable and leveled price. An e-business can easily alter the price of the products in one entry of the database, which is linked to its Web site. According to current inventories and demand, this type of ability allows an e-business to increase revenues by adjusting prices. Airline tickets are a good example where low-cost available tickets are shown on a Web site for flights with unsold seats. This can reduce the price war between competitors.
- **Reliable product:** By working together, it is also possible to sell reliable products to customers.
- **Globally available and less transportation cost:** Because organizations can work together globally, then the goods can also be available globally. For example, a customer in Thailand can place an order on the Internet. If there is a warehouse situated in Thailand for that item, then it is easier to get the item; otherwise the seller would need to ship the item. In the case of limited stock, it might not be profitable to have an item available globally when there are high transportation costs. Consequently, by glob-

ally working together, organizations can earn more profit and lower transportation costs.

- **Reduce operational cost:** Operating costs can also be decreased if a manufacturer is using e-business to sell directly to customers, as there are fewer supply chain stages for the product as it makes its way to a customer.

- **Reduce delivery time:** If a warehouse exists locally, then this will also lower the delivery time, in addition to delivery costs.

- **Enhanced customer service locally:** If a problem arises for the product, then it can be serviced locally. As a result, a customer will feel more confident in buying further products.

- **Fewer inventories:** E-business can reduce inventory levels and costs by improving supply chain coordination and creating a better match between supply and demand. For example, Amazon.com requires fewer inventories than local retail bookshops. As a result, e-business reduces inventory cost.

- **24-hour access from any location:** Customers are able to place their order any time day or night and from any location through the Internet. Therefore, it is possible for an organization to increase sales.

- **Maximum profit:** All of the above points have the potential to maximize profit for organizations.

- **Expansion of business:** By working together, large organizations have the opportunity to expand their business with the cooperation of SMEs. Thus, SMEs also have the opportunity to share tasks with large organizations. Ultimately, through collaboration, organizations can collectively increase their profits.

- **Duplication of work:** Reducing the duplication of work can save both time and money. For example, a pricing strategy for a product can be negotiated electronically,

and then can be used for the collaborating organizations.

In summary, factors in conducting e-business can be categorized as general problems associated with the operation of the organization (finances, pricing strategy, and customer service), and more specific problems in managing the supply chain (lack of information sharing and access, and lack of agreement to work together). To examine the supply chain further in an electronic context, the use of agent-based technology is investigated.

AGENT-BASED TECHNOLOGY

Agent-based technology has emerged as the preferred technology for enabling flexible and dynamic coordination of spatially distributed entities in a supply chain. Authors have defined agents from different perspectives. The main focus of this chapter is a discussion of software intelligent agents, and the definition presented is adapted and based on Wooldridge and Jennings (1995). An agent is a computer system that is situated in a particular environment, and is capable of *flexible autonomous actions* in that environment in order to meet its design objectives. Autonomy is a complicated concept, but it can be simply explained that the system should be able to perform without the direct intervention of humans (or other agents). At the same time, it should have control over its own actions and internal state. The meaning of *flexible actions* is that the system must be:

- **Responsive:** Agents should be able to perceive their environment, which may include the physical world, a user, a set of agents, or the Internet and can respond timely according to changes that occur in it.
- **Proactive:** Agents cannot only perform based on their environment, but should

also be able to exhibit opportunistic, goal-oriented behavior by taking the *initiative* according to their intention.

- **Social:** Agents should be able to interact with one another as humans do, based on their own problem solving ability to help others with their activities, as required.

Therefore, if the above characteristics exist in a single software entity, then we can consider it is an intelligent agent that provides the capability of the agent paradigm. This paradigm is different from the software paradigm, for instance, object-oriented systems, distributed systems, and expert systems.

Multi-Agent Systems

By using agent-based systems, the key abstraction used is that of an agent. It might be conceptualized in terms of an agent, but implemented without any software structures corresponding to agents at all. A situation exists with an agent-based system, which is designed and implemented in terms of agents. Again, a collection of software tools exist that allow a user to implement software systems as agents, and as societies of cooperating agents.

There is no such thing as a single agent system. Therefore, we should always consider the system of agents as a multi-agent system, where the agents will need to interact with each other and cooperate as required. Jennings (2000) illustrates the typical structure of a multi-agent system (see Figure 1). The system consists of a collection of agents that are able to interact with each other by communication. The agents perform their activities in the environment and different agents have different "spheres of influence," and have control over, or at least are able to, influence different parts of the environment. In some cases, the spheres of influence may coincide or may require dependency relationships between the agents. For instance, two robotic agents have the ability to

move through the door, but they may not be able to move simultaneously. Another example might be "power" relationships, where one agent is the "boss" of another agent.

Dependency Relations in Multi-Agent Systems

In multi-agent systems, the agents need to be dependent in some way to be able to perform their tasks. The basic idea of such dependency was identified by Sichman and Demazeau (1995) and Sichman (1994) and there are a number of possible dependency relations:

- **Independence:** In this case, no dependency exists between the agents.
- **Unilateral:** This type includes one agent depending on the other agent, but not vice versa.
- **Mutual:** Both agents depend on each other according to the same goal.
- **Reciprocal:** The first agent depends on the other for a goal, while the second agent depends on the first agent for another goal. These two goals may not be same, and mutual dependency implies reciprocal dependence.

The above dependency relations may also be qualified by whether or not they are *locally believed* or *mutually believed*. The locally believed dependency is when the agent believes the dependency exists, but may not believe that the other agent is aware of it. The mutual belief is when one agent believes that the dependency exists and the other agent is aware that this dependency exists.

The suppliers, manufacturers, retailers, and consumers are all in a supply chain related network, which needs proper, efficient, and timely coordination, cooperation, and negotiation. Therefore, overall benefits will be achieved when ap-

Figure 1. Typical structure of a multi-agent system (Jennings, 2000)

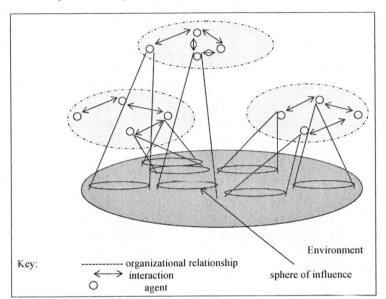

plying multi-agent systems to improve efficient performance among these entities.

In summary, the use of a multi-agent system has emerged as a flexible and dynamic method for coordination of spatially distributed entities in a supply chain. Efficient performance is possible between business partners in an online environment through coordination and cooperation.

DEFINITION/THEORY OF COORDINATION

We all have a common understanding about coordination and cooperation from our everyday lives. At times, we need to coordinate and cooperate with others for a variety of reasons. When we watch a winning soccer or cricket team or high-quality synchronized swimming, we notice how well the program is organized. In contrast, we could spend hours waiting to return something, or when we thought we had booked an airline ticket that had already been sold, or when a com-

pany repeatedly fails to make its expected profit, then we may become very aware of the effects of poor coordination. The dictionary definition of coordination is: *the act of working together harmoniously.* It is essential that an intention to work together "harmoniously" includes handling conflict as well as cooperation.

Malone and Crowston (1990) specified that computer science does not deal primarily with people; however different computational processes must certainly "work together harmoniously," and as numerous researchers have pointed out, certain kinds of interactions among computational processes resemble interactions among people (e.g., Fox, 1981; Hewitt, 1986; Huberman, 1988; Miller & Drexler, 1988; Smith & Davis, 1981). Malone and Crowston's (1990) observation is not completely correct, due to the fact that software developers implement computational processes according to user requirements. Therefore, it is possible to develop software agents, which will perform coordination tasks for human beings in order to facilitate e-business.

Literature Review: Cooperation and Coordination

Finnie, Berker, and Sun (2004) proposed a multi-agent architecture for cooperation and negotiation in supply networks (MCNSN), which incorporated a learning capability for some agents, and discusses the issues that need to be addressed for coordination, cooperation, and negotiation. They mainly concentrate on case-based reasoning (CBR) as a framework for learning the best strategy between buyers and suppliers and also focus on customer relationship management (CRM). They did not concentrate on business-to-business (B2B) cooperation and coordination.

Beck and Fox (1994) developed the mediated approach to coordinate the supply chain, which has a global perspective and gathers information on commitments from other agents when there is an event disrupting supply. They conducted an experiment, which showed that the mediated approach has a better performance than the negotiation approach. Although the multi-agent approach in SCM has received considerable attention, a number of unresolved questions remain in cooperation and negotiation in supply networks (Schneider & Perry, 2006). A multi-agent system (MAS) was considered by Finnie and Sun (2003) in such a way that only some agents had the CBR capability.

Several reasons have been identified for multiple-agent coordination (Jennings, 1990; Nwana, 1994):

- **Dependencies between agents' actions:** Interdependencies occur when goals undertaken by individual agents are related, either because local decisions made by one agent have an impact on the decisions of other community members (selling a commodity depends on a salesperson for customer service and customers), or because there is a possibility of a clash among the agents (two cars may simultaneously attempt to

pass on a narrow road, resulting in the risk of a collision). Ultimately, dependencies prevent anarchy or chaos and coordination is necessary among the agents to achieve common goals.

- **Meeting global constraints:** Commonly, some global constraints exist that a group of agents must satisfy if they agree to participate. For instance, a system of agents allocating components to organizations may have constraints of a predefined budget. Similarly, if one organization fails to sell their products for some reasons, then other organizations can coordinate to minimize the problem.

- **Distributed expertise, resources or information:** All agents may not have the same capability, but have different resources and specialized knowledge in various areas. For example, treating a patient in the hospital requires different expertise (anaesthetists, surgeon, heart specialist, neurologist, ambulance personnel, nurse, and so on), resources (equipment like an x-ray machine and ultra sound machine) and information (different reports) to diagnose the patient. In this type of case, it is not possible to work individually. Therefore coordination and cooperation are both necessary to solve the entire problem.

- **Efficiency:** When an individual agent works independently, time can be a factor. If another agent helps to finish that work, then it can be completed twice as fast. For instance, if two people plant 50 seedlings each, then 50% of the time is saved.

Nwana, Lee, and Jennings (1996) specified that coordination may require cooperation, but it would not necessarily need cooperation among all agents in order to get coordination. This could result in disjointed behavior, because for agents to cooperate successfully, they must maintain models of each other as well as develop and maintain models of future interactions. If an agent thinks that other agents are not functioning correctly, then disjointed behavior may still give a good result. Coordination may be completed without cooperation. For example, if somebody drives very close towards your lane, you might get out of the path, which coordinates your actions with the other person, without actually cooperating. To facilitate coordination, agents need to cooperate with others by sending communication messages. This results in agents having the opportunity to know the goals, intentions, outcomes, and states of other agents.

In summary, coordination and cooperation are practiced daily in physical world transactions, and the notion of creating a similar environment in the virtual world is not a trivial problem. Electronic cooperative problem solving using a multi-agent system is a complex challenge to address.

COOPERATIVE PROBLEM SOLVING

In the context of cooperation in multi-agent systems, Franklin and Graesser (1997) offer a cooperation typology (see Figure 2) with a number of characteristics. If each agent pursues its own agenda independently of the others, then it is termed an independent multi-agent system. There are two types of independent multi-agent systems: (a) discrete and (b) emergent cooperation. The discrete system involves agents with agendas that do not have any relation to each

other. Therefore, discrete systems do not have any cooperation. Becker, Holland, and Deneubourg (1994) specified that the puck gathering robots form an independent system, each moving in a straight line until an obstacle is encountered according to its agenda, it then backs up and goes in another direction. From an observer's point of view, this puck gathering is an emergent behavior of the system, as it looks like the agents are working together. However, from the agents' point of view, they are not working together. The agents only carry out their individual tasks.

On the other side of the independent system is the agent who is cooperating to its own agenda with other agents in the system (*cooperative systems*). This type of cooperation can be either communicative or noncommunicative. Communicative systems intentionally communicate with the other agents by sending and receiving messages or signals. The noncommunicative systems are those in which the agents coordinate their cooperative activity by observing and reacting to the behavior of the other agents, for example, lionesses on a hunt (Franklin, 1996). Intentional communicative systems are divided into two categories: (a) *deliberative*, where agents jointly plan their actions to achieve a particular goal; and such cooperation may, or may not entail coordination; and (b) *negotiating*, where agents act like deliberative systems, except that they have added challenge of competition.

Doran and Palmer (1995) offer a viewpoint that specifies cooperation as a property of the actions of

Figure 2. Cooperation typology (Adapted from Franklin & Graesser, 1997)

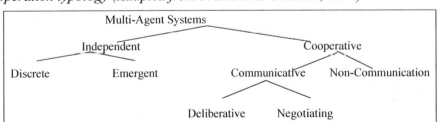

the agents involved. Thus, given a multiple-agent system in which the individuals and the various subgroups therein may be assigned one or more goals, possibly implicitly, then cooperation occurs when the actions of each agent satisfies either or both of the following conditions:

1. Agents have an implicit common goal (cannot be achieved in isolation) and actions tend towards that goal.
2. Agents carry out actions that enable or achieve their own goals, and also the goals of the other agents.

This definition does not require that the goals be explicit within the agents. For instance, two robots carrying a large object jointly, which is an example of the definition of the variant (1) assume that both have the goal of the moving object. If two robots are building two towers separately with different colored bricks, then if one of the robots finds colored bricks that match the other robot, it passes them to the other robot, which is an example of the variant (2). Therefore, agent developers need to know the more specific tasks and choices of actions to cooperate and achieve the intended goal.

The Cooperative Problem Solving Process

Wooldridge and Jennings (1999) developed a model that consists of four main stages:

a. **Recognition:** Where an agent is identified for potential cooperation.
b. **Team formation:** Where the agent applies for assistance.
c. **Plan formation:** Where the newly-formed collective agents attempt to prepare a joint contract.
d. **Execution:** When members of the team play out the roles they have negotiated.

Some questions arise in regard to the above stages:

1. Are the agents performing their task properly?
2. Has an agent left or decommitted in the middle of its task?
3. If it has, then who will complete that task?
4. Who will coordinate these tasks?

Gaps in the cooperation process have been recognized, and this research has identified that two more stages are necessary. The additional stages consist of *monitoring* and *post-execution evaluation* to support the completion of the cooperation activity. The monitoring stage will provide progress reports of the agents' tasks, and the evaluation stage will generate the overall result of the cooperative work. These six stages, four identified by Wooldridge and Jennings (1999) and two identified by this research, are discussed in the following section.

Recognition Stage

This stage commences when an agent in a multi-agent environment realizes that it has a common goal, and identifies the potential for cooperative action. Reasons for recognition include when an agent thinks that it is not able to complete the goal in isolation, or believes that cooperative actions can achieve that goal. For example, a supplier agent has excess goods in stock, but cannot sell these without the help of proper buyers. Therefore, cooperation is needed to achieve the goal. Alternatively, a large company may be able to achieve its goal but does not want to in isolation. This large company believes that if another company works with it, then it would be more beneficial. For example, a small company does not have enough capital to do business properly and a large company does, and wants to expand its business globally. This large company is looking for another company so that it can achieve its goal. Therefore, if the small company and large company work together, then the cooperative actions can provide good results for both companies more quickly and more accurately.

In regard to the above situation, the authors categorize the agents in the following manner:

Definition 1. Types of the agents
a. **Able agent:** Those agents that prefer to work with the group.
b. **Unable agent:** Any agent that does not prefer to work with a group.
c. **Partially able agent:** Those agents that prefer to cooperate and commence to do work, but cannot complete the task.

If an agent has the ability to do the task in the environment, then it is favorable to complete the task.

Theorem 1. An Able agent finishes its task if and only if the environment (*En*) is favorable, which can expressed from the definition as:

$$Able_{ag} \; Favourable \; En \rightarrow Achieve \; goal$$

Proof. Assume that an agent is going to do its task, which is possible if its surrounding environment is favorable to complete its task. On the other hand, because this agent has the ability to complete its task, it can complete it successfully. In the case of an Unable agent, we can introduce the following theorem:

Theorem 2. An Unable agent cannot finish its task even if its environment (*En*) is favorable, which can be expressed:

$$Unable_{ag} \; Favourable \; En \rightarrow \lnot Achieve \; goal$$

In regard to cooperation, a set of able agents will complete their task.

Theorem 3. A set of able agents finish its tasks if and only if the environment (*En*) is favorable, which can be formalized as:

$$Able \; ag_i \; Favourable \; En \rightarrow Achieve \; goal$$

Theorem 4. A set of able agents cannot finish their tasks although the environment (*En*) is favorable can formalized as:

$$Unable \; ag_i \; Favourable \; En \rightarrow \lnot Achieve \; goal$$

Therefore, it has been identified that agents are able and unable to have the potential for cooperative work. Then, it needs to go to the next stage of the cooperation process.

Team Formation Stage

After an agent identifies the potential for cooperative action with respect to one of its goal, what will the rational agent do? Wooldridge and Jennings (1999) proposed that an agent will attempt to *solicit assistance* from a group of agents that it believes can achieve the goal. If the agents are successful, then each member has a nominal commitment to collective action to achieve the goal. The agents have not undertaken any joint action in this stage; they are only aware of being able to act together. Actually, in this stage, there is no guarantee for successful forming of the team, only an attempt to form a team. The able agents will attempt to do some action α to achieve at least some goal. Therefore, it can be formalized as:

Theorem 5. Happens{Attempt *Able ag$_i$ α*} $\rightarrow$ Achieve goal

The characteristics of the team building can assume that it is mutually believed that:

1. The group can jointly achieve the goal.
2. Each agent in the group is individually committed to carry out its task towards the goal

or failing that, to at least cause the group to achieve the goal.

3. The individual agent has an individual goal.
4. There is a common goal which is jointly achievable.

The main assumption about team formation is that all agents attempt to form a group, and the group believes that they will have individual commitments and can jointly complete their task. If team building is successful, then it will proceed to the next step.

Plan Formation Stage

In this stage, after successfully attempting to solicit assistance, a group of agents have nominal commitment to collective action. This action will not be commenced until the group agrees on what they will actually do.

From the previous section, the authors have found that to perform collective action, it is assumed that the agents have a common belief that they can achieve their desired goal. The agents believe that there is at least one action known to the group, which will take them "closer" to the goal. Therefore, the possibility is many agents that know the actions of the group carry out the task in order to take them closer to the goal. In addition, in some cases, it is also possible in collective actions that some agents may not agree with one or more of these actions. Furthermore, in collective actions, agents will not simply perform an action because another agent wants them to (Wooldridge & Jennings, 1995). Therefore, it is necessary for the collective to make some agreement about what exactly needs to be done. This agreement is reached via *negotiation*.

Negotiation has long been recognized as a process of some multi-agent systems (Rosenschein & Zlotkin, 1994; Sycara, 1989). At the time of negotiation, the agents usually make reasoning arguments for and against particular courses of action, making proposals, counter proposals, suggesting modifications or amendments to plans. These continue until all the negotiators have agreed upon the final result. Negotiation is also an extremely complex issue. But in the case of joint negotiation, it is a bit simpler than self-interested individual agents.

In negotiating a plan, collective negotiation may also abort due to irrelevant circumstances. The minimum requirement to occur for negotiation is that *at least one* agent will propose a course of action, which is believed will take the collective closer to the goal. Therefore, negotiation may also be successful. Like team formation, we assume a group of agents also attempts to do something collectively. A group of agents g attempts to achieve a goal after performing mutual actions α which is completely or partially satisfied and can be formalized as:

$$\{\text{Attempt } g \ \alpha\} \rightarrow ?; \text{ Achieve goal}$$

The minimum condition to occur in negotiation is that the group will try to bring about a state in which all agents agree to a common plan, and intends to act on it. The authors assume that if any agent shows its preference, then it will attempt to bring this plan about. Similarly, if the plan has any objection, then it will attempt to prevent this plan from being carried out. In this way, the agents will agree on a plan to carry out their actions. If the plan formation stage is successful, then the team will have a full commitment to the joint goal and will proceed to execution phase.

Execution Stage

When the agents have a collective plan to do something, then they are ready to move to this phase, as the group knows what to do. That is, each agent has its own target and the group has its intention to perform actions to achieve the goal.

The group mutually believes that the action they intend to perform in order to achieve the goal can actually happen.

Monitoring Stage

How do we know that all the agents are performing their tasks according to the plans? What if an agent is unable to complete its task in the middle of the plan? Who will take this responsibility, or will another agent perform this task? How will it be solved? For these reasons, the authors identified that it is necessary to have a monitoring phase when the execution stage is carried out. An agent will need to monitor the execution phase; if something unusual occurs, it can be solved accordingly. For example, if an agent cannot finish its task, then the monitoring agent will request another agent to complete this task and the agent who could not finish its task can be defined as a partially able agent.

Evaluation Stage

This research identified some additional questions:

1. Which agent completed its task?
2. Which agent did not complete its task?
3. Which agent partially completed its task?
4. Which agents did extra tasks?
5. How do we know which agent performed what action?

Therefore, the authors recognized that it is also necessary to evaluate the execution stage by using an agent to evaluate and allocate reward benefits. From this evaluation, processes can be improved or updated according to necessity. After this stage, the agent can go back to the first stage to begin a new cooperative work. Therefore, we can consider it as *enhanced and effective cooperative stages*, as depicted in Figure 3.

In summary, the model developed by Wooldridge and Jennings (1999) has been extended by this research to include two more stages, the monitoring stage and the evaluation stage. The new model, shown in Figure 3, is applied to the TAC SCM game as a case study to investigate its potential performance.

Figure 3. Enhanced and effective cooperative processing stages

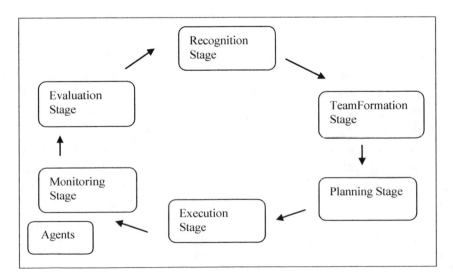

TAC SCM Game Overview

The TAC SCM is an international competition where six software agents are the manufacturers of personal computers (PC) in a simulated common market economy linked with two markets: the *component market* and the *product market*. The full specification can found at http://www.sics.se/tac/tac06scmspec_v16.pdf. TAC SCM is designed as a traditional supply chain model where supplier and end users (customers) are directly involved in an electronic market. Each manufacturing agent can manufacture 16 different types of computers, characterized by different *stock keeping units* (SKUs). SKUs consist of different combinations of components in 10 types.

During each TAC day of the game, customers send a set of request for quotes (RFQs) to the agents. Each RFQ contains a SKU, a quantity, due date, a penalty rate, and reserve price (the highest price that customers are willing to pay). Each agent responds to the RFQ by sending an offer that states a price less than the reserve price. The agent that sends the lowest price wins the bid. The winning agent delivers the entire order by the due date and is paid in full if it is delivered within five days of due date. If the order is not delivered by the due date, a penalty is incurred based on the number of late days. Consequently, if the agent cannot deliver the entire order within five days of its due date, then this order is canceled and the maximum penalty is incurred.

On the other hand, agents can send a RFQ to the suppliers for the required components and the expected delivery date. The suppliers can respond to the RFQ the next day with offers specifying the price per unit. Offers either have a delivery date on the day requested or a delivery date later than the requested day. The agent can accept or reject these offers according to their requirements and enter into an agreement with the supplier. The agent will be charged for the components on delivery. This simple negotiation mechanism must follow when agents purchase their components from suppliers. This mechanism only focuses on the accept or reject method.

Each agent must solve daily problems:

- Bidding problems for a customer's order of PCs.
- Negotiating a supply contract when the procurement problem deals with components that need to be purchased from the supplier.
- Production problems concerned with everyday scheduling.
- Allocation problems that deal with matching SKUs in the inventory to orders.

At the end of the game, the agents receive awards based on profits.

Product Market Performance

As we know, a pure competitor or monopolist can simply choose its price or output policy and directly calculate the resulting gain or loss. In an oligopoly market setting, the choice of a price, output, or other marketing policy does not uniquely determine profit, because the outcome for each firm depends on what its opponents decide to do. The Cournot and Chamberlin descriptions of oligopoly suggest the kind of interdependence that arises explicitly here, but do not take into account uncertainty about opponents' decisions (Meyer, 1976) .

The market price of PCs for all the agents depend on the quantity they produce. This means that the profit for each agent is linked directly to the profit of the other. Consequently, different agents have their own cost functions, which imply different payments for inputs. Therefore, each agent has its own policy to bid for a customer order, which it will enhance to win the bid.

The PC market is another vital part of TAC SCM in which agents are directly involved in winning. In the competition, the authors recognize the following critical questions to resolve or improve the agents' performance as price competition:

- How does the agent bid for a customer's reserve price for a PC?
- What strategies need to be adopted for this?
- How much does the agent need to reduce the price to win the bid?

To improve the performance of the agent, it is necessary to learn from the history of the game. For example, Figure 4 presents the average price of PC of the competition. The agents can learn from the chart when the market price of PCs are high, medium, and low. Equilibrium prices arise when supply equals demand: $Q^s_i = Q^d_i$ for product *i*. If $Q^s_i \geq Q^d_i$, agents will bid price P_i lower ; if $Q^s_i \leq Q^d_i$, agents will bid price P_i higher. Usually the price of the product increases at the beginning of game due to lack of supplies. Therefore,

the agents who supply the product at the time of low market supply can get a higher price and, as a result, can earn more market share with more profit. Consequently, the agent who can adopt this strategy of increased productivity, and bids according to the market situation will have a better opportunity to maximize profit.

Huq (2006) analyzed the product market of the TAC/SCM 2004 game and observes the lack of cooperation among agents involved in component purchasing and product selling. The average market demand for PCs in the semifinal and final round game can be depicted in Tables 1 and 2, where the second column is the average PCs delivered by the agents; the third column is the total average market demand. The authors subsequently find that the free agent bids on an average with a higher average price and a higher percentage of orders.

In summary, the TAC SCM has a distinct lack of cooperation among the agents involved in component purchasing and product selling, and this led the authors to conclude that the game was a likely case study to investigate modeling coordination and cooperation.

Figure 4. Market price of PC of the game 942–945

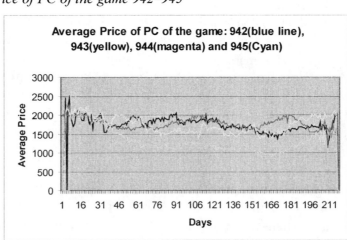

Table 1. Average total PCs delivered in semifinal of Gr-1 (TAC3 and TAC4)

Agents	Average Delivery	Total Average Price	% of Order	Average
FreeAgent	46882	291505	16	1656
SouthamptionSCM	61759		21	1508
Mr. UMBC	51587		18	1527
ScrAgent	40995		14	1504
KrokodilAgent	41551		14	1538
Socrates	48732		17	1323

Table 2. Average total PCs ordered by agents in final round

Agents	Average PC Delivery	Total Average Market Demand	% of Order Price	Average
FreeAgent	41659	201227	21	1842
SouthamptionSCM	45465		23	1670
Mr. UMBC	44665		22	1481
ScrAgent	13765		7	1434
KrokodilAgent	24487		12	1869
Socrates	31186		15	1764

MODELING COORDINATION AND COOPERATION IN TAC/SCM

According to the TAC/SCM, all manufacturer agents are rational or self-interested and their main focus is to maximize profit. If we assume that the agents cannot achieve their goal in isolation or that they would prefer to work with each other, then this has the potential for cooperation. In this context, all the manufacturer agents can work together towards their goal. On the one hand, manufacturer agents will be able to increase their production capacity and sell the final products to customers, and on the other hand, suppliers will benefit by supplying more components to the manufacturers, which will result in more profit. The following discussion proposes a theoretical model, which will be able to solve the coordina-

tion and cooperation problem of the TAC/SCM game.

The authors have found in the TAC/SCM competition that three or four agents always dominant the market of buying components or selling the products. Therefore, this research characterized these agents as big agents and the other agents as small/medium agents (SMAs). Again, it was also found that SMAs could not purchase enough share of the components to produce a final product to sell. This is a technical/strategic or financial problem for the SMAs. Consequently, if the SMAs purchase components from big agents and sell to customers, then it is possible to survive. Otherwise, the SMAs cannot compete with the big agents. In the real world, usually the intention of large organizations is to extend their business and make more profit. This increases production which ultimately leads to increased profit. Using this strategy, we assume that big agents want to extend their business, and at the same time, the SMAs would like to work with big agents. This way, big agents and SMAs can work together to achieve their common goals. As a result, every agent will be benefited by participating in shared activities. Therefore, to work together the agents need to follow the stages defined in the previous section, Cooperative Problem Solving Process. In this regard, the following characteristics can be defined (see the agent types in Definition 1 of the Recognition Stage of the Cooperative Problem Solving Process section):

Theorem 6.
a. There exist some group of agents g such that the individual agent i believes that the g can jointly achieve goal.
b. either:
c. An agent i cannot achieve goal individually.
d. an agent i believes for every action that could be performed to achieve the task, it has a goal of not performing the goal.

Theorem 7. The outcomes ensure their profit if and only if the *cooperative agents* complete their task successfully.

If the *cooperative agents* complete their task successfully, then all the participating agents will share the profit, otherwise it will be considered an incomplete task.

Theorem 8. The *cooperative agents* are those if and only if they agree to work together.

In the Cooperative Processing Stage, only those *able agents* that are determined to complete their tasks towards a common goal are considered *cooperative agents*.

Definition 2. A *decommitted agent* is an agent that started its task but did not complete that task, and therefore needs to be penalized.

Definition 3. Let a set of *able agents* that share their work to achieve a common goal be called *cooperative agents,* which is:

$$ag_i \in A = \{ ag_1, ag_2, \ldots\ldots ag_n \} = \phi \quad (1)$$

Definition 4. The accumulated task of the *cooperative agents A*, the utility u of that task can be considered as unique, and can be expressed as:

$$u(A) = \sum_{i=1} A_i = n1 \quad (2)$$

Definition 5. Profit allocation to the agents: The percentage of the utility of each agent can be worked out according to the contribution of each agent, which can be expressed as:

$$u(ag_i) = \frac{u(ag_i) \times 100}{u(A)} \quad (3)$$

Definition 6. Cooperative action takeover: If any agent fails to complete its task, then other agents will need to complete that task to achieve the goal.

If any agent is unable to finish its allocated tasks due to unavoidable circumstances, then the other agents will take over that unfinished task enthusiastically to achieve the goal.

Definition 7. The set of *cooperative agents A* are a finite set and said to be bounded.

The *cooperative agents* must be limited in number for efficiency in task allocation, as it is not possible to have an unlimited number of agents working together. The *cooperative agents* are bounded, for instance: (a) the agent who invests or sells the greatest is called the upper bounded; and (b) the agent who invests or sells the lowest is called the lower bounded.

Architecture of the Cooperative Processing Agents

Let us consider that a number of companies in different locations have agreed to sell some products to customers within a limited time frame. Assume that the agents are going to work together according to the Cooperative Processing Stages. A proposition for architecture of effective cooperative processing is shown in Figure 5. In this figure, there is a collection of manufacturer agents *n* in the domain. When these agents have agreed to perform tasks to achieve a specific goal, to complete the cooperative processing, other agents are needed. This research argues that these agents are Task Allocation Agent, Monitoring Agent, Evaluating Agent, Result Allocation Agent, and Coordination Manager Agent (CoManager).

When a problem is decomposed into smaller subproblems, the *Task Allocation Agent* is responsible for allocating tasks to the a*ble agents* in order to achieve the goal. The *Monitoring Agent* is responsible for monitoring the performance of the agents' tasks, that is, which agent is doing its task and which is not. Finally, this agent will produce a report to the *CoManager Agent*. According to this report, the *CoManager Agent* will reallocate the unfinished task to the agent that is willing to undertake that task.

The *Evaluating Agent* will evaluate all tasks from the *Monitoring Agent*. The *Evaluating Agent* will provide analytical and objective feedback on efficiency and effectiveness of the performance of agents. Finally, it will produce an overall final report including benefits of each agent to the *Result Allocation Agent*. Eventually, this final report allows the agents to learn lessons. The *Result Allocation Agent* then processes the benefits deserved by each agent, and finally produces a benefit report to the agents.

The contribution made by this research is the addition of the monitoring and evaluation stages for the Cooperative Problem Solving Process, and the results described in this section. The TAC/SCM was used as a case study to illustrate the concepts outlined in the theorems and definitions.

FUTURE TRENDS

The potential for B2B e-commerce is now projected to be much larger than that for consumer oriented e-commerce (Chan, Lee, Dillon, &

Figure 5. Architecture of effective cooperation model

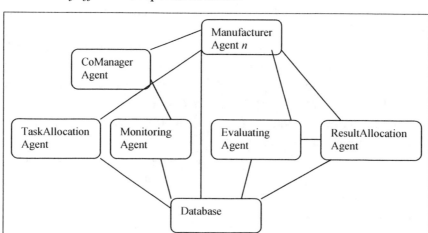

Chang, 2001). Conducting electronic B2B transactions is an emerging and potentially lucrative issue. For example, in the supply chain, manufacturer organizations or retailers are dependent on supplier organizations. There are many processes in selling and purchasing that can be conducted electronically. Particularly, at the time of purchasing, many processes are complex and involve negotiation, cooperation, and coordination. In the real world, these processes are very time consuming and complicated. Therefore, if we can utilize these processes electronically, we can avoid complexity and will be able to reduce costs and time taken. Figure 6 shows how B2B e-commerce has grown from 1998 to 2005.

As a result, we can predict that this trend in e-business utilization will increase into the future. As described in the previous sections, both large organizations and SMEs will be able to work together to conduct e-business on a global basis. In regards to implementing cooperative work utilizing multi-agent systems, the agents need to follow the stages defined in this chapter. As a result, all the participant organizations will benefit in overall

performance outcomes. In conclusion, the authors argue that team effort, rather than individual effort, will give more robust and sustainable results. The cooperation and coordination protocol, and information sharing among various agents can be future research areas, which will facilitate in building the software that enables coordination and cooperation activities.

CONCLUSION

This chapter identified problems in conducting e-business and managing the supply chain. It also identified expected benefits for supply chains with agents working together in coordinated and cooperative processes. The utilization of a multi-agent system in supply chain management and the *cooperative problem solving stages* have been presented and discussed. To apply these stages, the proposition for architecture of effective cooperative processing for agents and some characteristics in modeling coordination and cooperation for TAC/SCM have been outlined.

Figure 6. Projection growth of B2B e-commerce drawn from a report by Gartner Group

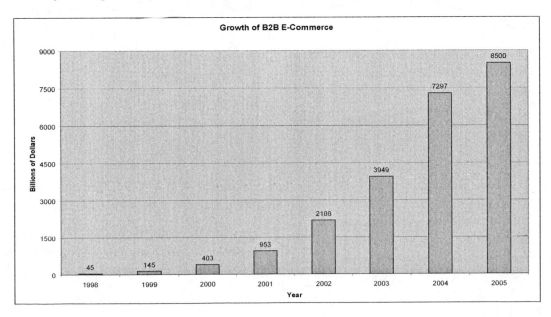

The ultimate goal is to develop the capability of organizations to work effectively together in online e-business transactions. In addition to this, large organizations can expand their businesses and SMEs can work with large organizations. Finally, it can reduce time for selling and buying activities and increase the total profits of the supply chain. In addition, it will facilitate the ability to cooperate and coordinate among multi-agents in e-commerce. Further, it will enhance customer satisfaction and streamline B2B transactions by reducing transaction costs of tasks at every stage of the supply chain. Therefore, it will increase trust and confidence in the component market and product market.

REFERENCES

Beck, J. C., & Fox, M. S. (1994, May 15). *Supply chain coordination via mediated constraint relaxation.* Paper presented at the First Canadian Workshop on Distributed Artificial Intelligence, Banff.

Becker, R., Holland, O. E., & Deneubourg, J. L. (1994). From local actions to global tasks: Stigmergy in collective robotics. In R. Brooks & P. Maes (Eds.), *Artificial Life IV.* Cambridge, MA: MIT Press.

Chan, H., Lee, R., Dillon, T., & Chang, E. (2001). *E-commerce fundamentals and applications.* West Sussex, England: John Wiley & Sons.

Chopra, S., & Meindl, P. (2003). *Supply chain management: Strategy, planning, and operation* (2nd ed.).

Collins, J., Arunachalam, R., Sadeh, N., Eriksson, J., Finne, N., & Janson, S. (2005). *The supply chain game for the 2006 trading agent competition, competitive benchmarking for the trading agent community.* Retrieved August 18, 2007, from http://www.sics.se/tac/tac06scmspec_v16.pdf

Doran, J. E., & Palmer, M. (1995). The EOS Project: Integrating two models of palaeolithic social change. In N. Gilvert & R. Conte (Eds.), *Artificial Societies: The Computer Simulation of Social Life* (pp. 103-105). London: UCL Press.

Finnie, G., Berker, J., & Sun, Z. (2004, August). *A multi-agent model for cooperation and negotiation in supply networks.* Paper presented at the Americas Conference on Information Systems, New York.

Finnie, G., & Sun, Z. (2003). *A knowledge-based model of multiagent CBR systems.* Paper presented at the International Conference on Intelligent Agents, Web Technologies, and Internet Commerce (IAWTIC'2003), Vienna, Austria.

Fox, M. S. (1981). An organizational view of distributed systems. *IEEE Transactions on Systems, Man and Cyvernetics, 11*(1), 70-79.

Franklin, S. (1996). *Coordination without communication.* Retrieved August 18, 2007 from http://www.msci.memphis.edu/~franklin/coord.html

Franklin, S., & Graesser, A. (1997). *Is it an agent, or just a program? A taxonomy for autonomous agents.* Paper presented at the Third International Workshop on Agent Theories, Architectures, and Languages.

Hewitt, C. (1986). Offices are open systems. *ACM Transactions on Office Systems, 4*(3), 271-287.

Holt, A. W. (1988). Diplans: A new language for the study and implementation of coordination. *ACM Transformations on Office Information Systems, 6*(2), 109-125.

Huberman, B. A. (1988). *The ecology of computation.* Amsterdam: North-Holland.

Huq, G. B. (2006, February 13-20). *Analysis, planning and practice of trading agent competition supply chain management (TAC/SCM).* Paper presented at the 2nd International Conference on Information Management Business, Sydney.

Jennings, N. R. (1990). Coordination techniques for distributed artificial intelligence. In G.M.P. O'Hare & N.R. Jennings (Ed.), *Foundations of Distributed Artificial Intelligence* (pp. 187-210). London: Wiley.

Jennings, N. R. (2000) On agent-based software engineering. *Artifical Intelligence, 117*(2), 277-296.

Melone, T. W., & Crowston, K. (1990, October 7-10). *What is coordination theory and how can it help design cooperative work systems?* Paper presented at the ACM Conference on Computer Supported Cooperative Work (CSCW), Los Angeles.

Meyer, R. A. (1976). *Microeconomic decisions.* Houghton Mifflin Company.

Miller, M. S., & Drexler, K. E. (1988). Markets and computation: Agoric open systems. Amsterdam: North-Holland. In B.A. Huberman (Eds.), *The Ecology of Computation* (pp. 133-176).

Nwana, H. S. (1994). *Negotiation strategies: An overview* (BT Laboratories internal report).

Nwana, H. S., Lee, L., & Jennings, N. (1996). Co-ordination in software agent systems. *British Telecom Technical Journal, 14*(4), 79-88.

Rosenschein, J. S., & Zlotkin, G. (1994). *Rules of encounter: Designing conventions for automated negotiation among computers.* Cambridge: MIT Press.

Schneider, P. G., & Perry, J. T. (2001). Electronic Commerce. In *Course Technology.* Canada.

Sichman, S. J. (1994). A social reasoning mechanism based on dependence networks. In *Proceedings of the 11th European Conference on Artificial Intelligence* (ECAI-94), Amsterdam.

Sichman, S. J., & Demazeau, Y. (1995). *Exploiting social reasoning to deal with agency level inconsistency.* Paper presented at the 1st International Conference on Multi-Agent Systems (ICMAS-95), San Francisco.

Smith, R. G., & Davis, R. (1981). Frameworks for cooperation in distributed problem solving. *IEEE Transactions on Systems, Man and Cyvernetics, 11*(1), 61-70.

Sycara, K. P. (1989). Multiagent compromise via negotiation. In L. Gasser & M. Huhns (Eds.), *Distributed Artificial Intelligence* (Vol. II, , pp. 119-138). London: Morgan Kaufmann/San Mateo, CA: Pitman Publishing.

Winogard, T., & Flores, F. (1986). *Understanding computers and cognition: A new foundation for design.* Noorwood, NJ: Ablex.

Wooldridge, M. (2002). *An introduction to multiagent systems.*: John Wiley & Sons.

Wooldridge, M., & Jennings, N. R. (1995). Intelligent agents: Theory and practice. *Knowledge Engineering Review, 2*(10), 115-152.

Wooldridge, M., & Jennings, N. R. (1999). The cooperative problem-solving process. *Journal of Logic Computation, 9*(4), 563-592.

Chapter VIII
An Agent–Based Framework for Emergent Process Management

John Debenham
University of Technology, Sydney, Australia

ABSTRACT

Emergent processes are business processes whose execution is determined by the prior knowledge of the agents involved and by the knowledge that emerges during a process instance. The amount of process knowledge that is relevant to a knowledge-driven process can be enormous and may include common-sense knowledge. If a process' knowledge cannot be represented feasibly, then that process cannot be managed, although its execution may be supported partially. In an e-market domain, the majority of transactions, including trading orders and requests for advice and information, are knowledge-driven processes for which the knowledge base is the Internet; therefore, representing the knowledge is not at issue. Multiagent systems are an established platform for managing complex business processes. What is needed for emergent process management is an intelligent agent that is driven not by a process goal but by an inflow of knowledge, where each chunk of knowledge may be uncertain. These agents should assess the extent to which they choose to believe that the information is correct, and thus, they require an inference mechanism that can cope with information of differing integrity.

INTRODUCTION

Emergent processes are business processes that are not predefined and are ad hoc. These processes typically take place at the higher levels of organizations (Dourish, 1998) and are distinct from production workflows (Fischer, 2003). Emergent processes are opportunistic in nature, whereas production workflows are routine. How an emergent process will terminate may not be known until the process is well advanced. The tasks involved in an emergent process typically are not predefined and emerge as the process develops. Those tasks may be carried out by collaborative groups as well as by individuals (Smith & Fingar, 2003) and may involve informal meetings, business lunches, and so forth. For example, in an e-market context, an emergent process could be triggered by "let's try to establish a business presence in Hong Kong." Further, the goal of an emergent process instance may mutate as the instance matures. So, unlike lower-order processes, the goal of an emergent process instance may not be used as a focus for the management of that instance.

Emergent processes contain knowledge-driven subprocesses but also may contain conventional goal-driven subprocesses. A knowledge-driven process is guided by its process knowledge and performance knowledge. The goal of a knowledge-driven process may not be fixed and may mutate. On the other hand, the management of a goal-driven process instance is guided by its goal, which is fixed. A multiagent system to manage the goal-driven processes is described in Debenham (2000). In that system, each human user is assisted by an agent that is based on a generic three-layer, BDI hybrid agent architecture. The term *individual* refers to a user/agent pair. The general business of managing knowledge-driven processes is illustrated in Figure 2.

Process management is an established application area for multiagent systems (Singh, 2004), although emergent processes typically are handled either manually or by CSCW systems rather than by process management systems. The use of these two technologies is not elegant and presents a barrier to a unified view of emergent process management.

In an experimental e-market, transactions include trading orders to buy and sell in an e-exchange, single-issue and multi-issue negotiations between two parties, and requests for information extracted from market data as well as from news feeds and other Internet data. In this e-market, every market transaction is managed as a business process. To achieve this, suitable process management machinery has been developed. To investigate what is suitable, the essential features

of these transactions are related to two classes of process that are at the high end of process management feasibility (Aalst & Hee, 2002). The two classes are goal-driven processes and knowledge-driven processes. The term *business process management* generally is used to refer to the simpler class of workflow processes (Fischer, 2003), although there are notable exceptions using multiagent systems (Singh, 2004).

The agent architecture described extends the simple, offer-exchange, bargaining agent described in Debenham (2004). The agent described here is driven by the contents of a knowledge base that represents the agent's world model in probabilistic first-order logic and manages emergent processes. Each message that the agent receives from another agent reveals valuable information about the sender agent's position. The agent aims to respond with messages that have comparable information revelation. In this way, it aims to gain the trust of its collaborating agents. The agent does not necessarily strive to optimize its utility and aims to make informed decisions in an information-rich but uncertain environment.

The emergent process management agent, Π, attempts to fuse the agent interaction with the information that is generated both by and because of it. To achieve this, it draws on ideas from information theory rather than game theory. Π decides what to do, such as what message to send, on the basis of its information that may be qualified by expressions of degrees of belief. Π uses this information to calculate and to continually recalculate probability distributions for what it does not know. One such distribution over the set of all possible actions expresses Π's belief in its own suitability in performing that action. Other distributions attempt to predict the behavior of another agent, Ω say, such as what proposals the agent might accept and of other unknowns that may effect the process outcome. Π makes no assumptions about the internals of the other agents in the system, including whether they have or even are aware of the concept of utility functions. Π is concerned purely with the other agents' behaviors—what they do—and not with assumptions about their motivations. This somewhat detached stance is appropriate for emergent process management in which each agent represents the interests of its owner while at the same time attempts to achieve the social goal of driving the processes toward a satisfactory conclusion.

As with the agent described in Debenham (2004), the process management agent makes assumptions about the way in which the integrity of information will decay and some of the preferences that another agent may have for some deals over others. It also assumes that unknown probabilities can be inferred using maximum entropy (ME) inference (MacKay, 2003), which is based on random worlds (Halpern, 2003). The maximum entropy probability distribution is "the least biased

estimate possible on the given information; i.e. it is maximally noncommittal with regard to missing information" (Jaynes, 1957,). In the absence of knowledge about the other agents' decision-making apparatuses, the process management agent assumes that the maximally noncommittal model is the correct model on which to base its reasoning.

PROCESS MANAGEMENT

Following Fischer (2003), a business process is "a set of one or more linked procedures or activities which collectively realise a business objective or policy goal, normally within the context of an organisational structure defining functional roles and relationships". Implicit in this definition is the idea that a process may be decomposed

Figure 1. Goal-driven process management

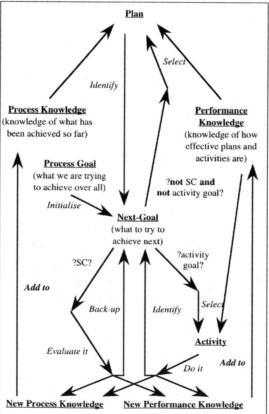

repeatedly into linked subprocesses, until those subprocesses are activities that are atomic pieces of work. "An activity is a description of a piece of work that forms one logical step within a process"(Fischer, 2003).

A particular process is called a (process) *instance*. An instance may require that certain things should be done; such things are called tasks. A *trigger* is an event that leads to the creation of an instance. The *goal* of an instance is a state that the instance is trying to achieve. The *termination condition* of an instance is a condition which, if satisfied during the life of an instance, causes that instance to be destroyed, whether its goal has been achieved or not. The *patron* of an instance is the individual who is responsible for managing the life of that instance. At any time in a process instance's life, the *history* of that instance is the sequence of prior subgoals and the prior sequence of knowledge inputs to the instance. The history is knowledge of all that has happened already. Three classes of business process are defined in terms of their management properties (i.e., in terms of how they may be managed).

- A *task-driven process* has a unique decomposition into a possibly conditional sequence of activities. Each of these activities has a goal and is associated with a task that always achieves this goal. Production workflows typically are task-driven processes.
- A *goal-driven process* has a process goal, and achievement of that goal is the termination condition for the process. The process goal may have various decompositions into sequences of subgoals, where these subgoals are associated with (atomic) activities and so with tasks. Some of these sequences of tasks may work better than others, and there may be no way of knowing which is which (Smith & Fingar, 2003). The possibility of task failure is a feature of goal-driven processes. In any case, a goal-driven process management system requires a mechanism for selecting plans

for subgoals. A simplified view of goal-driven process management is shown in Figure 1. In that figure, SC refers to the *success condition* that is a procedure at the end of each path through a plan that determines whether or not the plan has achieved its goal. *Activities* are atomic subprocesses.

- A *knowledge-driven process* may have a process goal, but the goal may be vague and may mutate (Dourish, 1998). Mutations are determined by the process patron, often in light of knowledge generated during the process. At each stage in the performance of a knowledge-driven process, the next goal is chosen by the process patron; this choice is made using general knowledge about the context of the process, called the process knowledge. The process patron also chooses the tasks to achieve that next goal; this choice may be made using general knowledge about the effectiveness of tasks, called the performance knowledge. So, insofar as the process goal gives direction to goal-driven and task-driven processes, the process knowledge gives direction to knowledge-driven processes. The management of knowledge-driven processes is considerably more complex than the other two classes of process. But knowledge-driven processes are not all bad—they typically have goal-driven subprocesses that may be handled in conventional ways. A simplified view of knowledge-driven process management is shown in Figure 2.

Managing knowledge-driven processes is rather more difficult than goal-driven processes (see Figure 2). The complete representation (never mind the maintenance) of the process knowledge may be an enormous job. But the capture of at least some of the knowledge generated during a process instance may not be difficult, if the tasks chosen used virtual documents such as workspace technology, for example. Some performance knowledge is not difficult to capture, represent,

Figure 2. Knowledge-driven process management

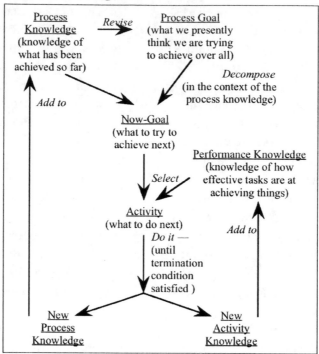

and maintain. For example, measurements of how long another agent took to complete a subprocess can be very useful. So, in the system described here, the process knowledge is left in the heads of the patron or nominated delegates, and the performance knowledge is captured by the system. The initial selection of the process goal is performed by the patron, and thus, this action is completely unsupported by the system (see Figure 2). The possible subsequent mutation of the process goal is performed by the agent using the process knowledge (see Figure 2). Task selection is supported by the agent for e-market processes, which can, for example, be given authority to withdraw a bid from two separate auctions and to negotiate for a package of goods from a single supplier. In this way, the system provides considerable assistance in the management of knowledge-driven processes. Further, if a now-goal is associated with a goal-driven, or task-driven, subprocess, then the

management system is given full responsibility for the management of that subprocess.

EMERGENT PROCESS AGENT

Π operates in an information-rich environment that includes the Internet. The integrity of Π's information, including information extracted from the Internet, will decay in time. The way in which this decay occurs will depend on the type of information and on the source from which it is drawn. Little appears to be known about how the integrity of real information such as news-feeds decays, although the effect of declining integrity has been analyzed. For example, Bernhardt and Miao (2004) consider how delays in the acquisition of trading data affect trading outcomes. One source of Π's information is the signals received from Ω. These include offers from Ω to Π the acceptance or rejection by Ω of Π's offers, and claims that Ω sends to Π. This information is augmented with sentence probabilities that represent the strength of Π's belief in its truth. If Ω rejected Π's offer of \$8 two days ago, then what is Π's belief now in the proposition that Ω will accept another offer of \$8? Perhaps it is around 0.1. A linear model is used to model the integrity decay of these beliefs, and when the probability of a decaying belief approaches 0.5[1], the belief is discarded. The model of decay could be exponential, quadratic, or whatever.

Interaction Protocol

A *deal* is a pair of commitments $\delta_{\Pi:\Omega}(\pi,\omega)$ between an agent Π and another agent Ω, where π is Π's commitment, and ω is Ω's commitment. $\mathcal{D} = \{\delta_i\}_{i=1}^{D}$ is the deal set (i.e., the set of all possible deals). If the discussion is from Π's point of view, then the subscript $\Pi : \Omega$ may be omitted. These commitments may involve multiple issues and not simply a single issue such as trading price. The set of terms, $\mathcal{T}$, is the set of all possible commitments that could occur in deals in the deal set. An agent may have a real-valued utility function: $\mathbf{U} : \mathcal{T}^{\circledR} \mathcal{R}$, that induces an ordering on $\mathcal{T}$. For such an agent, for any deal $\delta = (\pi,\omega)$, the expression $\mathbf{U}(\omega) - \mathbf{U}(\pi)$ is called the *surplus* of δ and is denoted by $\mathbf{L}(\delta)$, where $\mathbf{L} : \mathcal{T} \times \mathcal{T}^{\circledR} \mathcal{R}$. For example, the values of the function $\mathbf{U}$ may be expressed in units of money. It may not be possible to specify the utility function either precisely or with certainty.[2] This is addressed below where a predicate $\Omega Acc(.)$ represents the acceptability of a deal to Ω. The agents communicate using sentences in a first-order language $\mathcal{C}$. This includes the exchange, acceptance, and rejection of offers. $\mathcal{C}$ usual trading predicates including the following: *Offer*(δ), *Accept*(δ), *Reject*(δ), *Bid*(δ), and *Quit*(.), where *Offer*(δ) means "the sender is offering you a deal δ," *Accept*(δ) means "the sender accepts your deal δ," *Reject*(δ) means "the sender rejects your deal δ," *Bid*(δ) means "the sender submits the

bid δ," and *Quit*(.) means "the sender quits—the negotiation ends."

Agent Architecture

Π uses the language $\mathcal{C}$ for external communication and the language $\mathcal{L}$ for internal representation. Two predicates in $\mathcal{L}$ are $\Pi Acc(.)$ and $\Omega Acc(.)$. The proposition $(\Pi Acc(\delta) \mid I_t)$ means "Π will be comfortable accepting the deal δ given that Π knows information I_t at time t." The idea is that Π will accept deal δ, if $\mathbf{P}(\Pi Acc(\delta) \mid I_t) \geq \alpha$ for some threshold constant α. The precise meaning that Π gives to $\Pi Acc(.)$ is described in the following. The proposition $\Omega Acc(\delta)$ means "Ω is prepared to accept deal δ." The probability distribution $\mathbf{P}(\Omega Acc(.))$ is estimated in the following.

Each incoming message M from source S received at time t is time-stamped and source-stamped, $M_{[S,t]}$, and placed in an *inbox*, X, as it arrives. Π has an *information repository I*, a *knowledge base $\mathcal{K}$*, and a *belief set $\mathcal{B}$*. Each of these three sets contains statements in a first-order language $\mathcal{L}$. I, contains statements in $\mathcal{L}$ together with sentence probability functions of time. I_t is the state of I at time t and may be inconsistent. At some particular time, t, $\mathcal{K}_t$ contains statements that Π believes are true at time t, such as $\forall x(Accept(x) \leftrightarrow \neg Reject(x))$. The belief set $\mathcal{B}_t = \{\beta_i\}$ contains statements that each is qualified with a *given sentence probability*, $\mathbf{B}(\beta_i)$, that represents Π's belief in the truth of the statement at time t. The distinction between the knowledge base $\mathcal{K}$ and the belief set $\mathcal{B}$ is simply that $\mathcal{K}$ contains unqualified statements, and $\mathcal{B}$ contains statements that are qualified with sentence probabilities. $\mathcal{K}$ and $\mathcal{B}$ play different roles; $\mathcal{K}_t \cup \mathcal{B}_t$ is required to be consistent. Π's actions are determined by its strategy. A *strategy* is a function $\mathbf{S} : \mathcal{K} \times \mathcal{B}^{\circledR} \mathcal{A}$, where $\mathcal{A}$ is the set of actions. At certain distinct times, the function $\mathbf{S}$ is applied to $\mathcal{K}$ and $\mathcal{B}$, and the agent does something. The set of actions, $\mathcal{B}$, includes sending *Offer*(.), *Accept*(.), *Reject*(.), *Quit*(.) messages and claims to Ω. The way in which $\mathbf{S}$ works is described in Secs. Two instants of time before the **S** function is activated, an import function and a revision function are activated. The import function $\mathbf{I} : (X \times I_t-) \circledast I_t$ clears the inbox, using its import rules. An import rule takes a message M, written in language C, and from it derives sentences written in language L, to which it attaches decay functions and adds these sentences together with their decay functions to I_t- to form I_t. These decay functions are functions of the message type, the time the message arrived, and the source from which it came (an illustration is given in the following). An *import rule* has the form $\mathbf{P}(S \mid M_{[\Omega,t]}) = f(M,\Omega,t) \in [0,1]$, where S is a statement, M is a message, and f is the decay function. Then, the belief revision function $\mathbf{R} : I_t- \circledast (I_t \times K_t \times B_t)$ deletes any statements in I_t- whose sentence probability functions have a value that is ≈ 0.5 at time t. From the remaining statements, $\mathbf{R}$ selects a consistent set of statements and instantiates its sentence probability functions to time t and places the unqualified statements from that set in K_t and the qualified statements, together with their sentence probabilities, in B_t.

An example now illustrates the ideas in the previous paragraph. Suppose that the predicate $\Omega Acc(\delta)$ means that "deal δ is acceptable to Ω." Suppose that Π is attempting to trade a good "g" for cash. Then, a deal $\delta(\pi,\omega)$ will be $\delta(g,x)$, where x is an amount of money. If Π assumes that Ω would prefer to pay less than more, then I_t will contain: $\iota_0 : (\forall gxy)((x \geq y) \circledast (\Omega Acc(g,x)) \circledast \Omega Acc(g,y))$. Suppose Π uses a simple linear decay for its import rules: $f(M,\Omega,t_i) = trust(\Omega) + (0.5 - trust(\Omega)) \times \dfrac{t - t_i}{decay(\Omega)}$, where $trust(\Omega)$ is a value in $[0.5,1]$ and $decay(\Omega) > 0$.[3] $trust(\Omega)$ is the probability attached to S at time $t = t_i$, and $decay(\Omega)$ is the time period taken for $\mathbf{P}(S)$ to reach 0.5 when S is discarded. Suppose at time $t = 7$, Π receives the message $Offer(g,\$20)_{[\Omega,7]}$ and has the import rule $\mathbf{P}(\Omega Acc(g,x) \mid Offer(g,x)_{[\Omega,t_i]}) = 0.8 - 0.025 \times (t - t_i)$, (i.e., *trust* is 0.8 and *decay* is 12). Then, in the absence of any other information, at time $t =$

11, $K_{t_{11}}$ contains i_0, and $B_{t_{11}}$ contains $\Omega Acc(g,\$20)$ with a sentence probability of 0.7.

Π uses three things to make offers: (1) an estimate of the likelihood that Ω will accept any offer; (2) an estimate of the likelihood that Π will, in hindsight, feel comfortable accepting any particular offer; and (3) an estimate of when Ω may quit and leave the negotiation (Debenham, 2004). Π supports its negotiation with claims with the aim of either improving the outcome (i.e., reaching a more beneficial deal) or improving the process (i.e., reaching a deal in a more satisfactory way).

Random Worlds

Let G be the set of all positive ground literals that can be constructed using the predicate, function, and constant symbols in L. A *possible world* is a valuation function $V: G \circledast \{\top,\bot\}$. $\mathcal{V}$ denotes the set of all possible worlds, and $\mathcal{V}_K$ denotes the set of possible worlds that are consistent with a knowledge base K (Halpern, 2003). A *random world* for K is a probability distribution $W_K = \{p_i\}$ over $\mathcal{V}_K = \{V_i\}$, where W_K expresses an agent's degree of belief that each of the possible worlds is the actual world. The *derived sentence probability* of any $\sigma \in L$, with respect to a random world W_K, is ($\forall \sigma \in L$):

$$\mathbf{P}_{W_K}(\sigma) \triangleq \sum_n \{p_n : \sigma \text{ is } \top \text{ in } V_n\} \qquad (1)$$

A random world W_K is consistent with the agent's beliefs B if: $(\forall \beta \in B)(\mathbf{B}(\beta) = \mathbf{P}_{W_K}(\beta))$. That is, for each belief, its derived sentence probability as calculated using equation (1) is equal to its given sentence probability.

The *entropy* of a discrete random variable X with probability mass function $\{p_i\}$ is (MacKay, 2003) $\mathbf{H}(X) = -\sum_n p_n \log p_n$ where: $p_n \geq 0$ and $\sum_n p_n = 1$. Let $W_{\{K,B\}}$ be the maximum entropy probability distribution over $\mathcal{V}_K$ that is consistent with B. Given an agent with K and B, its *derived sentence probability* for any sentence, $\sigma \in L$, is:

$$(\forall \sigma \in \mathcal{L})\mathbf{P}(\sigma) \triangleq \mathbf{P}_{W_{\{K,B\}}}(\sigma) \tag{2}$$

Using equation (2), the derived sentence probability for any belief, bi, is equal to its given sentence probability. So the term *sentence probability* is used without ambiguity.

If X is a discrete random variable taking a finite number of possible values $\{x_i\}$ with probabilities $\{p_i\}$, then the *entropy* is the average uncertainty removed by discovering the true value of X and is given by $\mathbf{H}(X) = -\sum_n p_n \log p_n$. The direct optimization of $\mathbf{H}(X)$ subject to a number, q, of linear constraints of the form $\sum_n p_n g_k(x_n) = g_k$ for given constants g_k, where $k = 1, \ldots, q$, is a difficult problem. Fortunately, this problem has the same unique solution as the *maximum likelihood problem* for the Gibbs distribution (Pietra, Pietra & Lafferty, 1997). The solution to both problems is given by:

$$pn = \frac{\exp(-\sum_{k=1}^{\theta} \lambda_k g_k (x_n))}{\sum_m \exp(-\sum_{k=1}^{\theta} \lambda_k g_k (x_m))} \tag{3}$$

$n = 1, 2$, where the constants $\{\lambda_i\}$ may be calculated using equation (3), together with the three sets of constraints: $p_n \geq 0$, $\sum_n p_n = 1$, and $\sum_n p_n g_k(x_n) = g_k$. The distribution in equation (3) is known as *Gibbs distribution*.

If X is a discrete random variable taking a finite number of possible values $\{x_j\}$ with probabilities $\{p_j\}$, then the *entropy* is the average uncertainty removed by discovering the true value of X and is given by $\mathbf{H}(X) = -\sum_n p_n \log p_n$. The maximum entropy distribution: argmax $\underline{p}\mathbf{H}(\underline{p})$, $\underline{p} = (p_1, \ldots, p_N)$, subject to $M+1$ linear constraints:

$$g_j(\underline{p}) = \sum_{i=1}^{N} c_{ji} p_i - \mathbf{B}(\beta_j) = 0, j = 1, \ldots, M$$

$$g_0(\underline{p}) = \sum_{i=1}^{N} p_i - 1 = 0 \tag{4}$$

where $c_{ji} = 1$ if β_j is $\top$ in v_i and 0 otherwise, and $p_i \geq 0$, $i = 1, \ldots, N$, is found by introducing Lagrange multipliers and then obtaining a numerical solution using the multivariate Newton-Raphson method. In the subsequent subsections, we will see how an agent updates the sentence probabilities, depending on the type of information used in the update.

Given a prior probability distribution $\underline{q} = (q_i)_{i-1}^{n}$ and a set of constraints, the *principle of minimum relative entropy* chooses the posterior probability distribution $\underline{p} = (p_i)_{i-1}^{n}$ that has the least relative entropy with respect to $\underline{q}$:

$$arg \ min \sum_{i=1}^{n} p_i \log \frac{p_i}{q_i} \tag{5}$$

and that satisfies the constraints. The principle of minimum relative entropy is a generalization of the principle of maximum entropy. If the prior distribution $\underline{q}$ is uniform, the relative entropy of $\underline{p}$ with respect to $\underline{q}$ differs from $-\mathbf{H}(\underline{p})$ only by a constant. So, the principle of maximum entropy is equivalent to the principle of minimum relative entropy with a uniform prior distribution.

SUITABILITY OF AN ACTION

The proposition $(\Pi Acc(\delta) \mid I)$ was introduced previously. This section describes how the agent estimates its beliefs of whether this proposition is true for various δ.

An Exemplar Application

An exemplar application follows. Π is placing bids in an e-market, attempting to purchase with cash a particular second-hand motor vehicle with some period of warranty. So, the two issues in this negotiation are the period of the warranty and the cash consideration. A deal δ consists of this pair of issues, and the deal set has no natural ordering. Suppose that Π wishes to apply *ME* to estimate values for $\mathbf{P}(\Omega Acc(\delta))$ for various δ. Suppose that the warranty period is simply 0, four years, and that the cash amount for this car certainly will be at least \$5,000 with no warranty and is unlikely to be more than \$7,000 with four years' warranty.

In what follows, all price units are in thousands of dollars. Suppose, then, that the deal set in this application consists of 55 individual deals in the form of pairs of warranty periods and price intervals: { $(w, [5.0, 5.2))$, $(w, [5.2, 5.4))$, $(w, [5.4, 5.6))$, $(w, [5.6, 5.8))$, $(w, [5.8, 6.0))$, $(w, [6.0, 6.2))$, $(w, [6.2, 6.4))$, $(w, [6.4, 6.6))$, $(w, [6.6, 6.8))$, $(w, [6.8, 7.0))$, $(w, [7.0, \infty))$ }, where $w = 0, 4$. Suppose that Π has previously received two offers from Ω. The first is to offer 6.0 with no warranty, and the second is to offer 6.9 with one year's warranty. Suppose Π believes that Ω still stands by these two offers with probability 0.8. Then, this leads to two beliefs: $\beta_1 : \Omega Acc(0, [6.0, 6.2))$; $\mathbf{B}(\beta_1) = 0.8$, $\beta_2 : \Omega Acc(1, [6.8, 7.0))$; $\mathbf{B}(\beta_2) = 0.8$. Following the previous discussion, before switching on *ME*, Π should consider whether it believes that $\mathbf{P}(\Omega Acc(\delta))$ is uniform over δ. If it does, then it includes both β_1 and β_1 in B and calculates $W_{\{K, B\}}$ that yields estimates for $\mathbf{P}(\Omega Acc(\delta))$ for all δ. If it does not, then it should include further knowledge in K and B. For example, Π may believe that Ω is more likely to bid for a greater warranty period the higher her bid price is. If so, then this is a multi-issue constraint that is represented in B and is qualified with a sentence probability.

Estimation of Beliefs

Here, agent, Π is attempting to buy a second-hand motor vehicle with a specific period of warranty, as described previously. This section describes how Ω estimates: $\mathbf{P}(\Pi Acc(\delta) | I_i)$. This involves the introduction of four predicates into the language L: *Me(.)*, *Suited(.)*, *Good(.)*, and *Fair(.)*.

General information is extracted from the World Wide Web using special-purpose bots that import and continually confirm information. These bots communicate with Π by delivering messages to Π's inbox X using predicates in the communication language C in addition to those described previously. These predicates include *IsGood*(Γ, Ω, r) and *IsFair*(Γ, δ, s), meaning,

respectively, that according to agent Γ, agent Ω is a good person to deal with certainty r; and, according to agent Γ, δ is a fair market deal with certainty s. The continual inflow of information is managed as described in Debenham (2003). Import functions are applied to convert these messages into beliefs. For example:

$$\mathbf{P}(Good(\Omega) | (\Gamma, \Omega, r)_{[\Theta, t_i]}) = f(IsGood, r, \Gamma, t),$$

where $Good(\Omega)$ is a predicate in the agent's internal language L, meaning Ω will be a good agent to do business with. Likewise, *IsFair*(.) messages in C are imported to I as *Fair*(.) statements in L, where *Fair*(δ) means δ is generally considered to be a fair deal at least.

With the motor vehicle application in mind, $\mathbf{P}(\Pi Acc(\delta) | I)$ is derived from conditional probabilities attached to four other propositions: *Suited*(ω), *Good*(Ω), *Fair*(δ), and *Me*(δ), where *Suited*(ω) means terms w are perfectly suited to Π's needs, and *Me*(δ) means on strictly subjective grounds, the deal δ is acceptable to Π. These four probabilities are $\mathbf{P}(Suited(\omega) | I)$, $\mathbf{P}(Good(\Omega) | I)$, $\mathbf{P}(Fair(\delta) | I_t \cup \{Suited(\omega), Good(\Omega)\})$, and $\mathbf{P}(Me(\delta) | I_t \cup \{Suited(\omega), Good(\Omega)\})$. The last two of these four probabilities factor out both the suitability of ω and the appropriateness of the other Ω. The third captures the concept of a fair market deal and the fourth a strictly subjective what ω is worth to Π. The *Me*(.) proposition is related closely to the concept of a private valuation in game theory. This derivation of $\mathbf{P}(\Pi Acc(\delta) | I)$ from the four other probabilities may not be suitable for assessing other types of deals. For example, in eProcurement, some assessment of the value of an ongoing relationship with a collaborating agent may be a significant issue. Also, for some low-value trades, the inclusion of *Good*(.) may not be required.

The whole estimation of beliefs apparatus is illustrated in Figure 3.

INTERACTION

Π engages in bilateral bargaining with another agent Ω. Π and Ω each exchanges offers alternately at successive discrete times (Kraus, 2001). They enter into a commitment, if one of them accepts a standing offer. The protocol has three stages:

1. Simultaneous, initial binding offers from both agents.
2. A sequence of alternating offers.
3. An agent quits and walks away from the negotiation.

In the first stage, the agents simultaneously send *Offer*(.) messages to each other that stand for the entire negotiation. These initial offers are taken as limits on the range of values that are considered possible. This is crucial to entropy-based inference, when there are domains that would otherwise be unbounded. The exchange of initial offers "stakes out the turf" on which the subsequent negotiation will take place. In the second stage, an *Offer*(.) message is interpreted as an implicit rejection, *Reject*(.), of the other agent's offer on the table. Second stage offers stand only if accepted by return—Π interprets these offers as indications of Ω's willingness to accept. They are represented as beliefs with sentence probabilities that decay in time. The negotiation ceases either

in the second round, if one of the agents accepts a standing offer, or in the final round, if one agent quits and the negotiation breaks down.

To support the offer-exchange process, Π has to do two things. First, it must respond to offers received from Ω. Second, it must send offers and possibly information to Ω. This section describes machinery for estimating the probabilities $\mathbf{P}(\Omega Acc(\delta))$, where the predicate $\Omega Acc(\delta)$ means Ω will accept Π's offer δ. In the following, Π is attempting to purchase with cash a particular second-hand motor vehicle with some period of warranty from Ω, as described previously. So, a deal δ will be represented by the pair (w, p), where w is the period of warranty in years, and $\$p$ is the price.

Π assumes the following two preference relations for Ω, and K contains:

$$\kappa_{11} : \forall\ x,y,z((x<y) \circledast (\Omega Acc(y, z) \circledast \Omega Acc(x, z)))$$
$$\kappa_{12} : \forall\ x,y,z((x<y) \circledast (\Omega Acc(z,x) \circledast \Omega Acc(z,y)))$$

These sentences conveniently reduce the number of possible worlds. The two preference relations κ_{11} and κ_{12} induce a partial ordering on the sentence probabilities in the $\mathbf{P}(\Omega Acc(w, p))$ array from the top left, where the probabilities are ≈ 1, to the bottom right, where the probabilities are ≈ 0. There are 51 possible worlds that are consistent with $\mathcal{K}$.

Suppose that the offer exchange has proceeded as follows: Ω asked for \$6,900 with a one-year warranty, and Π refused; then, Π offered \$5,000 with two years' warranty, and Ω refused; and then, Ω asked for \$6,500 with three years' warranty, and Π refused. Then, the next time, step B contains $\beta_{11} : \Omega Acc(3, [6.8,7.0))$, $\beta_{12} : \Omega Acc(2, [5.0,5.2))$, and $\beta_{13}: \Omega Acc(1, [6.4,6.6))$, and with a 10% decay in integrity for each time step: $\mathbf{P}(\beta_{11}) = 0.7$, $\mathbf{P}(\beta_{12}) = 0.2$, and $\mathbf{P}(\beta_{13}) = 0.9$, equation (3) is used to calculate the distribution $\mathbf{W}\{\mathcal{K},\mathcal{B}\}$ which shows that there are just five different probabilities in it. The probability matrix for the proposition $\Omega Acc(w, p)$ is:

Figure 3. Estimating Π's beliefs

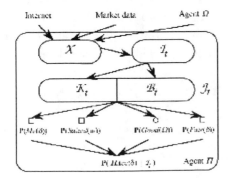

p_w 0 1 2 3 4

[7.0,∞) 0.9967 0.9607 0.8428 0.7066 0.3533
[6.8,7.0) 0.9803 0.9476 0.8330 **0.7000** 0.3500
[6.6,6.8) 0.9533 0.9238 0.8125 0.6828 0.3414
[6.4,6.6) 0.9262 **0.9000** 0.7920 0.6655 0.3328
[6.2,6.4) 0.8249 0.8019 0.7074 0.5945 0.2972
[6.0,6.2) 0.7235 0.7039 0.6228 0.5234 0.2617
[5.8,6.0) 0.6222 0.6058 0.5383 0.4523 0.2262
[5.6,5.8) 0.5208 0.5077 0.4537 0.3813 0.1906
[5.4,5.6) 0.4195 0.4096 0.3691 0.3102 0.1551
[5.2,5.4) 0.3181 0.3116 0.2846 0.2391 0.1196
[5.0,5.2) 0.2168 0.2135 **0.2000** 0.1681 0.0840

In this array, the derived sentence probabilities for the three sentences in B are shown in bold type; they are exactly their given values.

Π's *negotiation strategy* is a function $\mathbf{S} : \mathcal{K} \times \mathcal{B} \circledast \mathcal{A}$, where $\mathcal{A}$ is the set of actions that send *Offer*(.), *Accept*(.), *Reject*(.), and *Quit*(.) messages to Ω. If Π sends *Offer*(.), *Accept*(.), or *Reject*(.) messages to Ω, then P is giving Ω information about himself or herself. In an infinite-horizon bargaining game, where there is no incentive to trade now rather than later, a self-interested agent will sit and wait and do nothing except, perhaps, ask for information. The well-known bargaining response to an approach by an interested party, "Well, make me an offer," illustrates how a shrewd bargainer may behave in this situation.

An agent may be motivated to act for various reasons; three are mentioned: (1) if there are costs involved in the bargaining process due either to changes in the value of the negotiation object with time or to the intrinsic cost of conducting the negotiation itself; (2) if there is a risk of breakdown caused by another agent walking away from the bargaining table; and (3) if the agent is concerned with establishing a sense of trust (Ramchurn, Jennings, Sierra & Godo, 2003) with another agent (this could be the case in the establishment of a business relationship). Of these three reasons, the last two are addressed here. The risk of breakdown may be reduced, and a sense of trust may be established, if the agent

appears to the other agent to be approaching the negotiation in an even-handed manner. One dimension of appearing to be even-handed is to be equitable with the value of information given to the other agents. Various bargaining strategies, both with and without breakdown, are described in Debenham (2004), but they do not address this issue. A bargaining strategy is described here that is founded on a principle of equitable information gain; that is, Π attempts to respond to Ω's messages so that Ω's expected information gain is similar to that which Π has received.

Π models Ω by observing his or her actions and by representing beliefs about his or her future actions in the probability distribution **P**(Ω*Acc*). Π measures the value of information that it receives from Ω by the change in the entropy of this distribution as a result of representing that information in **P**(Ω*Acc*). More generally, Π measures the value of information received in a message, μ, by the change in the entropy in its entire representation, $\mathcal{I}_t = \mathcal{K}_t \cup \mathcal{B}_t$, as a result of the receipt of that message; this is denoted by: $\Delta_\mu |\mathcal{I}_t^\Pi|$, where $|\mathcal{I}_t^\Pi|$ denotes the value (as negative entropy) of Π's information in $\mathcal{I}$ at time t. Although both Π and Ω will build their models of each other using the same data, the observed information gain will depend on the way in which each agent has represented this information. To support its attempts to achieve equitable information gain, Π assumes that Ω's reasoning apparatus mirrors its own, and so is able to estimate the change in Ω's entropy as a result of sending a message μ to Ω: $\Delta_\mu |\mathcal{I}_t^\Omega|$. Suppose that Π receives a message $\mu = $ *Offer*(.) from Ω and observes an information gain of $\Delta_\mu |\mathcal{I}_t^\Pi|$. Suppose that Π wishes to reject this offer by sending a counter-offer, *Offer*(δ), that will give Ω expected equitable information gain. $\delta = \{\operatorname{argmax}_\delta \mathbf{P}(\Pi Acc(\delta) \mid I_t) \geq \alpha \mid (\Delta_{Offer(\delta)} |\mathcal{I}_t^\Omega| \approx \Delta_\mu |\mathcal{I}_t^\Pi|)\}$. That is, Π chooses the most acceptable deal to himself or herself that gives his or her bargaining partner expected equitable information gain, provided that there is such a deal. If there is not, then Π chooses the best available compromise

$\delta = \{argmax_\delta(\Delta_{Offer(\delta)}|\mathcal{J}_t^\Omega|) \mid \mathbf{P}(\Pi Acc(\delta) \mid I_t) \geq \alpha\}$, provided there is such a deal. If there is not, then Π does nothing.

COLLABORATION

The mechanism that Π uses for managing process delegation is described in full. Join(·) messages are managed similarly. This next section discusses the sorts of payoff measures and estimates that are available and that are combined to give a value for the expected payoff vector $\underline{v}_i$ for each agent. Let $\mathbf{P}(A \gg)$ denote that A is the best choice in terms of some combination of the parameter estimates described in the following. These measurements then are used by agent Π to determine $\mathbf{P}(X_i \gg)$ and then, in turn, to determine the delegation strategy $(p_i)_{i=1}^n$.

The Performance Parameters

Agent Π continually measures the performance of itself and of other agents in the system using four measures. Three are: *time, cost,* and *likelihood of success,* which are attached to all of its delegations-in and delegations-out. The last one is a *value* parameter that is attached to other agents. Time is the total time taken to termination. Cost is the actual cost of the resources allocated. For example, the time that the agent (possibly with a human assistant) actually spent working on that process. The likelihood of success is the probability that an agent will deliver its response within its constraints. The value parameter is the value added to a process by an agent. Unfortunately, value is often very difficult to measure. It is treated here by a subjective estimate delivered by users of the system.

The three parameters *time, cost,* and likelihood of *success,* are observed and recorded every time an agent, including Π, delivers or fails to deliver its commitments. This generates a large amount of data whose significance reasonably can be

expected to degrade over time. So, a cumulative estimate only is retained. The integrity of information evaporates as time goes by. If we have the set of observable outcomes as $O = \{o_1, o_2, \ldots, o_m\}$, then complete ignorance of the expected outcome means that our expectation over these outcomes is $\frac{1}{m}$ — (i.e., the unconstrained maximum entropy distribution). This natural decay of information integrity is offset by new observations.

Given one of the parameters, u, with m possible outcomes[4], suppose that $\mathbf{P}^t(u\mathfrak{c} \mid \delta)$ is the estimate at time t of the probability that the actual outcome $u\mathfrak{c}$ will be observed, given that the agent being observed has committed to δ. Suppose that Π observes the actual outcome r, and on the basis of this outcome, Π believes that the probability of r being observed at the next time is g_r. Then let $\mathbf{P}_{gr}^t(u\mathfrak{c} \mid \delta)$ be the posterior minimum relative entropy distribution calculated using equation (5) with prior distribution $\mathbf{P}^t(u\mathfrak{c} \mid \delta)$ and satisfying the constraint that $\mathbf{P}_{gr}^t(r \mid \delta) = g_r$. Then, update $\mathbf{P}^t(u\mathfrak{c} \mid \delta)$ with:

$$\mathbf{P}^{t+1}(u' \mid \delta) = \frac{1-\rho}{n} + \rho \cdot \mathbf{P}_{gr}^t(u' \mid \delta) \qquad (6)$$

This equation determines the development of $\mathbf{P}^t$ $(u\mathfrak{c} \mid \delta)$ for some large $\rho \in [0,1]$.

Π uses the method in equation (6) to update its estimates for all probability distributions representing each of the agents with which it deals (e.g., if $\mathbf{P}^t(\cdot)$ is Π's estimate of the time that Ω will take to deliver on a particular type of agreement). Suppose that at time t, Ω delivers his or her response after having taken time u. Then, Π attaches a belief (i.e., a sentence probability) to the proposition that this is how Ω will behave at time $t+1$. This becomes the constraint in the minimum relative entropy calculation, and then, equation (6) gives $\mathbf{P}^{t+1}(\cdot)$. The *process delegation problem* belongs to the class of resource allocation games, which are inspired by the El Farol Bar problem (see Galstyan, Kolar, and Lerman [2003] for recent work).

Choosing the "Best" Collaborator

The probability distributions previously described may be used to determine the probability that one agent is a better choice than another by calculating the probability that one random variable is greater than another in the usual way. This method may be extended to estimate the probability that one agent is a better choice than a number of other agents. For example, if there are three agents to choose from (A, B, and C), then:

$$\mathbf{P}(A\gg) = \mathbf{P}((A\gg B)\wedge(A\gg C))$$
$$= \mathbf{P}(A\gg B) \times \mathbf{P}((\gg\_C)\mid(A\gg B))$$

The difficulty with this expression is that there is no direct way to estimate the second conditional probability. This expression shows that:

$$\mathbf{P}(A\gg B) \times \mathbf{P}(A\gg C) \le \mathbf{P}(A\gg) \le \mathbf{P}(A\gg B)$$

By considering the same expression with B and C interchanged:

$$\mathbf{P}(A\gg B)\times\mathbf{P}(A\gg C) \le \mathbf{P}(A\gg) \le \mathbf{P}(A\gg C)$$

and so:

$$\mathbf{P}(A\gg) \le \min[\mathbf{P}(A\gg B), \mathbf{P}(A\gg C)]$$

So, for some $\tau_A \in [0,1]$:

$$\mathbf{P}(A\gg) = \mathbf{P}(A\gg B)\times\mathbf{P}(A\gg C) +$$
$$\tau_A\times[\min[\mathbf{P}(A\gg B),\mathbf{P}(A\gg C)] -$$
$$\mathbf{P}(A\gg B)\times\mathbf{P}(A\gg C)]$$

Similar expressions may be constructed for the probabilities that B and C are the best agents, respectively. This is as far as probability theory can go without making some assumptions. To proceed, assume that $\tau_A = \tau_B = \tau_C = \tau$; this assumption is unlikely to be valid, but it should not be too far from correct. Either A or B or C will be the best plan, so the sum of the three expressions for the probabilities of A, B, and C being the "best" plan will be unity. Hence:

$$t = \frac{1-d}{q-d}\text{ where:}$$

$$d = [(\mathbf{P}(A\gg B) \times \mathbf{P}(A\gg C))+$$
$$(\mathbf{P}(B\gg C) \times \mathbf{P}(B\gg A)) + (\mathbf{P}(C\gg A) \times \mathbf{P}(C\_B))]$$
$$q = [\min[\mathbf{P}(A\gg B),\mathbf{P}(A\gg C)]+$$
$$\min[\mathbf{P}(B\gg C),\mathbf{P}(B\gg A)]+$$
$$\min[\mathbf{P}(C\gg A),\mathbf{P}(C\gg B)]]$$

This expression for τ is messy but easy to calculate. The probability that each of the three agents (A, B, and C) is the "best" choice is $\mathbf{P}(A\gg)$, $\mathbf{P}(B\gg)$ and $\mathbf{P}(C\gg)$.

An alternative to the previous is simply to use equation (2) to estimate the probability of the propositions that each of the agents is the "best" collaborator; that is, for agent A:

$$\mathbf{P}(A\gg) = \mathbf{P}_{W_{\{\mathcal{K},\mathcal{B}\}}}(A\gg)$$

To calculate this probability, then, requires the calculation of the maximum entropy distribution that is consistent with $\mathcal{K}$ and $\mathcal{B}$, and then simply the adding up of the probabilities in that distribution that are associated with possible worlds in which agent A is the "best." This alternative approach involves a maximum entropy calculation, whereas the previous approach does not.

Delegation Strategy

A delegation strategy is a probability distribution $\{p_i\}_{i=1}^n$ that determines to whom from $\{X_i\}_{i=1}^n$ to offer responsibility for doing what. The delegation strategy determines who does what stochastically by determining the $\{p_i\}_{i=1}^n$, where p_i is the probability that the i'th agent will be selected. The choice of the agent to which to delegate, then, is made with these probabilities. The expression of the delegation strategy in terms of probabilities enables the strategy to balance conflicting goals, such as achieving process quality and process efficiency.

A greedy strategy *best* picks the agent that promises greatest returns:

$$p_i = \begin{cases} \frac{1}{m} & \text{if } X_i \text{ is such that } \mathbf{P}(X_i \gg) \text{ is} \\ 0 & \text{maximal 0 otherwise} \end{cases}$$

where m is such that there are m agents for whom $\mathbf{P}(X_i \gg)$ is maximal. This strategy attempts to maximize expected payoff, but it is shortsighted in that it rewards success with work, although it is not uncommon in practice. Another strategy *prob* also favors high payoff but gives all agents a chance to prove themselves, sooner or later, and is defined by $p_i = \mathbf{P}(X_i \gg)$. The strategy *random* is equitable and picks agents by: $p_i = \frac{1}{n}$.

An *admissible* delegation strategy has the properties:

if $\mathbf{P}(X_i \gg) > \mathbf{P}(X_j \gg)$ then $p_i > p_j$
if $\mathbf{P}(X_i \gg) = \mathbf{P}X_j \gg$ *then* $p_i = p_j$
$(\forall i)p_i > 0$ and $\sum_i p_i = 1$

So, *best* and *random* are not admissible strategies, but *prob* is admissible.

$\mathbf{P}(X_i \gg)$ is the probability that Ω_i is the "best" choice. The strategy *best* that continually chooses the "best" on the basis of historic data is flawed, because an agent who "goes through a bad patch" may never be chosen, which this means that if an agent wants "the quiet life," then all it would have to do is to make a series of mistakes. The delegation strategy *prob* is a compromise between being equitable and utility optimization; it chooses agents with probability $p_i = \mathbf{P}(X_i \gg)$. That is, the probability that Π will attempt to delegate a process to Ω_i is equal to the probability that Π estimates Ω_i to be the "best" choice for the job.

CONCLUSION

Emergent processes are business processes whose execution is determined by the prior knowledge of the agents involved and by the knowledge that emerges during a process instance. The establishment of a sense of trust (Ramchurn et al., 2003) contributes to the establishment of business relationships and to preventing breakdown in one-off negotiation. The agent architecture is based on a first-order logic representation.

Emergent processes are managed by these agents that extract the process knowledge from the Internet using a suite of data mining bots. The agent achieves this by using ideas from information theory and by using maximum entropy logic to derive integrity estimates for knowledge about which it is uncertain.

The agents make no assumptions about the internals of the other agents in the system, including their motivations, logic, and whether they are conscious of a utility function. These agents focus only on the information in the signals that they receive.

REFERENCES

Bernhardt, D., & Miao, J. (2004). Informed trading when information becomes stale. *The Journal of Finance, LIX*(1).

Debenham, J. (2000). Supporting strategic process. *Proceedings of the 5th International Conference on the Practical Application of Intelligent Agents and Multi-Agents* (pp. 237-256).

Debenham, J. (2003). An eNegotiation framework. *Proceedings of the 23rd International Conference on Innovative Techniques and Applications of Artificial Intelligence, AI'2003* (pp. 79-92).

Debenham, J. (2004). Bargaining with information. *Proceedings of the 3rd International Conference on Autonomous Agents and Multi Agent Systems AAMAS-2004* (pp. 664-671).

Dourish, P. (1998). Using metalevel techniques in a flexible toolkit for CSCW applications. *ACM Transactions on Computer-Human Interaction (TOCHI), 5*(2), 109-155.

Fischer, L. (2003). *The workflow handbook 2003.* Future Strategies Inc.

Galstyan, A., Kolar, S., & Lerman, K. (2003). Resource allocation games with changing resource capacities. *Proceedings of the 2nd International Joint Conference on Autonomous Agents and Multiagent Systems Aamas-03*, 145-152.

Halpern, J. (2003). *Reasoning about uncertainty.* MIT Press.

Jaynes, E. (1957). Information theory and statistical mechanics: Part I. *Physical Review, 106*, 620-630.

Kraus, S. (2001). *Strategic negotiation in multiagent environments.* MIT Press.

MacKay, D. (2003). *Information theory, inference and learning algorithms.* Cambridge University Press.

Pietra, S. D., Pietra, V. D., & Lafferty, J. (1997). Inducing features of random fields. *IEEE Transactions on Pattern Analysis and Machine Intelligence, 19*(2), 380-393.

Ramchurn, S., Jennings, N., Sierra, C., & Godo, L. (2003). A computational trust model for multi-agent interactions based on confidence and reputation. *Proceedings of the 5th International Workshop on Deception, Fraud and Trust in Agent Societies.*

Singh, M. (2004). Business process management: A killer ap for agents? *Proceedings of the Third International Conference on Autonomous Agents and Multi Agent Systems AAMAS-2004,* (pp. 26-27).

Smith, H., & Fingar, P. (2003). *Business process management (bpm): The third wave.* Meghan-Kiffer Press.

van der Aalst, W., & van Hee, K. (2002). *Workflow management: Models, methods, and systems.* MIT Press.

ENDNOTES

[1] A sentence probability of 0.5 represents null information (i.e., "maybe, maybe not").

[2] The often-quoted oxymoron "I paid too much for it, but it's worth it" attributed to Samuel Goldwyn, movie producer, illustrates that intelligent agents may negotiate with uncertain utility.

[3] In this example, the value for the probability is given by a linear decay function that is independent of the message type, and *trust* and *decay* are functions of Ω only. There is scope for using learning techniques to refine the *trust* and *decay* functions in the light of experience.

This work was previously published in International Journal of Intelligent Information Technologies, Vol. 2, Issue 2, edited by V. Sugumaran, pp. 30-48, copyright 2006 by IGI Publishing, formerly known as Idea Group Publishing (an imprint of IGI Global).

Chapter IX
Beyond Intelligent Agents:
E–Sensors for Supporting Supply Chain Collaboration and Preventing the Bullwhip Effect

Walter Rodriguez
Florida Gulf Coast University, USA

Janusz Zalewski
Florida Gulf Coast University, USA

Elias Kirche
Florida Gulf Coast University, USA

ABSTRACT

This chapter presents a new concept for supporting electronic collaboration, operations, and relationships among trading partners in the value chain without hindering human autonomy. Although autonomous intelligent agents, or electronic robots (e-bots), can be used to inform this endeavor, the chapter advocates the development of e-sensors, i.e., software based units with capabilities beyond intelligent agent's functionality. E-sensors are hardware-software capable of perceiving, reacting and learning from its interactive experience through the supply chain, rather than just searching for data and information through the network and reacting to it. E-sensors can help avoid the "bullwhip" effect. The chapter briefly reviews the related intelligent agent and supply chain literature and the technological gap between fields. It articulates a demand-driven, sense-and-response system for sustaining e-collaboration and e-business operations as well as monitoring products and processes. As a proof of concept, this research aimed a test solution at a single supply chain partner within one stage of the process.

INTRODUCTION: FROM E-BOTS TO E-SENSORS

As e-business and e-commerce has grown, so has the need to focus attention on the: (1) electronic communications between e-partners; (2) operational transactions (e.g., sales, purchasing, communications, inventory, customer service, ordering, submitting, checking-status, and sourcing, among others); and (3) monitoring improvements in the supply (supply, demand, value) chain of products, systems, and services (Gaither & Fraizer, 2002).

Integrating continuous communication protocols and operational and supply chain management (SCM) considerations, early on in the enterprise design process, would greatly improve the successful implementation of the e-collaboration technologies in the enterprise. It is particularly important to examine the resources and systems that support the electronic communications, and relationships among partners, in the supply chain.

In addition, there is a need for obtaining (sensing) real time data for managing (anticipating, responding) throughout the supply chain. Typically companies need to synchronize orders considering type, quantity, location, and timing of the delivery in order to reduce waste in the production and delivery process. The data collection and availability provided by the e-sensing infrastructure/architecture discussed later in this chapter will allow for a collaborative environment, improve forecast accuracy, and increase cross-enterprise integration among partners in the supply chain.

Current supply chain information technologies (IT) allow managers to track and gather intelligence about the customers purchasing habits. In addition to point-of-sale Universal Product Code (UPC) barcode devices, the current IT infrastructure may include retail radio frequency identification (RFID) devices and electronic tagging to identify and track product flow. These technologies aid mainly in the marketing and re-supply efforts. But, how about tracking partners' behaviors throughout the chain in real time?

Artificial intelligent agents (or e-bots) can be deployed throughout the supply chain to seek data and information about competitive pricing, for instance, e-bots can search for the cheapest supplier for a given product and even compare characteristics and functionality. For this reason, the concept of an *agent* is important in both the Artificial Intelligence (AI) and the e-operations fields.

The term "intelligent agent" or "e-bot" denotes a software system that enjoys at least one of the following properties: (1) Autonomy; (2) "Social" ability; and (3) Reactivity (Wooldridge & Jennings, 1995). Normally, agents are thought to be autonomous because they are capable to operate without direct intervention of people and have some level of control over their own actions (Castelfranchi, 1995). In addition, agents may have the functionality to interact with other agents and automated systems via an agent-communication language (Genesereth & Ketchpel, 1994). This agent attribute is termed here *e-sociability* for its ability to interact with either people, or systems (software).

The next evolution of the intelligent agent concept is the development of integrated hardware/software systems that may be specifically designed to sense (perceive) and respond (act) within certain pre-defined operational constrains and factors, and respond in a real time fashion to changes (not a just-in-time fashion) occurring throughout the supply chain. These integrated hardware-software systems are termed *e-sensors,* in this chapter. Indeed, there is a real opportunity for process innovation and most likely organizations will need to create new business applications to put e-sensors at the centre of a process if they want to be competitive in this new supply chain environment. Aside from asset tracking, each industry will have specialized applications of e-sensors that cannot be generalized. Before getting into the e-sensors details, let us review some key supply chain management (SCM) issues relevant to this discussion.

SUPPLY CHAIN MANAGEMENT IN THE E-COLLABORATION CONTEXT

SCM is the art and science of creating and accentuating synergistic relationships among the trading partners in supply and distribution channels with the common shared objective of delivering products and services to the 'right customer' at the 'right time.' (Vakharia, 2002)

In the e-collaboration/e-business context, supply chain management (SCM) is the operations management discipline concerned with these synergistic communications, relationships, activities and operations in the competitive Internet enterprise. SCM involves studying the movement of physical materials and electronic information and communications—including transportation, logistics and information-flow management to

improve operational efficiencies, effectiveness and profitability. SCM consists in the strategies and technologies for developing and integrating the operations, communications and relationships among the e-trading partners (producers, manufacturers, services providers, suppliers, sellers, wholesalers, distributors, purchasing agents, logisticians, consultants, shipping agents, deliverers, retailers, traders and customers) as well as improving their operations throughout the products' or services' chain.

Integrated e-business SCM can enhance decision making by collecting real time information as well as assessing and analyzing data and information that facilitate collaboration among trading partners in the supply chain.

To achieve joint optimization of key SCM decisions, it is preferable that there be a free flow of all relevant information across the entire chain leading to a comprehensive analysis. (Vakharia, 2002)

As shown in Figure 1, IT systems, such as, enterprise resource planning (ERP), point of sale

Figure 1. Information flow using electronic information technologies in the supply chain (after Burke & Vakharia, 2002; Vakharia, 2002)

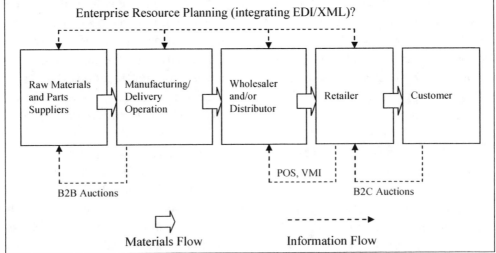

(POS), and vendor managed inventory (VMI) systems permit and, to some extend, automate information sharing.

The advent of reliable communication technologies has forced business partners throughout the supply chain to rethink their strategies as well as change the nature of the relationships with suppliers and customers. Companies that have made the shift have benefited from: "Reduced operating expenses, increased revenue growth, and improved customer levels," according to IBM ERP/Supply Management Division (Cross, 2000). According to the same source, the companies that have implemented supply chain improvement projects have been able to increase forecast accuracy and inventory reduction (up to 50% in overall improvement!). Some of the newer activities being implemented include: Supply-and-demand auctions, integrated collaborative product design (CAD/CAM), cross-enterprise workflow processes, demand management collaboration. In addition, some companies are even deploying SCM as an offensive tactic to gain a competitive edge (Cross, 2000).

Meixell's "Collaborative Manufacturing for Mass Customization" (2006)site, at http://www.som.gmu.edu/faculty/profiles/mmeixell/collaborative%20Planning%20&%20Mass%20Customization.pdf, provides extensive information about the use of collaborative technologies in the supply chain. The same author recently compiled a literature review; particularly, on decision support models used for the design of global supply chains (Meixell & Gargeya, 2005). This, however, does not mean that there are no strategic and technological gaps in the supply chain.

PARADIGM SHIFT: FROM 'PUSH' (SCM) TO 'PULL' (SRS)

We are not smart enough to predict the future, so we have to get better at reacting to it more quickly. (GE saying quoted by Haeckel, 1999)

E-business forces have shifted both the enterprise landscape and the competitive power from the providers of goods and information (makers, suppliers, distributors and retailers) to the purchasers of goods and information (customers). For this reason, e-businesses must collaborate electronically and sense-and-respond very quickly to the individual customer's needs and wants. So, rather than considering SCM analysis from the "supply" perspective, some researchers and practitioners advocate analyzing the market operations from the "demand" perspective: Sensing-and-responding to the consumer changing needs and wants by quickly collaborating and communicating in real-time throughout the chain. Researchers argue that e-businesses should measure and track customers' demands for products and services, rather than relying solely on demand forecasting models.

Fisher (1997) studied the root cause of poor performance in supply chain management and the need to understand the demand for products in designing a supply chain. Functional products with stable, predictable demand and long lifecycle require a supply chain with a focus almost exclusively on minimizing physical costs—a crucial goal given the price sensitivity of most functional products. In this environment, firms employ enterprise resource planning systems (ERP) to coordinate production, scheduling, and delivery of products to enable the entire supply chain to minimize costs and maximize production efficiency. The crucial flow of information is internal within the supply chain. However, the uncertain market reaction to innovation increases the risk of shortages or excess supplies for innovative products. Furthermore, high profit margins and the importance of early sales in establishing market share for new products, the short product lifecycles increasing the risk of obsolescence, and the cost of excess supplies require that innovative products have a responsive supply chain that focuses on flexibility and speed of response of the supplier. The critical decision to be made about inventory and capacity is not about minimizing costs, but where in the chain to position inventory and available production capacity in order to hedge against uncertain demand. The crucial flow of information occurs not only within the chain, but also from the market place to the chain.

While Selen and Soliman (2002) advocate a demand-driven model, Vakharia (2002) argues that push (supply) and pull (demand) concepts apply in different settings. That is, since businesses offering mature products have developed accurate demand forecasts for products with predictable lifecycles, they may rely more heavily on forecasting models. While businesses offering new products, with unpredictable short cycles, are better off operating their chains as a pull (demand) system, because it's harder to develop accurate demand forecasts for these new (or fluctuating demand) products.

The difficulty in synchronizing a supply chain to deliver the right product at the right time is caused by the distortion of information traveling upstream the supply chain. One of the most discussed phenomena in the e-operations field is called the Forrester (1958) or "bullwhip" effect which portrays the supply chain's tendency to amplify or delay product demand information throughout the chain (Sahin & Robinson, 2002). For instance, a particular supplier may receive a large order for their product and then decide to replenish the products sold. This action provides the quantity to restock the depleted products, plus some additional inventory to compensate for potential variability in demand. The overstated order and adjustments are passed throughout the supply chain causing demand amplification. At some point, the supply chain partners loose track of the actual customer demand.

Lee et al., (1997) proved that demand variability can be amplified in the supply chain as orders are passed from retailers to distributors and producers. Because most retailers do not know their demand with certainty, they have to make their decisions based on demand forecast.

When it is not very accurate, the errors in the retailers forecast are passed to the supplier in the form of distorted order. They found that sharing information alone would provide cost savings and inventory reduction. Other factors that contribute to the distortion of information is over reliance on price promotion, use of outdated inventory models, lack of sharing information with partners, and inadequate forecasting methods.

An important question in supply chain research is whether the bullwhip effect can be preventable. Chen et al., (2000) quantified the bullwhip effect for a multi-stage system and found that the bullwhip effect could be reduced but not completely eliminated, by sharing demand among all parties in the supply chain. Zhao et al., (2002) also studied the impact of the bullwhip effect and concluded that sharing information increases the economical efficiency of the supply chain. In a later study, Chen (2005) found that through forecast sharing the bullwhip effect can be further reduced by eliminating the need for the supplier to guess the retailer's underlying ordering policy.

The causes of uncertainty and variability of information leading to inefficiency and waste in the supply chain can be traced to demand forecasting methods, lead-time, batch ordering processes, price fluctuation, and inflated orders. One of the most common ways to increase synchronization among partners is to provide at each stage of the supply chain with complete information on the actual customer demand. Although this sharing of information will reduce the bullwhip effect, it will not completely eliminate it (Simchi-Levy et al., 2003). Lee et al., (1997a, 2004) suggests a framework for supply chain coordination initiatives which included using electronic data interchange (EDI), internet, computer assisted ordering (CAO), and sharing capacity and inventory data among other initiatives. Another important way to achieve this objective is to automate collection of point of sale data (POS) in a central database and share with all partners in a real time e-business environment. Therefore, efficient information acquisition and

Figure 2. SRS framework for integrating communication, information and materials flow and monitoring the e-business supply/demand chain

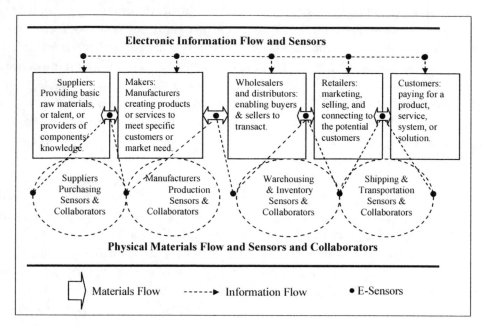

sharing is the key to creating value and reducing waste in many operations. A specially designed adaptive or sense-and-response system may help provide the correct information throughout the supply chain. The proposed system would have two important system functions—maintaining timely information sharing across the supply chain and facilitating the synchronization of the entire chain.

Haeckel (1999) indicates that "unpredictable, discontinuous change is an unavoidable consequence of doing business in the information age." And, since this "intense turbulence demands fast—even instantaneous—response," businesses must manage their operations as adaptive systems. Adaptive (sense-and-response) models may help companies systematically deal with the unexpected circumtances, particularly, e-businesses need to be able to anticipate and preempt sensed problems.

SENSE-AND-RESPONSE SYSTEM (SRS) MODEL AND FRAMEWORK

Figure 2 shows the proposed SRS model and framework for integrating real-time electronic communications, information sharing, and materials flow updating as well as monitoring the e-supply/demand/value chain—towards a new e-collaboration paradigm.

The "e-sensors" in the diagram are computer programs (software code) and its associated data and information collection devices (hardware), and communication interfaces. These sensors are designed for e-collaboration, data capturing (sensing), and information sharing, monitoring and evaluating data (input) throughout the value chain. Ultimately, this approach would result in semi-automated analysis and action (response) when a set of inputs are determined (sensed) without hindering human autonomy. That is, the sensors will gather the data, monitor, and evaluate the exchange in information between designated servers in the e-partners (suppliers and distribution channel) networks. Sensors will adjust plans and re-allocate resources and distribution routes when changes within established parameters are indicated. In addition, sensors will signal human monitors (operations or supply chain managers) when changes are outside the established parameters. The main advantage of this approach is that sensors will be capable of assessing huge amounts of data and information quickly to respond to changes in the chain environment (supply and demand) without hindering human autonomy. Particularly, e-sensors can provide the real-time information needed to prevent the bullwhip effect.

Companies like Cisco, Dell, IBM and Wal-Mart have led the development of responsive global supply chains. These companies and a few others have discovered the advantages of monitoring changes in near real-time. By doing so, they have been able to maintain low inventories, implement lean production and manufacturing operations, and even defer building and assembly resulting in lower costs and increase responsiveness to variable customer demands. This practice can be extended to incorporate e-sensors and human collaborators throughout the value chain and perceive and react to the demands.

SYSTEM ARCHITECTURE AND IMPLEMENTATION

To develop the implementation of the entire framework outlined in Figure 2 one faces involvement of multiple supply chain partners and months, if not years, of work just to develop a reliable communication infrastructure. In order to provide an immediate viable solution to test the concepts, in this research, the authors aimed at a single supply chain partner/company at only one stage

illustrated in Figure 2, to provide interfaces to the immediate preceding and the immediate succeeding stage (Kirche et al., 2005). Choosing a wholesaler/distributor (the middle box in Figure 2) as the company to automate its information flows and material flows with e-sensors and e-controls interfacing to the manufacturers and retailers, as well as to internal storage and distribution centers, we developed the overall design architecture as illustrated in Figure 3.

The selected communication architecture is based on CORBA (common object request broker architecture), a standard solution available from multiple vendors (Bolton, 2002). CORBA is an open system middleware with high scalability and potentially can serve an unlimited number of players and virtually any number of business processes and partners in the supply chain environment. As a communication infrastructure, it enables an integrated view of the production and distribution processes for an efficient demand management. Other benefits include continuous availability, business integration, resources availability on demand, and worldwide accessibility. The architecture presented in Figure 3 gives the wholesaler/distributor direct access to the assembly lines of the manufacturers and their shipping/transportation data via the operational data server. Full communication with the retailers is available. The wholesaler/distributor company does have itself full control over their financial data server and optimization server. The detailed functions of this architecture are described in (Kirche et al., 2005).

The goal of the real time system based on this architecture is to dynamically integrate end-to-end processes across the organization (key partners, manufacturers and retailers) to respond with speed to customer changes and market requirements. The real time CORBA framework enables employees to view current process capability and load on the system and provide immediate information to customers, by enabling tuning of

Figure 3. Architecture of distributed services for the wholesaler or distributor (after Kirche et al, 2005)

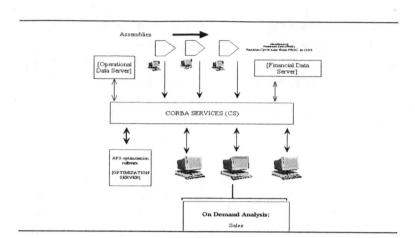

resources and balancing workloads to maximize production efficiency and adapt to dynamically changing environment.

A sample implementation of the system architecture from Figure 3 is presented in the form of a context diagram in Figure 4. To achieve the project's objective, that is, remote data access to enterprise networks with e-sensors/e-controls, we provide the capability of accessing enterprise-wide systems from a remote location or a vehicle, for both customers and employees.

The overall view of the system is as follows:

- When access to manufacturers from Figure 2 is considered, the focus can be on *plant access* for immediate availability of data and functions of the system; in that case, a remote *e-sensor/e-control* application using LabVIEW data acquisition software (Sokoloff, 2004) comes into play, with graphical user interface capable of interacting with remote users connected via the Internet.
- When access to warehousing from Figure 2 is considered, the focus can be on *business integration* via a multi-purpose enterprise-wide network; in that case, a CORBA based

framework is employed for a remote access to data objects identified as *e-sensors*, that can be stored on typical SQL database servers (Kirche et al., 2005).

From the network operation and connectivity perspective, e-sensors and e-controls provide business services, so they play the role of servers. Access to servers in this system is implemented via two general kinds of clients:

- When focus is on the *customer access* to obtain services, a cell phone location-aware application for business transactions has been developed, using order services as an example
- When focus is on the *employee access* to obtain services, such as conducting business on the road, a wireless PDA application for remote vending machine access has been developed, using the IEEE Std 802.11 wireless network protocol.

Several tests have been conducted to check behavior and performance of all four applications listed above and presented in Figure 4. For

Figure 4. Context diagram of the system being implemented (DAQ stands for data acquisition and control, 802.11 stands for an IEEE Std 802.11 for wireless networks, SQL stands for standard query language)

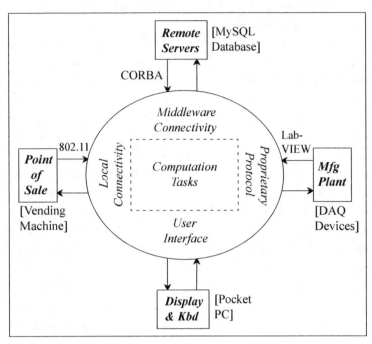

Figure 5. PDA client connectivity/performance test

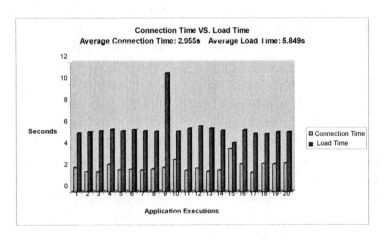

concision, it shows only a sample behavior of a PDA client via connectivity/performance test, in Figure 5. The graph shows how long it takes for the server to receive the connection request from the client application after the application was started. It is marked "Connection time." Another bar on the same chart shows how long the program itself took to load completely after being started (marked "Load time"). The connection graph was created to give an indication of how long, on average, one can expect for requests to be acknowledged and accepted by the server. Since all requests are handled the same way as the initial connection, this average connection time reflects sending and receiving of data to and from the client application. The load time is just a measure of performance for the application on the PDA itself. The data collected that way show the feasibility of all applications built within the SRS framework, as presented in Figure 2, for the architecture outlined in Figure 3.

CONCLUSION

This chapter briefly reviewed the current intelligent agent and supply chain paradigm and presented a conceptual framework for integrating e-collaboration tools in the operation and monitoring of products and services across value chain networks without hindering human autonomy. The demand-driven, sense-and-response framework model incorporates e-sensors and e-collaborators (humans using communication tools, computer software programs and its associated data-capturing hardware devices) throughout the supply chain. In practice, these e-sensors would be designed for data-capturing (sensing), monitoring and evaluating data (input) throughout the value chain, while humans collaborate and communicate in real-time, as tested in the above solution.

The implications of this new framework are that it contributes to the enhancement of the current SCM/DCM systems (such as Manugistics'

demand planning system) that analyzes manufacturing, distribution and sales data against forecasted data. The addition of SRS sensors would signal human monitors (operations or supply chain managers) when changes are outside the established parameters. The main advantage of this approach is that sensors would be capable of assessing huge amounts of data and information quickly to respond to changes in the chain environment (supply and demand) without hindering human autonomy.

Ultimately, this approach would result in the semi-automated analysis and action (response) when a set of inputs are determined (sensed) without hindering human autonomy. That is, the e-sensors would gather the data and monitor and evaluate the exchange in information between designated servers in the e-partners (suppliers and distribution channel) networks. E-sensors would adjust plans and re-allocate resources and distribution routes when changes within established parameters are indicated. Particularly, the new approach will aid managers in the prevention of the bullwhip effect.

Having real time data is critical in managing supply chain efficiently. Typically companies need to synchronize orders considering type, quantity, location and timing of the delivery in order to reduce waste in the production and delivery process. The data collection and availability provided by the e-sensing infrastructure/architecture will allow for a collaborative environment, improve forecast accuracy and increase cross-enterprise integration among partners in the supply chain. E-sensors will also offer a more proactive solution to current ERP systems by giving them the ability to process in real time relevant constraints and simultaneously order the necessary material type and quantities from multiple sources.

This e-sensor concept opens additional research opportunities within the boundaries of the operations management and information technology fields, particularly in the development of new software-hardware interfaces, real-time data capturing devices and other associated technologies. Finally, it leads to future 'automated decision-making' where IT/operations managers can "embed decision-making capabilities in the normal flow of work" (Davenport and Harris, 2005).

REFERENCES

Burke, G., & Vakharia, A. (2002). Supply chain management. In H. Bidgoli (Ed.), *Internet Encyclopedia*, New York: John Wiley.

Bresnahan, J. (1998). Supply chain anatomy: The incredible journey. *CIO Enterprise Magazine*, August 15. Retrievedon March 12, 2006 from http://www.cio.com site

Bolton, F. (2002). Pure CORBA: A code intensive premium reference. Indianapolis: Sams Publishing.

Castelfranchi, C. (1995). Guarantees for autonomy in cognitive agent architecture. In Wooldrige, M. and Jennings, N. R. (Eds.), *Intelligent Agents: Theories, Architectures, and Languages*, 890, pp. 56-70. Heidelberg, Germany: Springer-Verlag.

Chen, L. (2005). *Optimal information acquisition, inventory control, and forecast sharing in operations management*. Dissertation thesis. Stanford, CA: Stanford University.

Cheng, F., Ryan, J.K., & Simchi-Levy, D. (2000). Quantifying the 'bullwhip effect' in a supply chain: The impact of forecasting, lead times, and information. *Management Science, 46*(3), 436-444.

Cross, Gary J. (2000). How e-business is transforming supply chain management. *Journal of Business Strategy, 21*(2), 36-39.

Davenport, T.H., & Harris, J.G., (2005). Automated decision making comes of age. *MIT Sloan Management Review, 46*(4), 83-89.

Fisher, M. (1997). What is the right supply chain for you? *Harvard Business Review,* March-April, 105-117.

Forrester, J. W. (1958). Industrial dynamics. *Harvard Business Review*, July-August, 37-66.

Frohlich, M.T. (2002). E-integration in the supply chain: Barriers and performance, *Decision Sciences, 33*(4), 537-556.

Gaither, N. & Frazier, G. (2002). *Operations management*, 6th Edition, Cincinnati: Southwest.

Genesereth, M. R. & Ketchpel, S.P. (1994). Software agents. *Communications of the ACM, 37*(7), 48-53.

Haeckel, S.H. (1999). *Adaptive enterprise: Creating and leading sense-and-response organizations*. Boston: Harvard Business School Press.

Kirche, E., Zalewski, J., & Tharp, T. (2005). Real-time sales and operations planning with CORBA: Linking demand management and production Planning. In C.S. Chen, J. Filipe, I. Seruca, J. Cordeiro (Eds.), *Proceedings of the 7th International Conference on Enterprise Information Systems* (pp. 122-129). Washington, DC: ICEIS, Setubal, Portugal.

Lee, H., Padmanabhan, V., & Whang, S. (1997). The bullwhip effect. *Sloan Management Review, 38*(3), 93-103.

Lee, H., Padmanabhan, V., & Whang, S. (1997a). Information distortion in a supply chain: The bullwhip effect. *Management Science,43*, 546-548.

Lee, H., Padmanabhan, V., & Whang, S. (2004). Information distortion in a supply chain: The bullwhip effect/comments on "information distortion in a supply chain: The bullwhip effect." *Management Science, 50*(12), 1875-1894.

Mcixell, M.J. (2006). *Collaborative manufacturing for mass customization*. George Manson University. Retrieved February 15,2006 http://www.som.gmu.edu/faculty/profiles/mmeixell/

collaborative%20Planning%20&%20Mass%20 Customization.pdf

Meixell, M.J. & Gargeya, V.B. (2005). Global supply chain design: A literature review and critique. Transportation Research, *41*(6), 531- 550 Science Direct. Retrieved February 15, 2006 http://top25. sciencedirect.com/index.php?subject_area_id=4 .]

Sahin, F. & Powell Robinson, E.P. (2002). Flow coordination and information sharing in supply chains: Review, implications, and directions for future research. *Decision Sciences, 33*(4), 505-536.

Selen, W., & Soliman, F. (2002). Operations in today's demand chain management framework. *Journal of Operations Management, 20*(6), 667-673.

Schneider, G.P., & Perry, J.T. (2000). *Electronic Commerce*. Cambridge, MA: Course Technology.

Simch-Levy, D., Kaminsky, P., & Simchi-Levy, E. (2003). *Designing and managing the supply chain— concepts, strategies and case studies, Second Edition*. New York: McGraw-Hill.

Sokoloff, L. (2004). *Applications in LabVIEW*. New Jersey: Prentice Hall.

Vakharia, A.J. (2002). E-business and supply chain management. *Decision Sciences, 33*(4), 495-504.

Wooldridge., M. & Jennings, N.R. (1995). Intelligent agents: Theory and practice. GRACO. Retrieved on February 15, 2006 at http://www.graco.unb. br/alvares/DOUTORADO/disciplinas/feature/ agente_definicao.pdf .]

This work was previously published in International Journal of e-Collaboration, Vol. 3, Issue 2, edited by N. Kock, pp. 1-15, copyright 2007 by IGI Publishing, formerly known as Idea Group Publishing (an imprint of IGI Global).

Section IV
Agent Technologies in E–Business Infrastructure

Chapter X
An Automated Negotiation Mechanism for Agent Based on International Joint Ventures

Yee Ming Cheng
Yuan Ze University, Taiwan

Pei-Ni Huang
Yuan Ze University, Taiwan

ABSTRACT

One consequence of market globalization has been the growing incidence of collaborative ventures among companies from different countries. Small and large, experienced and novice, companies increasingly are choosing partnerships as a way to compete in the global marketplace. International joint ventures have emerged as the dominant form of partnership in light of intense global competition and the need for strategic organizational viability. The success of international joint ventures depends on many factors, but the most critical is vendors selection from among many suppliers based on their ability to meet the quantity requirements, delivery schedule, and the price limitation. The supplier selection negotiation mechanism is often the most complex, since it requires evaluation and decision making under uncertainty, based on multiple attributes (criteria) of quantitative and qualitative nature, involving temporal and resource constraints, risk and commitment problems, varying tactics and strategies, domain specific knowledge, information asymmetries, and so forth. In this chapter, we propose a negotiation mechanism employing fuzzy logic to evaluate different quantitative/qualitative scale of each attribute, generating similarity matching with bilateral alternatives offered by buyer and seller agents, and then modeling some constraint-based rules for sellers when receiving counter-proposals from buyers, consequently proceeding to trade-off mechanisms between both sides to gain an agreement. The negotiation mechanism is mainly classified into five parts. We first define negotiation parameters set and iso-curve computation in the preliminary setting. Second, negotiation alternative processing service will be proposed to select buyer's alternative based on iso-curve. After selecting negotiation alternatives, the buyer agent will send its alternative (counter-proposal) to seller agents to determine if it satisfies the seller's constraints, then decide iso-curve relaxation which is buyer's subjective behavior. Consequently, we use trade-off to find out buyer's partner and determine which attributes need to change along with the iso-curve. An example application to negotiating a supplier selection among agents to demonstrate the how the agents negotiation attribute parameters and reach the agreement. In the last part, post-negotiation analysis, we will compare two results from the preceding trade-off strategies (i.e., risk-seeking and risk-aversion) to inform decision making on how to make the most beneficial decision for a company. In our proposed negotiation mechanism scheme, agents autonomously negotiate multi-attribute fuzzy values of trade-off in an international joint venture selection tested with a notebook computer manufacturing company scenario.

INTRODUCTION

International joint ventures (IJV) are an increasingly important way for organizations to expand internationally. There is no apparent reason for this trend not to continue with pressures from global competition. Therefore, IJV becomes a major trend in cooperative business. The concept of IJV has been applied to many forms of cooperative business relations, like outsourcing, supply chains, or temporary consortiums. Specialization and flexibility are some of the key aspects of an everyday more dynamic and global market. The success of IJV depends on many factors, but the most critical include recognition of cultural differences, specified **workflow, information-sharing** through electronic data interchange and the Internet, and joint planning and other models that facilitate successful supply chain management. Technological support to the creation of such relationships is arising in many forms. The most ambitious ones intend to automate (part of) the process of creation and negotiation of IJV, mainly through **multi-agent** technology approaches, where each agent can represent each of the different enterprises. In fact, research on multi-agent technology addresses issues that fit the IJV buyer/seller relationship scenario. Agents are autonomous, interact with other agents, and enable

approaching inherently distributed problems with negotiation and coordination capabilities. The **negotiation** mechanism is often the most complex, since it requires evaluation and decision making under uncertainty, based on multiple **attributes (criteria)** of quantitative and qualitative nature, involving temporal and resource constraints, risk and commitment problems, varying tactics and strategies, domain specific knowledge, information asymmetries, and so forth. The negotiation cycle typically involves a sequence of interdependent activities (evaluation and decision making)—from suppliers' selection to enter the negotiation, through the negotiation per se to the execution of the agreed deal. Supplier selection and negotiation then are of a special importance for supply chain management. Thus, the objective of this chapter is to develop an agent-based cooperative negotiation mechanism which can be seen as a **decision-making** process of automatically resolving a conflict involving many parties over mutual goals.

In this chapter, we propose a set of negotiation mechanism schemes through employing **fuzzy logic** to evaluate different scales of each attribute, generating similarity matching with bilateral alternatives offered by buyer agent and seller agents, and then modeling some constraint-based rules for sellers when receiving **counter-proposal** by buyer, consequently proceeding to trade-off mechanism between both sides to gain an agreement. The negotiation mechanism is mainly classified into five parts. We first defined negotiation parameters and iso-curve computation in a *preliminary setting*. Second, *negotiation alternative processing service* will be proposed to select buyer's alternative based on iso-curve. After selecting a negotiation alternative, the buyer agent will send its alternative (counter-proposal) to seller agents to determine if it satisfies seller's constraints or not decide *iso-curve relaxation* which is buyer's subjective behavior. Consequently, we use *trade-off* to find out buyer's partner and determine which attributes need to change along with the **iso-curve**. An example application to negotiate a supplier selection among agents to demonstrate for the agents how to adjust the negotiation attribute parameters and reach an agreement. At the last part, *postnegotiation analysis*, we will compare two results from the preceding trade-off strategies (i.e., **risk-seeking** and **risk-averse**) to let decision-making know how to make the most benefit decision for the company.

BACKGROUND

The shift, in recent decades from an industrial economy (based on mass production models) to an information economy associated with the globalization of markets, has brought an enormous increase in competitiveness, leading to the need for new organizational models. Enterprise cooperation models have emerged, where different enterprises coordinate the necessary means to accomplish shared activities or reach common goals. This association of strengths enables enterprises to build privileged relationships, based on an increase of advantages through resource and competence sharing and risk minimization. Cooperation arrangements are particularly relevant in small and medium enterprises due to their reduced size and high specialization and flexibility. These kinds of enterprises have been

adopting new strategies that enable them to adapt to a constantly changing market, organizing themselves in strategic partnerships. Furthermore, many large companies are isolating parts of their businesses, making them autonomous in order to increase the overall flexibility and achieve greater performance. Outsourcing models are also becoming dominant, enabling enterprises to concentrate on their core competencies. Thus, there is an increasing emphasis in cooperation and coordination of small and medium enterprises. The concept of an IJV arose from this trend and has been defined as "a temporary consortium of autonomous, diverse and possibly geographically dispersed organizations that pool their resources to meet short-term objectives and exploit fast-changing market trends" (Davulcu, Kifer, Pokorny, Ramakrishnan, Ramakrishnan, & Dawson, 1999). The creation of the IJV starts with the definition of the business to be developed; this process may initiate because of a client need or because of a market opportunity detected by an enterprise. The formation phase typically includes the definition of goals, the selection of participants through negotiation, and the definition of their roles and respective obligations. There are many interesting works on automated IJV negotiations. In this survey, we focus on multi-agent technology approaches.

Software agent technologies are promising great advantages to the way we do business (Jennings, Faratin, Norman, O'Brien, Odgers, & Alty, 2000). Systems that use **software agent** technologies are proving to be effective in helping users make better decisions when buying or selling through the Internet (Bailey & Bakos, 1997). Software agents can also play an important role in providing automation and support for the negotiation stage of online trading (Maes, Guttman, & Moukas, 1999). They vary from decision-making models of negotiation to learning methods for supporting the negotiation, based on a variety of approaches including **game theory**, **heuristics**, and **machine learning**. Game theory, the first approach, is the study of conflicts and cooperation between people using mathematical models (Binmore & Vulkan, 1999). Since negotiation can resolve conflicts (Pruitt, 1981), game theory has been applied to analyze negotiation processes. There are two methods in game theory. The first method is the prediction of the possible outcome using axioms (Nash, 1950), where the axioms reflect the desirable properties of the solution. The other method is the design of the negotiation mechanism. A "mechanism design" is a protocol that restricts the number of strategies that can be used by players to achieve the best outcome for all players (Faratin, 2000). However, game theory makes a number of assumptions including knowledge of circumstances (Jennings et al., 2000). This means that we should know rules of the encounter, specify our preferences, and know our partners' preferences or at least be able to formulate beliefs about their preferences. Another assumption of game theory is the full rationality of negotiators, which means that agents have sufficient reasoning and computational capacity to maximize their expected payoffs given their beliefs. Because of these shortcomings, we do not consider game theoretic approaches in the remainder of this chapter.

Negotiation is defined as a process by which a joint decision is made by two or more parties; the parties first verbalize contradictory demands and then move towards agreements (Pruitt, 1981; Sierra, Faratin, & Jennings, 2000). Most current digital marketplaces systems use predefined and non-adaptive negotiation strategies in the generation of offers and counter-offers during the course of negotiation (Wong, Zhang, & Kara-Ali, 2000). For example, Kasbah is an electronic marketplace populated by selling and buying software agents who engage in a single issue negotiation (Chavez & Maes, 1996.). Experiments with Kasbah led to a design of Tête-à-Tête, a system capable of handling multi-issue negotiations (Maes et al., 1999). Based on the users' issue weights, it constructs a rating function to evaluate offers made by other agents. Users may also specify bounds on the issue values which describe their reservation levels.

A successful negotiation occurs when the two opposing offers meet, so the negotiation process consists of a number of decision-making processes, each of which is characterized by evaluating an offer, determining strategies, and generating a counter-offer (Bertsekas, 1995; Cyert & DeGroot, 1987; Zeng & Sycara, 1998). Negotiation is a form of decision making where

Figure 1. Negotiation system architecture

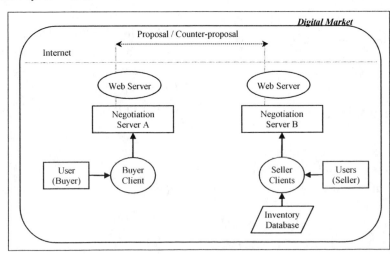

two or more parties jointly explore possible solutions in order to reach a consensus (Rosenschein & Zlotkin, 1994). In common, negotiation can be classified according to the number of parities involved which can be one-to-one, one-to-many, or many-to-many negotiation, and the number of attributes negotiated such as single-attribute or multiple-attribute (Rahwan, Kowalczyk, & Pham, 2002). Our research supports one-to-many multi-attribute negotiation and its negotiation system architecture is shown in Figure 1 which illustrates two agents negotiate on the Internet through a Web server in the digital market. The detail of negotiation process will be described in the following section.

ONE-TO-MANY NEGOTIATION MECHANISM SCHEME

In this section, we propose a set of negotiation mechanism schemes to execute how to proceed to one-to-many negotiation. In the preliminary setting, we negotiate parameterst and iso-curve computation before starting negotiation. When entering negotiation, we proceed to select buyer's alternative based iso-curve from seller's proposal and consider seller's constraints to choose the buyer's alternative. Consequently, using the attributes trade-off makes an agreement between buyer agent and seller agents. At the end, we compare and analyze the results from two trade-off strategies to let decision makers decide which strategy will be used in different conditions based on seller's constraints.

Negotiation Parameters Set

i suppliers will be chosen from the buyer (an assembling company) for the key component. We discuss n attributes (a_j) about this key component.

Definition 1. Buyer generates several feasible alternatives ($a_{a_j}^B$) by its own set of attributes (

$[a_{a_1}^B, a_{a_2}^B, \cdots, a_{a_n}^B]$), which include quantitative and qualitative attributes, at the same iso-curve value that is shown in Definition 5. And buyer's range of attribute ($a_{a_1}^B\_lower$, $a_{a_1}^B\_upper$), seller's range of attribute ($a_{a_1}^{S_i}\_lower$, $a_{a_1}^{S_i}\_upper$), and the negotiation times are all determined by themselves:

$$a_{a_j}^B = [a_{a_1}^B, a_{a_2}^B, \cdots, a_{a_n}^B]$$

At first, each seller (S_i) will also offer their own set of attributes ($a_{a_j}^{S_i}$) to the buyer:

$$a_{a_j}^{S_i} = [a_{a_1}^{S_i}, a_{a_2}^{S_i}, \cdots, a_{a_n}^{S_i}]$$

Definition 2. The weights determined by buyer ($W_{a_j}^B$) and sellers ($W_{a_j}^{S_i}$) for n attributes (a_j); weights are from 0 to 1:

$$W_{a_j}^B : [W_{a_1}^B, W_{a_2}^B, \cdots, W_{a_n}^B]$$

$$W_{a_j}^{S_i} : [W_{a_1}^{S_i}, W_{a_2}^{S_i}, \cdots, W_{a_n}^{S_i}]$$

$$\sum_{j=1}^{n} W_{a_j}^B = 1, \quad \sum_{j=1}^{n} W_{a_j}^{S_i} = 1$$

Evaluating attributes in negotiation is a key role when solving process. However, in evaluating process, each attribute has its different scale to describe and divides into quantitative and qualitative attributes. To solve this problem, we applied membership functions in fuzzy logic which are determined by buyer and sellers, and used fuzzy value to normalize different scales between all attributes and make them have the same scale.

Definition 3. Fuzzy value can be perceived as the means to measure the degree of compatibility of a quantitative attribute value to a fuzzy set. A triangular number (a, b, c) is a **fuzzy set** that has a **membership function** of the following form:

$$\mu(x) = \begin{cases} \dfrac{x-a}{b-a} , & if \ x < a \\ \dfrac{c-x}{c-b} , & if \ a \le x < b \\ 1 , & if \ x \ge b \end{cases} \quad (1)$$

In addition, triangular numbers also provide qualitative attributives that may be given by means of the **fuzzy singleton** (Castro-Schez1, Jennings, Luo, & Shadbolt, 2004) shown in Figure 2a c. For instance, whether the outsourced item is with service or not is a Boolean type attribute, and specification of the item can be ranked by buyer/seller preference which is ranking by order type attribute. Each value is expressed by the fuzzy singleton that is determined from formula (1).

In order to better facilitate interaction between the user and software agent from a practical viewpoint, we then employed triangular membership function to determine the user linguistic value of each quantitative/qualitative attribute. When the fuzzy membership function of each attribute is decided, we can then convert these human linguistic values of attributes into fuzzy value ($FV_{a_1}^B$ ($FV_{a_1}^{S_i}$)). And Table 1 shows the notations of fuzzy value of buyer (seller) through fuzzy membership function.

Computing Iso-Curve Model

In negotiation, buyer agent and seller agent can not only reject or accept a message of proposal (alternative), but also evaluate their alternatives ($a_{a_j}^B$ ($a_{a_j}^{S_i}$)). In this subsection, we use the iso-curve model to evaluate total scores of each alternative formed by a set of attributes.

Table 1. Fuzzy value for different scale of each attribute

attribute	Attribute value of Buyer (Seller)	Fuzzy value of Buyer (Seller)
a_j	$a_{a_j}^B$ ($a_{a_j}^{S_i}$)	$FV_{a_1}^B$ ($FV_{a_1}^{S_i}$)

Definition 4. Total scores (TS_{a_j}) of an alternative made up of weights W_{a_j} ($W_{a_1}^B$ ($W_{a_1}^{S_i}$)) and fuzzy value FV_{a_j} ($FV_{a_1}^B$ ($FV_{a_1}^{S_i}$)) for attribute a_j is formulated as:

$$TS = (\sum_{j=1}^{n} W_{a_j} \times (FV_{a_j}))^{\frac{1}{r}} \qquad (2)$$

where r represents the value from $-\infty$ to ∞, a spectrum total score models values such as *minimum, weighted arithmetic mean, maximum*, and so forth (Dujmovic, 1975; Su, Huang, Hammer, Huang, Li, Wang, et al., 2001). Some commonly used functions are given in Table 2. The total score models can be selected by a user to suit different decision situations and for the selection of different attributes. The total score models represent different degrees of conjunction and disjunction of negotiation data conditions. They can be selected by a user to suit different decision situations and for the selection of different products and services.

Figure 2. Various continuous/discrete attribute types that triangular membership numbers can represent

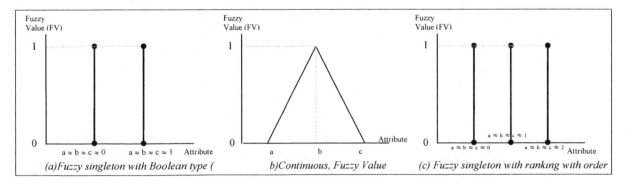

(a)Fuzzy singleton with Boolean type (b)Continuous, Fuzzy Value   (c) Fuzzy singleton with ranking with order

For example, the maximum model is suitable when one or more of the negotiation conditions are acceptable to the user. In that case, the maximal value among all the preference scores derived for the negotiation conditions will be used as the global preference score. For example, if CPU speed is most important to a client and she is 95% satisfied (i.e., preference score of 0.95) with the speed of the computer under consideration, the preference scores of the rest of the attributes can be ignored. In this case, the total score is 0.95. On the other end of the spectrum, the minimum model would use the minimal score among all the preference scores as the total score. In the above example, if a client is only 10% satisfied with the speed of the CPU, the total score is 0.1 even though he may be totally satisfied with all other attribute values. As pointed out, the total score models defined by negotiation agents represent different degrees of conjunction/disjunction in the spectrum $[-\infty, \infty]$.

The main idea of this tactic is to find an alternative with the same total score as the previous one offered, but expect it to be more acceptable for the client's opponent. Intuitively, a trade-off is where one partner lowers the scores on some negotiation attributes. The trade-off is intended to generate an offer that, although of the same value to the alternative, may benefit the negotiation opponent and hence increase the overall gains between both seller and buyer. An agent will decide to make a trade-off action when it does not wish to decrease its aspirational level for a given benefit-oriented negotiation. Thus, the agent first needs to generate some/all of the potential alternatives for which it receives the total score. Technically, it needs to generate alternatives that lie on the iso-value (Faratin, Sierra, & Jennings, 2002). Because all these potential alternatives have the same value for the agent, it is indifferent among them. Given this fact, the aim of the trade-off mechanism is to find the alternative that are most preferable (and hence acceptable) to the negotiation opponent since this maximizes the total score. In the first place, therefore, the buyer agent will generate some potential alternatives that are the same total score value as the previous one offered, but expect it to be more acceptable than the other alternatives with different attribute combinations for its seller agents. The generating method is initiated by first generating new alternatives that lie on what is called the iso-value curves which derive from the next definition (Raiffa, 1982).

Definition 5. An iso-curve is defined as the curve formed by all the proposals with the same total

Table 2. Total score model types

Total Score Models	Function	r
Minimum model	$TS = \min(FV_{a_1}, FV_{a_2}, \cdots, FV_{a_n})$	$r = -\infty$
Harmonic mean model	$TS = 1 / \sum_{j=1}^{n}(W_{a_j} / FV_{a_j})$	$r = -1$
Geometric mean model	$TS = (FV_{a_1})^{W_{a_1}} \cdot (FV_{a_2})^{W_{a_2}} \cdots (FV_{a_n})^{W_{a_n}}$	$r = 0$
Weighted arithmetic mean model	$TS = \sum_{j=1}^{n} W_{a_j} \times FV_{a_j}$	$r = 1$
Square mean model	$TS = \sqrt{\sum_{j=1}^{n} W_{a_j} \times (FV_{a_j})^2}$	$r = 2$
Maximum model	$TS = \max(FV_{a_1}, FV_{a_2}, \cdots, FV_{a_n})$	$r = \infty$

score values for the buyer agent. Given a score d, the iso-curve set at degree d_k (in the k level) for buyer's alternative $a_{a_j}^B$ is defined as:

$$iso(d_k) = \{a_{a_j}^B \mid TS_{a_j}^B = d_k\} \tag{3}$$

Similarity Matching

In this research, we use **similarity matching** to support sellers easily finding more suitable alternative of buyer because of different cognitions for attributes between both sides. There are two kinds of similarity evaluation: *united similarity* for each alternative and *individual similarity* for each attribute.

United similarity is used in two parts. When the buyer agent receives the seller's proposals first, it will use *united similarity* to match its own and sellers' alternatives to find out one set of attributes which has the largest similarity with each seller. And the other part is used to choose the buyer's partner which has the most similar alternative between buyer and sellers.

Definition 6. *United similarity* between buyer and sellers for alternative $(a_{a_j}^B \ (a_{a_j}^{S_i}))$ is defined as:

$$Sim_{a_j}^{B,S_i} = \frac{\sum W_{a_j}^{B,S_i} \times S_{a_j}^{B,S_i}}{\sum W_{a_j}^{B,S_i}} \tag{4}$$

Where $W_{a_j}^{B,S_i} = \sqrt{W_{a_j}^B \times W_{a_j}^{S_i}}$ represents joint weights of buyer and sellers for attribute a_j, and $S_{a_j}^{B,S_i}$ represents the similarity degree of buyer and sellers for attribute a_j which can be divided into computing quantified attributes,

$$\left(S_{a_j}^{B,S_i} = 1 - \frac{|a_{a_j}^B - a_{a_j}^{S_i}|}{\max\{a_{a_j}^B, a_{a_j}^{S_i}\}}\right)$$

and qualified attributes (see Box 1).

The buyer will offer a counter-proposal to each seller; at this time, each seller will check and match each attribute in buyer's alternative offered. And they will use *individual similarity* to do the matching.

Definition 7. *Individual similarity* between each attribute a_j of buyer and sellers' alternatives is defined as:

$$Sim_{a_j}^{B,S_i} = \frac{W_{a_j}^{B,S_i} \times S_{a_j}^{B,S_i}}{W_{a_j}^{B,S_i}} \tag{5}$$

Negotiation Procedure

In this subsection, we present how to execute our negotiation procedure, mainly classified into five parts (Figure 3). First, we defined negotiation parameters set and iso-curve computation in the *preliminary setting*. Second, the *negotiation alternative processing service* will be proposed to select the buyer's alternative based on iso-curve. After selecting the negotiation alternative, the buyer agent will send its alternative (counter-proposal) to seller agents to determine if it satisfies seller's constraints or not; then it decides the *iso-curve relaxation*. Consequently, we use *trade-off* to find the buyer's partner and determine which attributes need to change. Last, in *postnegotiation analysis*, we will compare two results from the

Box 1.

$$S_{a_j}^{B,S_i} = \begin{cases} 1, & \text{if } (u_{a_j}^{S_i} a_{a_j}^{S_i}) \le u_{a_j}^{S_i}(a_{a_j}^B) \\ 1 - [u_{a_j}^{S_i}(a_{a_j}^{S_i}) - u_{a_j}^{S_i}(a_{a_j}^B)], & \text{otherwise} \end{cases}, \text{where } 0 \le u \le 1$$

preceding trade-off strategies to allow the most beneficial decision making.

Step 1: Start Negotiation

Each seller agent proposes an alternative $(a_{a_j}^{S_i})$ to buyer agent.

Step 2: Selecting Buyer's Alternative Based Iso-Curve from Sellers' Proposal

When receiving each seller agent's alternative, the buyer agent starts to proceed to united similarity matching using formula (4) and finds an alternative for the buyer agent $(a_{a_j}^B)$ which has the largest similarity with all seller agents.

Figure 3. Scheme for negotiation mechanism

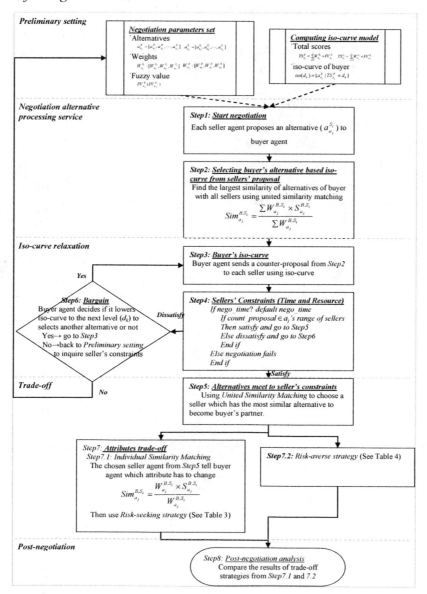

Step 3: Buyer's Iso-Curve Values

In this step, the buyer agent sends a counter-proposal (alternative $a_{a_j}^B$ from *Step 2*) back to each seller agent using iso-curve.

Step 4: Sellers' Constraints (Time and Resource)

When each seller agent receives the counter-proposal (alternative $a_{a_j}^B$) from the buyer agent, it uses mathematical constraints to decide if the alternative $a_{a_j}^B$ satisfies the seller's limitations of time and resources. About time limitation, if negotiation times between buyer and seller agents exceed default negotiation times, this negotiation fails. Moreover, about resource limitation. If the attributes in the buyer agent's alternative satisfies seller's constraints, then it goes to *Step 5*; if dissatisfied, it goes to *Step 6*.

> *If nego_time□default nego_time*
> If count_proposal $\in$ a$_j$'s range of sellers($\forall a_j$)
> Then satisfy and go to Step 5
> Else dissatisfy and go to Step 6
> End if
> Else negotiation fails
> End if

Step 5: Alternatives Meet to Seller's Constraints

Choose a seller which has the largest similarity as buyer's partner based on *united similarity matching* between buyer agent and seller agents.

Step 6: Bargain

When buyer agent receives dissatisfy message from *Step 4*, it will decide to whether alternative relax iso-curve or not. If yes, then go to *Step 3* to lower iso-value to the next level (d_k) and select another alternative. If not, go back to *preliminary setting* to inquire about the seller's constraints; then proceed with the following steps.

Step 7: Attributes Trade-Off

The trade-off rules are according to two kinds of strategies toward risk (risk-seeking and risk-averse), which are determined by buyer's attitude toward risk. And each strategy is also subdivided into *cost-oriented* and *benefit-oriented*, respectively. For instance, the lower the attribute *price* for the buyer agent, the better, so it is classified as *cost-oriented*. However, attribute *Quantity* is better for the buyer agent, so it is sorted as *benefit-oriented*.

Step 7.1: Risk-Seeking Strategy

In this step, the chosen seller agent first uses individual similarity matching which lets the buyer agent know which attributes have to change. After individual similarity matching, we use the risk-seeking strategy (shown in Table 3) to cope with the attributes trade-off corresponding to the largest unsimilarity between buyer agent and seller agent.

Cost-oriented: (Buyer agent and seller agents all have their own ranges for attribute a_j).If one of the attributes in buyer's alternative $(a_{a_j}^{B'})$ is within the range

Table 3. Risk-seeking strategy in attributes trade-off

Cost-oriented for buyer	Benefit-oriented for buyer
When $a_{a_1}^{S_i}\_upper \geq a_{a_j}^B\_lower$	When $a_{a_1}^B\_lower \geq a_{a_1}^{S_i}\_upper$
If $a_{a_1}^{S_i}\_lower \leq a_{a_j}^{B'} \leq a_{a_1}^B\_upper$	If $\underline{a}_{a_1}^B\ lower \leq a_{a_j}^{B'} \leq a_{a_1}^{S_i}\_upper$
then accept $a_{a_j}^{B'}$	then accept $a_{a_j}^{R'}$
else deal $\leftarrow a_{a_1}^{S_i}\_lower$	else deal $\leftarrow a_{a_1}^{S_i}\_upper$
end if	end if

between the seller's lower bound ($a_{a_1}^{S_i}$ _ *lower*) and the buyer's upper bound ($a_{a_1}^{B}$ _ *upper*) for attribute a_j, this value ($a_{a_j}^{B}$) is accepted, or else, seller's lower bound is a deal.

Benefit-oriented: If one of the attributes in buyer's alternative ($a_{a_j}^{B}{}'$) is within the range between the buyer's lower bound and the seller's upper bound ($a_{a_1}^{S_i}$ _ *upper*) for attribute a_j, this value ($a_{a_j}^{B}{}'$) is accepted, or else, the seller's upper bound is a deal.

Step 7.2: Risk-Averse Strategy

We already know the alternative of the chosen seller, and then we use the risk-averse strategy to proceed with attributes trade-off. The trade-off strategies are determined by the buyer as shown in Figure 4. And buyer's attitudes to trade-off can be classified as "anxious," "cool-headed," and "frugal" (Maes et al., 1999).

Because the buyer agent wants to keep the same level d_k of iso-curve ($iso(d_k)$), we next used mathematical programming to calculate the buyer's new alternative based on buyer and seller's initial requirements ($a_{a_j}^{B}$, $a_{a_j}^{S_i}$). Table 4 shows the risk-averse strategy in attributes trade-off.

Step 8: Postnegotiation Analysis

Compare two results of trade-off strategies in *Step 7*.

RELATED WORK

The proposed negotiation mechanism scheme is generic and handles one-to-many and multi-attribute negotiation. The key features of the scheme are (1) the iso-curve model based on weighted fuzzy values, (2) attributes trade-off using risk-seeking and risk-averse strategies, and (3) a set

Figure 4. Trade-off strategies

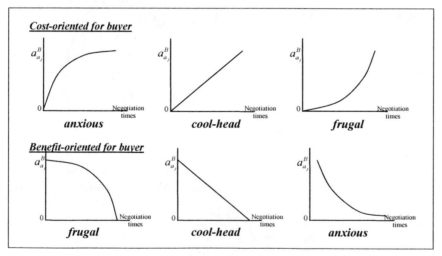

Table 4. Risk-averse strategy in trade-off

Cost-oriented for buyer	Benefit-oriented for buyer
Objective: ($iso\ d_k$) *keeps the same*	*Objective*: ($iso\ d_k$) *keeps the same*
*Constra*int: $a_{a_j}^{B} \le a_{a_j}^{B}{}' \le a_{a_j}^{S_i}$	*Constra*int: $a_{a_j}^{S_i} \le a_{a_j}^{B}{}' \le a_{a_j}^{B}$

of negotiation tactics. Software agent researchers have developed a number of computational models of negotiation to date. Generally speaking, they have been more inclined to autonomous agents, which characterize the scheme along a set of dimensions. Table 5 summarizes the key features of four representative schemes along a set of dimensions. The models handle two-participator (i.e., one-to-one) or multiple participators (one-to-many) negotiation (dimension 1). The model accounts for quantitative and qualitative attributes (dimension 2). Some schemes handle only competitive behavior and other schemes handle cooperative/competitive behavior (dimension 3). Finally, two schemes adopt integration of negotiation strategies from risk-averse/seeking and various tactics (dimension 4).

IJV NEGOTIATION SCENARIO

To demonstrate this proposed negotiation mechanism scheme, we used a notebook computer manufacturing company by assembling company (*AC*) as our sample scenario. In order to maintain the confidentiality of the firm utilized in the case illustration, the notebook computer production company is referred to as Company *AC*. It assembles various functional components on the motherboard to its notebook computer. The key components will be outsourced in our suppliers.

It is assumed that only two potential sellers are qualified to supply the outsourced motherboard components. So, in this study, two international suppliers of Company *AC* will be evaluated and

Table 5. Comparison of our scheme with related work

	Chavez et al.	**Jennings et al.**	**Madhu et al.**	**Our scheme**
1. Number of participators	2	n	n	n
2. Quantitative/qualitative attributes	Quan.	Quan. ./Qual.	Quan./Qual.	Quan./Qual.
3.Cooperative/competitive behavior	Comp.	Comp.	Coop./ Comp.	Coop./ Comp.
4. Strategies and tactics	Strat.	Strat./tact.	Strat.	Strat./tact.

Table 6. Buyer/sellers' weights and their fuzzy value for all attributes

attributes	Buyer (Company AC)			Seller A			Seller B		
	Range	Fuzzy Value	Weights	Range	Fuzzy Value	Weights	Range	Fuzzy Value	Weights
Specification of CPU	Intel® Celeron™	0	0.35	Intel® Celeron™	0	0.3	Intel® Celeron™	0	0.3
	Intel® Pentium® 4	1		Intel® Pentium® 4	2		Intel® Pentium® 4	1	
	AMD Athlon™ XP	2		AMD Athlon™ XP	1		AMD Athlon™ XP	2	
Delivery Time (Days)	[1, 6]	[0.9, 0.4]	0.2	[2,6]	[0.5, 0.7]	0.5	[4,8]	[0.6, 0.8]	0.3
Price ($)	[2000, 4000]	[0.8, 0.4]	0.45	[4000,5750]	[0.45, 0.8]	0.2	[3250,3750]	[0.3, 0.5]	0.4

named as Supplier A (from China) and Supplier B (from Malaysia). With this basic background about the IJV, we start our negotiation mechanism process step by step.

Table 6 shows the set of attributes involved in negotiation for the buyer (*AC*), Seller A, and Seller B, respectively. It details the ranges of acceptable values, fuzzy value, and the weights used to signify the level of importance for attributes, which are *Specification of CPU*, *Delivery Time*, and *Price*. The attribute *Specification of CPU* is qualitatively subdivided into three standards which are Intel® Celeron™, Intel® Pentium® 4, and AMD Athlon™ XP. These three standards are selected because they are commonly used. Moreover, we consider another two quantitative attributes, *Delivery Time* and *Price*, which are of the most concern to buyers.

After we know the basic information, the buyer agent (Company *AC*) and the seller agents start to negotiate with each other. The negotiation

mechanism of agents for Company *AC* and Sellers A and B is illustrated in Figure 5.

The current implementation of the proposed negotiation mechanism has been made within the JADE 3.1 agent platform. The main reason for this selection was the fact that **JADE** is one of the best modern agent environments. JADE is open-source, **FIPA** compliant, and runs on a variety of operating systems, including Windows and Linux. JADE architecture matches well with our requirements. Negotiations between seller and buyer agents take place in JADE containers. There is one Main container that is responsible for performing all database operations (updates and queries). Users can create as many containers they need to hold their Seller agents. Figure 6 presents the mapping of the negotiation procedure (Figure 5) onto JADE. Figure 6 presents message exchanges captured in the experiment with the help of a JADE provided sniffer agent. Figure 6 shows (1) seller and buyer agents subscribing to

Figure 5. Negotiation mechanism of agents for Company AC and Seller A, B

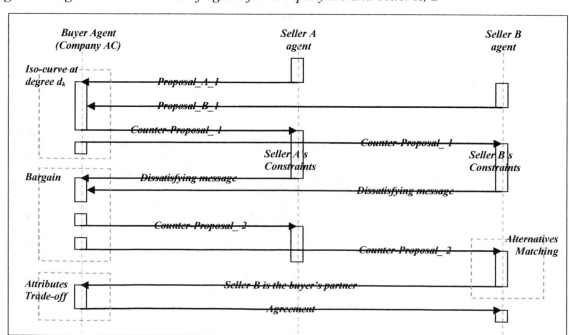

Figure 6. Screen captures showing our negotiation mechanism in action

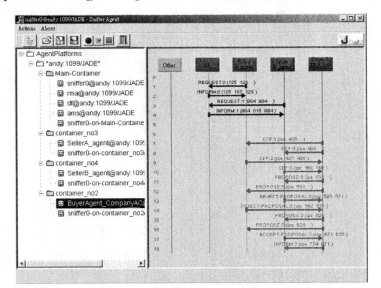

the Main container; (2) seller agents' proposal of alternatives; (3) negotiation alternative processing service proposed to select buyer's alternative based on iso-curve; (4) the buyer agent sending its alternative (counter-proposal) to seller agents to determine if it satisfies seller's constraints or not, then decide on iso-curve relaxation.

Here, we introduce a case experiment to illustrate the main features of our implementation. First of all, each seller agent sends the buyer agent an alternative; Seller A offers a CPU with its specification AMD Athlon™ XP and specifies delivering within four days for a price of $5500 [AMD Athlon™ XP, 4, 5500] which is named *Proposal_A_1*. Seller B offers *Proposal_B_1* [Intel® Pentium® 4, 5, 3750] to the buyer agent. And the total scores of the two alternatives for agents of Seller A and Seller B, which chooses a weighted arithmetic mean model from Table 2, are 0.56 and 0.415, respectively.

Then, the buyer agent generates a set of iso-curve alternatives and finds *Counter-Proposal_1* [AMD Athlon™ XP, 6, 2000] which has the most united similarity with two seller agents using formula (4). So the buyer agent sends this

alternative back (also named counter-proposal) to each seller agent.

Because buyer's *Price* ($2000) falls in neither Seller A's nor Seller B's range for the attribute *Price*, the buyer agent must lower the iso-curve value to the next level and determine another alternative until satisfying the seller's constraints. Consequently, the buyer agent offers each seller agent *Counter-Proposal_2* [AMD, 5, 3300] and then does united similarity matching to choose Seller B as its partner, which has the most similarity to the buyer agent.

In the next step, the agent of Seller B uses individual similarity matching to let the buyer agent know the attribute *Price* has to change. And the buyer agent first uses risk-seeking as its changing strategy. Because *Price* is the lower the better for buyer, it chooses cost-oriented as its attributes trade-off strategy. In the end, a final agreement [AMD, 5, 3300] is achieved through trade-off between Company *AC* and Seller B.

Then, the buyer agent uses risk-averse with cool-headed attitude as its trade-off strategy. In the end, a final agreement [AMD, 4, 3000] is achieved through trade-off between Company *AC*

Figure 7. Negotiation process between Company AC and Seller B using Risk-averse strategy

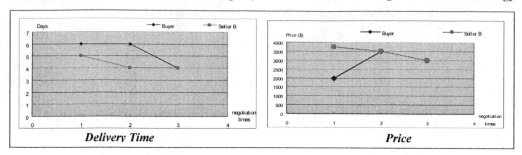

and Seller B. The negotiation process is shown in Figure 7. We can see that they get a consensus the third time.

As we know, the two results above are different; using risk-averse strategy gets the greater agreement. However, we can try to change the total score model chosen then carry out the following steps. If we choose *maximum model* and *Specification of CPU* is the most important to the buyer and sellers, the buyer's partner will become Seller A. Using the trade-off strategy of risk-seeking, the final agreement is [AMD Athlon™ XP, 3, 4000]. But when the buyer agent uses risk-averse as its trade-off strategy, the final agreement [AMD Athlon™ XP, 3, 4250] is achieved through trade-off between Company *AC* and Seller A. We find risk-seeking strategy is better when the buyer cooperates with Seller A. As we know, using different total score models will cause totally distinct results. Because of the buyer's subjective views on deciding total score, or if it is a lower iso-curve to the next level, it will cause distinct trade-off strategies used even though it has to choose a different partner. According to decision-maker subjective views, this negotiation mechanism will lead to the most beneficial and effective outcome for the company.

CONCLUSION

Negotiation recently was developing dramatically and very important to the success of a manufac-

turing firm. This is because the cost and quality of goods and services are usually of concern to buyers who need to negotiate. Therefore, we developed an automated negotiation mechanism for facilitating multi-agent technology approaches in IJV scenarios. A mechanism of one-to-many multi-attribute negotiation was proposed in this research. Sellers and buyers in Internet-based supply chain through negotiations have significant impact on supplier selection and partners' profit. Using the fuzzy value in fuzzy logic transfers linguistic value into the same scale in order to evaluate sequent process. Then through the communications between buyer's and sellers' agents, both sides know opponents' behavior and decide which trade-off strategy can be implied. Among the negotiation, similarity matching applied in two parts; one is to find the most similar alternative as the counter-proposal of buyer; the other is for buyer to choose a business partner and know which attributes have to change and make a consensus with seller. This negotiation contributed more practical approach because it included fuzzy logic to represent the attributes and jointed buyer's behavior within negotiation, therefore, this research real world negotiation. In future work, we will make more effort in post-negotiation to let the decision maker make more accurate decisions.

FUTURE RESEARCH DIRECTIONS

This chapter aims at improving computerized negotiation support through the use of software agents. The primary objective is the design of the **Web-based** multi-agent negotiation mechanism. The negotiation agent is simple at this stage, providing a proof of concept. But the idea offers a very rich set of research issues. In the future, we would like to extend our work by implementing different trade-off strategies in more detail, by conducting a formal or empirical study. We believe there is a chance for exploring a large number of trade-off strategies. For example, the negotiation agent may issue commands to change the subnegotiators' individual strategies according to what is happening in the big picture. More sophisticated risk-seeking/averse strategies and knowledge sharing between agents may open new possibilities. One of the most important issues for the success of e-negotiation is trust. Trust is an important factor of social behavior. Although it is of high practical relevance, the production of trust in **e-negotiation** settings requires further studies and will therefore be integrated in our research framework. Another important issue of research direction could be investigating the use of learning techniques in order to allow agents to reuse their negotiation experience to improve the final outcomes.

REFERENCES

Bailey, J., & Bakos, Y. (1997). An exploratory study of the emerging role of electronic intermediaries. *International Journal of Electronic Commerce, 1*(3), 7-20.

Bertsekas, D. P. (1995). *Dynamic programming and optimal control.* Belmont, MA: Athena Scientific.

Binmore, K., & Vulkan, N. (1999). Applying game theory to automated negotiation. *Netnomics, 1*(1).

Castro-Schezl, J. J., Jennings, N. R., Luo, X., & Shadbolt, N. R. (2004). Acquiring domain knowledge for negotiating agents: a case of study. *International Journal of Human-Computer Studies, 61*, 3-31.

Chavez, A., & Maes, P. (1996). *Kasbah: An agent market-place for buying and selling goods.* Paper presented at the Conference on Practical Applications of Intelligence Agents and Multi-Agent Technology.

Cyert, R. M., & DeGroot, M. H. (1987). *Bayesian analysis and uncertainty in economic theory.* New York: Rowman & Littlefield.

Davulcu, H., Kifer, M., Pokorny, L. R., Ramakrishnan, C. R., Ramakrishnan, I. V., & Dawson, S. (1999). Modeling and analysis of interactions in virtual enterprises. In *Proceedings of the 9th International Workshop on Research Issues on Data Engineering: Information Technology for Virtual Enterprises (RIDE 1999)* (pp. 12-18). IEEE Computer Society.

Dujmovic, J. J. (1975). Extended continuous logic and the theory of complex criteria. *Series on Mathematics and Physics, 537*, 197-216.

Faratin, P., Sierra, C., & Jennings, N. R. (2002). Using similarity criteria to make issue trade-offs in automated negotiations. *Artificial Intelligence, 142*, 205-237.

Jennings, N. R., Faratin, P., Norman, T. J., O'Brien, P., Odgers, B., & Alty, J. L. (2000). Implementing a business process management system using ADEPT: A real-world case study. *International Journal of Applied Artificial Intelligence, 14*(5), 421-465.

Maes, P., Guttman, R. H., & Moukas, A. G. (1999). Agents that buy and sell. *Communications of the ACM, 42*(3), 81-91.

Nash, J. F. (1950). The bargaining problem. *Econometrica, 18*(2), 155-162.

Pruitt, D. G. (1981). *Negotiation behavior.* Academic Press.

Rahwan, I., Kowalczyk, R., & Pham, H. H. (2002). Intelligent agents for automated one-to-many e-commerce negotiation. In *Proceedings of the 25th Australasian Conference on Computer Science* (pp. 197-204).

Raiffa, H. (1982). *The art and science of negotiation.* Cambridge: Harvard University Press.

Rosenschein, J. S., & Zlotkin, G. (1994). *Rules of encounter.* MIT Press.

Sierra, C., Faratin, P., & Jennings, N. R. (2000). Deliberative automated negotiators using fuzzy similarities. In *Proceedings of EUSFLAT* (pp. 155-158).

Su, Y. W., Huang, C., Hammer, J., Huang, Y., Li, H., Wang, L., Liu, Y., Pluempitiwiriyawej, C., Lee, M., & Lam, H. (2001). An internet-based negotiation server for e-commerce. *The VLDB Journal, 10*, 72-90.

Wong, W. Y., Zhang, D. M., & Kara-Ali, M. (2000). Negotiating with experience. In *Proceedings of KBEM-2001*, Austin, TX.

Zeng, D., & Sycara, K. (1998). Bayesian learning in negotiation. *International Journal of Human Computer Studies, 48*, 125-141.

ADDITIONAL READING

This section offers this list of works for further reading on negotiation agents relevant to this chapter.

General Background

Arunachalam, R., & Sadeh, N. (2004). *The 2003 Supply Chain Management Trading Agent Competition* (pp. 113-120). ACM Press.

Axelrod, R. (1984). *The evolution of cooperation.* New York: Basic Books.

Cohen, R. (1997). *Negotiating across cultures: International communication in an interdependent world.* Washington, DC: United States Institute of Peace Press.

Cohen, R. (2001). Resolving conflict across languages. *Negotiation Journal,* pp. 17-34.

Cook, K., Hardin, R., & Levi, M. (2005). *Trust without cooperation.* New York: Russell Sage Foundation.

Davis, A. (1989). An interview with Mary Parker Follett. *Negotiation Journal,* p. 235.

Deutsch, M. (2003). Cooperation and conflict: A personal perspective on the history of the social psychological study of conflict resolution. In M.A. West, D.J. Tjosvold, & K.G. Smith (Eds.), *International Handbook of Organizational Teamwork and Cooperative Working.* Hoboken, NJ: John Wiley & Sons. Retrieved August 18, 2007, from http://www.apec.umn.edu/faculty/spolasky/Deutsch.pdf

Dickinson, I. (2004). The Semantic Web and software agents: Partners, or just neighbours? *Agentlink News, 15*, 3-6.

Flowerdew, J. (1999). Face in cross-cultural political discourse. *Text, 19*, 3-23.

Follett, M. (1973). Constructive conflict. In E. Fox & L. Urwick (Eds.), *Dynamic administration: The collected papers of Mary Parker Follett.* Pitman: London.

Kinsella, W. (1999). Discourse, power, and knowledge in the management of "big science":

The production of consensus in a nuclear fusion research laboratory. *Management Communication Quarterly, 13,* 171-208.

Ramchurn, S. D., Huynh, D., & Jennings, N. R. (2004). Trust in multiagent systems. *Knowledge Engineering Review, 19,* 1-25.

Susskind, W. (1985). Scorable games: A better way to teach negotiation? *Negotiation Journal, 1*(3), 205-209.

Methodology

Axelrod, R. (2000). *On six advance in cooperation theory* (pp. 1-39). Unpublished manuscript.

Babanov, A., Collins, J., & Gini, M. (2004). Harnessing the search for rational bid schedules with stochastic search and domain-specific heuristics. *IEEE Computer Society,* pp. 269-276.

Bohman, J. F. (1995). Public reason and cultural pluralism political liberalism and the problem of moral conflict. *Political Theory, 23*(2), 253-279.

Child, J. (1995). *Mary Parker Follett: Prophet of management: A celebration of writings from the 1920s* (p. 88). Cambridge University Press.

Davis, A. (1991). Follett on facts: Timely advice from an ADR pioneer. *Negotiation Journal,* pp. 131-138.

Doshi, P., Goodwin, R., Akkiraju, R., & Verma, K. (2004). *Dynamic workflow composition using Markov decision processes* (pp. 576-582).

Feldman, S. (2004). The culture of objectivity: Quantification, uncertainty, and the evaluation of risk at NASA. *Human Relations, 57*(6), 691-718.

Fisher, R. (1985). Beyond YES. *Negotiation Journal, 1,* 67-70.

Fisher, R., Ury, W., & Patton, B. (1991). *Getting to YES: Negotiating agreement without giving in* (2nd ed.). New York: Penguin.

Johnson, D. (1971). Role reversal: A summary and review of the research. *International Journal of Group Tensions, 1,* 318-334.

Kahane, D. (2003). Dispute resolution and the politics of cultural generalization. *Negotiation Journal, 19*(1), 5-27.

Kolb, D. (1995). The love for three oranges or: What did we miss about Ms. Follett in the library? *Negotiation Journal, 11,* 339-348.

Lax, D., & Sebenius, J. (1986). Interests: The measure of negotiation. *Negotiation Journal, 2,* 73-92.

McCarthy, W. (1985). The role of power and principle in getting to YES. *Negotiation Journal, 1,* 59-66.

Patel, J., Teacy, W. T. L., Jennings, N. R., & Luck, M. (2005). *A probabilistic trust model for handling inaccurate reputation sources* (pp. 193-209).

Tonn, J. (2003). *Mary P. Follett: Creating democracy, transforming management.* New Haven, CT: Yale University Press.

Winham, G. (1987). Multilateral economic negotiation. *Negotiation Journal, 3*(2), 175-189.

Agent Techniques

Adair, W., Brett, J., Lempereur, A., Okumura, T., Shikhirev, P., Tinsley, C., & Lytle, A. (2004). Culture and negotiation strategy. *Negotiation Journal, 20*(1), 87-111.

Avruch, K. (2000). Culture and negotiation pedagogy. *Negotiation Journal, 16*(4), 339-346.

Axelrod, R. (1997). *The complexity of cooperation: Agent-based models of competition and collaboration.* Princeton University Press.

Belecheanu, R. A., Munroe, S., Luck, M., Payne, T. R., Miller, T., McBurney, P., & Pechoucek, M.

(2006). *Commercial applications of agents: Lessons, experiences and challenges.*

Cardoso, J., Sheth, A., Millerb, J., Arnoldc, J., & Kochutb, K. (2004). Quality of service for workflows and Web service processes. *Journal of Web Semantics: Science, Services and Agents on the World Wide Web, 1,* 281-308.

Deutsch, M., & Krauss, R. (1960). The effect of threat upon interpersonal bargaining. *Journal of Abnormal and Social Psychology, 61,* 181-189.

Deutsch, M., & Krauss, R. (1962). Studies of interpersonal bargaining. *Journal of Conflict Resolution, 6*(1), 52-76.

Foster, I., Jennings, N. R., & Kesselman, C. (2004). *Brain meets brawn: Why grid and agents need each other* (pp. 8-15).

Klusch, M., Gerber, A., & Schmidt, M. (2005). *Semantic Web service composition planning with OWLS-XPlan.*

Luck, M., McBurney, P., Shehory, O., & Willmott, S. (2006). *Agent technology: Computing as interaction (a roadmap for agent-based computing).* AgentLink.

Maximilien, E. M., & Singh, M. P. (2004). A framework and ontology for dynamic Web services selection. *IEEE Internet Computing, 8,* 84-93. IEEE Educational Activities Department.

Maximilien, E., & Singh, M. (2005). *Multiagent system for dynamic Web services selection.*

Norman, T. J., Preece, A., Chalmers, S., Jennings, N. R., Luck, M., Dang, V. D., Nguyen, T. D., Deora, V., Shao, J., Gray, A. W., & Fiddian, N. J. (2004). Agent-based formation of virtual organisations. *Knowledge-Based Systems, 17,* 103-111.

Senger, J. (2002). Tales of the bazaar: Interest-based negotiation across cultures. *Negotiation Journal, 18*(3), 233-250.

Singh, M. P., & Huhns, M. N. (2005). *Service-oriented computing: Semantics, processes, agents.* John Wiley & Sons.

Sirin, E., Parsia, B., & Hendler, J. (2005). *Template-based composition of Semantic Web services.*

Sreenath, R. M., & Singh, M. P. (2004). Agent-based service selection. *Journal of Web Semantics: Science, Services and Agents on the World Wide Web, 1,* 261-279.

Implementation

Byde, A. (2006). *A comparison between mechanisms for sequential compute resource auctions.*

Chaisiri, J., & Flax, J. (2004). Cross-cultural issues in a life sciences company. *Negotiation Journal, 20*(1), 79-86.

Dang, V. D. (2004). *Coalition formation and operation in virtual organisations.* University of Southampton.

Eiseman, J. (1978). Reconciling "incompatible" positions. *The Journal of Applied Behavioral Science,* pp. 133-150.

Irwin, D. E., Grit, L. E., & Chase, J. S. (2004). *Balancing risk and reward in a market-based task service.*

Kuhn, T., & Poole, M. S. (2000). Do conflict management styles affect group decision making? Evidence from a longitudinal field study. *Human Communication Research, 26*(4), 558-590.

Winham, G., & Bovis, H. (1978). Agreement and breakdown in negotiation: Report on a state department training simulation. *Journal of Peace Research, 4*(15), 285-303.

Chapter XI
An Agent–Mediated Middleware Service Framework for E–Logistics

Zongwei Luo
The University of Hong Kong, Hong Kong

Minhong Wang
The University of Hong Kong, Hong Kong

William K. Cheung
Hong Kong Baptist University, Hong Kong

Jiming Liu
Hong Kong Baptist University, Hong Kong

F. Tong
The University of Hong Kong, Hong Kong

C.J. Tan
The University of Hong Kong, Hong Kong

ABSTRACT

Service-oriented computing promises an effective approach to seamless integration and orchestration of distributed resources for dynamic business processes along the supply chains. In this chapter, the integration and adaptation needs of next generation e-logistics, which motivates the concept of a middleware integration framework, are first explained. Then, an overview of a service oriented intelligent middleware service framework for fulfilling the needs is presented with details regarding how one can embed the autonomy oriented computing (AOC) paradigm in the framework to enable autonomous service brokering and composition for highly dynamic integration among heterogeneous middleware systems. The authors hope that this chapter can provide not only a comprehensive overview on technical research issues in the e-logistics field, but also a guideline of technology innovations which are vital for next generation on-demand e-logistics applications.

INTRODUCTION

Logistics, according to the Council of Supply Chain Management Professionals,[1] is defined as the process of planning, implementing, and controlling the efficient, effective flow and storage of goods, services, and related information from point of origin to point of consumption for the purpose of conforming to customer requirements. It includes inbound, outbound, internal, and external movements, and return of materials. How to add further values to such kinds of business activities is especially important for regions with strong manufacturing economies, such as Pearl-River-Delta (PRD). PRD is a region inside Guangdong province, benefiting from China's economics boom in the recent 30 years, and has made itself one of the largest manufacturing bases in the world. The signing of the Mainland/Hong Kong Closer Economic Partnership Arrangement 2 (CEPA 2) further enlarges the PRD into nine Pan-PRD provinces/autonomous regions in China. According to an official newsletter published by the Hong Kong Trade Development Council,[2] the total import and export value of the nine Pan-PRD regions reached US$433.44 billion in 2004, with a year-on-year increase of 27.5%. The eight provinces/autonomous regions other than Guangdong that imported and exported goods were up 43.4%. Guangdong's import and export trade is growing at an annual rate of over 20%. The volume of logistics in the Pan-PRD is predicted to grow at an annual rate of nearly 20% over the next 10 years.

Such logistics booming has pushed for the need of more advanced logistics management and opens up opportunities for this research. In modern logistics management, tremendous coordination and scheduling effort is typically needed, and yet the respective processes may not be properly optimized. Achieving real-time management of business processes like logistics is generally considered to be important and challenging as there exist at least the needs of (1) managing items with their geographic context (e.g., location, time zone) changing from time to time, (2) streamlining the information exchange among widely distributed heterogeneous partners, and (3) optimizing the logistics efficacy in an on-demand manner.

The recent advent of wireless computing has opened up new opportunities for tackling the challenges. For instance, inbound and outbound logistics activities used to be difficult to maintain but can be seamlessly recorded by tagging goods items with passive sensors like radio frequency identification (RFID) tags. That implies that up-to-minute status information of inventory and goods delivery could become highly available which can thus support better logistics services and planning. Also, delivery vehicles installed with global positioning system (GPS) sensors allow their locations to be accurately tracked for more effective fleet management. While the involved sensor technologies have been getting mature in recent years, integrating them together for building e-logistics infrastructure and networks demands middleware frameworks to enable different parties to communicate and collaborate in an efficient and effective manner. For smooth logistics operations with optimal performance, a well designed e-logistics framework and methodology must be in place to support agile and responsive logistics planning and business decisions based on timely information available from logistics participants' information systems.

In this chapter, an intelligent e-logistics middleware service framework (*i*MSF) for addressing the challenges is described. The *i*MSF adopts the concept of a service-oriented *meta-middleware* and is designed with event-processing and service provisioning components for providing management support for dynamic communication and interconnections among heterogeneous logistics participating middleware systems. The additional tier of abstraction can improve and streamline logistics operations through orchestrating various middleware systems for integrating business functions and systems, leading to, for example, shorter lead times, reduced working capital needs, and closer customer relationships through enhanced middleware integration capabilities. Furthermore, *i*MSF can be designed with the autonomy oriented computing (Liu, Jing, & Tsui, 2005) paradigm embedded to allow dynamic and cost effective creation of interconnections of disparate logistics middleware systems and processes to go autonomous, and thus resulting in improved efficiency and visibility throughout the supply chains, enabling just-in-time business planning, decision making, and evaluation of the results. The underlying concept is illustrated using our recent work on agent-mediated bidding and constraint coordination for dynamic service/resource management. Generally speaking, *i*MSF is designed with the *flexibility* to support time-varying and context-dependent needs of processing events triggered by different distributed logistics middleware systems and at the same time with the *autonomy* to assist the users in managing the flexibility provided. It is believed that this can, in turn, allow more effective exploration of the full value of logistics IT, the middleware in particular, and investment within and beyond the enterprise domain.

The organization of the chapter is as follows. First, the problems typically faced in the e-logistics environment are analyzed. Then, the proposed service-oriented approach to e-logistics middleware services framework addressing the challenges are briefly stated, which is followed by two sections dedicated to the details regarding the flexible architecture and the autonomous service provisioning capability. The last section concludes the chapter by analyzing the limitations and providing future directions for extending the proposed framework.

APPLICATION DOMAIN AND CHALLENGES

In recent years, technologies like RFID (Shepard, 2005) and Electronic Product Code (EPC)[3] have emerged to enable global information visibility for supply chain management/logistics participants. Global track and trace, business activity monitoring, and sense and respond along whole supply chains are made possible via these revolutionary technologies to enable the next generation e-logistics. The conventional approach to tackle the sensing devices integration is to simply put the right middleware in place. Such per device middleware linkage with the sensing devices no longer works in dynamic and changing e-logistics infrastructure and network development (Mahmoud, 2004). Today's enterprises may deploy many different sensing devices which in turn require different middleware systems (Langendoerfer, Maye, Dyka, Sorge, Winker, & Kraemer, 2004). The proliferation of different middleware approaches and technologies has complicated matters for IT asset management.

Reactivity Challenges

RFID and EPC technologies have promised real time global information visibility for e-logistics participants. To benefit from such visibility, the e-logistics participants have to be able to identify the interested situations and react to such situations when they happen. Such reactivity throughout the global supply chain shall be able to model the situations and abstract them. These situations would be linked to simple and/or composite events that their occurrence could then be the triggers that the e-logistics participants can react on. The reaction on the triggers has to respond in a timely manner. The events associated with the triggers have to be reported and notification has to be sent to interested e-logistics participants. Moreover, the reactive capability has to scale up to thousands of such triggers due to the scale of sensor deployment and global supply chain networks. This requires that we have to design middleware systems, managing the sensing devices and their data streams to scale up and provide quality of service (QoS) guarantees.

Proliferation and Heterogeneity Challenges

Sensing devices are available in the forms of RFIDs, bar code prints, mobile cell phones, geographic information system/global positioning system (GIS/GPS) sensors, and so forth. Each type of sensing device will have a middleware system in between to interface with business systems to enable location based services (Langendoerfer et al., 2004; Mahmoud, 2004). Moreover, each monitoring and sensing device requires its specific and unique middleware functions. For example, different sensing devices may have their own data stream formats and semantics, leading to the proliferation and heterogeneity of middleware systems for sensing devices. Furthermore, the change or removal of one sensing device usually requires removal of the associated middleware systems and puts it into another one. Thus, the system integration and maintenance burden increases with the number of choices of the middleware systems available to interface with the e-logistics systems. The challenges posed by such middleware proliferation and heterogeneity as well as dynamic e-logistics environment impose certain architectural and functional requirements on developing e-logistics middleware systems (Luo & Li, 2005; Luo, Zhang, Cai, & Kun 2005).

Adaptability Challenge

The proliferation of sensing and other logistics related services also calls for new solutions to be able to handle resources or services which are highly dynamic. In e-logistics, the relationship establishment among the participants is expected to be highly dynamic, and optimizing the way to manage the supply network in a global sense is by no means easy to achieve. For instance, logistics participants could have time-varying properties like availability, resource capacity, performance, charge, and so forth. The uncertainties resulting from these time-varying properties makes brokering among the participants challenging, let alone achieving some form of optimality. Nonstationary nature of the uncertainties is being one of the causes of difficulty. The huge number of potential participants that could be involved in an e-logistics environment further forbids any existing centralized methodology for handling uncertainties to be applicable. What is being described in the context of logistics in fact is just one of the many examples causing a recent demand for scalable computational means to support distributed middleware service composition/configuration and adaptive resource management in a highly dynamic networked service environment.

A MIDDLEWARE SERVICE FRAMEWORK: *i*MSF

To help tackle the challenges of reactivity, proliferation, and heterogeneity, as well as adaptability for sensing device middleware systems, we adopt the service oriented architecture (SOA) approach and propose an intelligent middleware service framework (*i*MSF). *i*MSF is an Enterprise Service Bus (Chappell, 2004) for e-logistics infrastructure and networks implementations based on global standards, best practices, and methodologies provided in SOA for services provisioning, solution development, and technology adoption and migration. The middleware service framework (*i*MSF) will promote adoption of the e-logistics related standards, prepared for adaptation for new middleware components, technologies, and services, while leveraging existing middleware components and services (Luo & Li, 2005; Luo et al., 2005). Also, *i*MSF adopts the AOC paradigm and provides intelligent service provisioning capability in finding, using, and composing middleware services to scale up and prepare for large-scale e-logistics sensing devices deployment (Cheung, Liu, Tsang, & Wong, 2004; Wang, Cheung, Liu, & Luo, 2006a; Wang, Cheung, Liu, Xie, & Luo, 2006b).

An Overview of *i*MSF's Flexible Architectural Design

Middleware has been at the core of system integration (Langedoerfer et al., 2004; Vinoski, 2002). With the introduction and growing popularity of SOA, the loosely coupled nature of distributed service applications in many cases changed the way of middleware systems development. *i*MSF based on SOA targets to streamline the e-logistics middleware development and delivery, e-logistics technology deployment, and adoption. Benefits inherited from SOA include building upon a foundation of global standards, and allowing more flexible yet loosely coupled middleware systems to be developed and extended.

The *i*MSF is built to realize a middleware service framework based on SOA and to embrace different middleware models which will be captured in *i*MSF logics. A middleware service can support multiple communication protocols, each intended to perform a different role that a middleware service can play in the framework. The *i*MSF makes it possible that application service developers can use these middleware services when building applications. As depicted in Figure 1, *i*MSF includes the following functional modules:

- **_i_MSF coordinator:** The *i*MSF coordinator shown in Figure 1 will provide middleware service provisioning, life cycle management, and scheduling functionalities. The

behaviors of each of these services may vary according to their own middleware logics.

- **_i_MSF participant:** The *i*MSF participant shown in Figure 1 will provide middleware service information services, that is, publication, discovery, and notification.

- **_i_MSF repository:** *i*MSF will have middleware resource management functions to manage the middleware resources federated by the *i*MSF coordinator. It also helps manage the middleware service context to establish the necessary relationships between participants to help fulfill the coordination role of the middleware services. Meanwhile, an associated middleware service context will be created to maintain the relationship among the *i*MSF coordinator and participants.

- **Service logic:** In *i*MSF, the middleware service logics are used to model different middleware systems for coordinating their services.

- **_i_MSF adapter:** The *i*MSF will use *i*MSF adapters for protocol exchange to interface with the different e-logistics system through *i*MSF messaging.

Note that the design of *i*MSF makes the development of new middleware by composition of existing middleware services possible. The *i*MSF will enable dynamic middleware orchestration through provisioning middleware services in the *i*MSF repository. *i*MSF will be able to interpret the middleware messages through the messaging services and broker the messages to coordinate the corresponding middleware transactions.

Middleware Services for Event Processing

The middleware service framework will provide services for event processing to support streaming capabilities. The middleware services for event processing are intended to address the unique

Figure 1. Service oriented middleware service framework architecture (Luo, Li, Tan, Tong, Kwok, Wong, & Wang, 2006b)

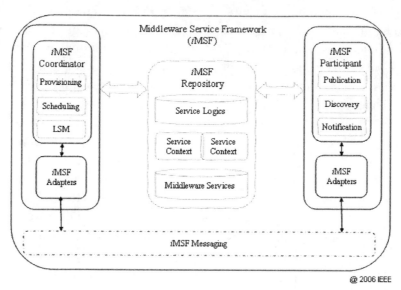

computational requirements presented by different sensing devices messaging and streaming. The *i*MSF abstracts the event processing functions into processing messaging, processing logics, and processing context. These functions are concerned with data reduction operations such as filtering, aggregation, and counting. Each of them will be the features supported by the middleware published in the middleware service framework.

The processing messaging mainly concerns the message streaming protocol. The process logics contain algorithms to handle the message streams for filtering, aggregation, counting, and so forth. The process context establishes the relationships for the streaming patterns among the event processing modules. The context can also provide session support.

The middleware services for event processing will interface with various sensing devices supported by the middleware services in the *i*MSF. The interface will be described in Web service description language (WSDL). The transformation may be necessary for the interface WSDL description and description of the messaging format and streaming processing logics supported by particular middleware services.

The middleware services for event processing also interface with the middleware service framework modules like *i*MSF coordinator and participant. The *i*MSF participant will allow the event processing middleware services to register themselves and the *i*MSF coordinator can manage their life cycles in the middleware service framework.

An XML-based language for specifying the event processing processes as Event-Condition-Action rules has been incorporated in the *i*MSF, which was called JECA rules in Luo, Sheth, Miller, and Kochut (1998). Such a construct for the specification is especially useful in e-logistics application. For example, in modern logistics applications, there are many sensors deployed and events are streamed out from those sensors. Tempted to analyze those events to track and trace the goods movement, it is necessary to apply filtering rules to the events. It can be assumed that

Figure 2. Architecture for middleware service event processing (Luo et al., 2006b)

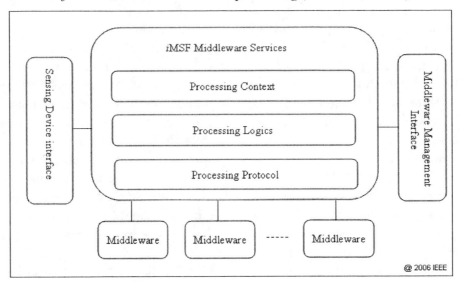

each sensor could have its own logics to filter out those un-needed events. However, each such sensor has limited knowledge about what to filter out. Its filtering rules are local. Globally, collaborative filtering rules have to be applied to coordinate several or a network of sensors. Assumptions can also be made that communication among sensors is constrained and sometimes quite expensive. This is often true in many logistics applications which have to leverage sensor network technologies. The use of event-condition-action rules proposed in Luo et al. (1998) is a good candidate for describing those filtering rules. Part of the filtering knowledge about the filtering, usually the local knowledge, can then be modeled as rules with events and conditions. Other filtering knowledge, usually the context dependent knowledge, will be modeled as rules with context and conditions. In a sensor network, evaluation of a context dependent collaborative knowledge is an expensive process. In such a network, if a sensor cannot obtain its neighboring sensors' information, it has to make assumptions in order to carry out the collaborative rules evaluation. If the assumptions cannot

be overturned, then the sensor can go ahead with collaborative rules evaluation.

Middleware Service Brokering

In *i*MSF, functions are necessary in finding and using services provided by the middleware framework. To meet the challenges like heterogeneity and usability for service brokering, we use service description technologies and annotation languages like WSDL[4] to provide an information abstraction layer for the e-logistics middleware services brokerage:

- To describe and correlate middleware services provided by different middleware through WSDL and annotation.
- To map, transform, and assemble middleware services using XML transformation and service orchestration technologies.
- To ensure the functional consistence of the middleware services provided by different system developers.

A brokering mechanism should be established in middleware service provisioning, that is, selection, discovery, and composition (see Figure 4). Similar to the approaches proposed for adaptive and dynamic service selection and compositions (Cai, Luo, Qian, & Gao, 2005; Luo, Zhang, & Badia, 2006a; Mahmoud & Zahreddine, 2005; Ran, 2003), the broker in *i*MSF provides functions to manage a network of middleware resources as an e-logistics application support platform for middleware service sharing. The service broker will have its interfaces published with WSDL descriptions. Interfaces include WSDL descriptions for queuing client access and management and implementing the *i*MSF participant and co-ordinator functions. The service broker can plug in prebuilt suitable algorithms in service brokering with information contained in the brokering context such as middleware invocation path and network models. Also, it can be embraced with autonomous agent techniques to improve the overall quality of service (QoS) of the composed process, with more details to be provided in the next section. Service brokers can connect to other service brokers as well and have the capability to

access functions provided by various middleware systems.

The middleware service brokers handle the process of making decisions involving service resources over multiple middleware systems. The core module is the brokering engine driving the orchestration engine to execute the brokering plans with broking logics (algorithms) and context. Depending upon different nature of the middleware service request, the service broker in turn invokes one of the specific brokers to help fulfill the service request. These specific types of service brokers include license broker, publication broker, discovery broker, and notification broker.

- The license broker manages licenses based on license usage analysis and prediction, as well as considerations on service provisioning policies (algorithms). It ensures license availability for sharing to enable organizations to maximize and optimize the utilization of the software licenses.
- The publication broker provides capabilities for invoking service registration services to describe services and register services.

Figure 3. Architecture diagram for middleware service brokering (Luo et al., 2006b)

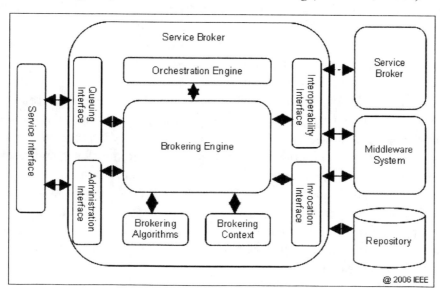

Necessary security is enforced in order to publish and modify the description of middleware services.

- The discovery broker provides standard interfaces to locate service discovery facility to help select the middleware services to fulfill the service requests.

- The notification broker allows subscriptions for the notifications of situations. If service requestors subscribe to a situation and specify the notification contract including an initial lifetime for the subscription, a stream of notification messages will flow to all the subscribers when corresponding events occur about the situation. The message may trigger actions on the service requestors if action receivers are specified in the notification contract.

AUTONOMOUS SERVICE PROVISIONING IN *i*MSF

The flexible architecture provided by *i*MSF opens up the opportunity for effectively reconfiguring and then sewing up distributed middleware services at a metalevel for supporting more agile and scalable development of distributed systems. For instance, the discovery broker in *i*MSF allows it to be able to select from a set of service candidates (after the rule-based filtering) a single or a composition of optimal middleware services for fulfilling the user request. However, there still remain the issues of what optimality to be achieved and how it can be achieved given the real-time requirement.

The optimality criteria for service selection in the discovery broker could take into the accounts of performance or cost related issues. Also, it could take the forms of objective functions (soft constraints) to be optimized and/or constraints to be satisfied. In addition, for the decision model, the consideration for picking a particular service could be as simple as being independent of other service

selection decisions, or could be as complicated as being affected by how the other component services within a middleware service composition are selected. As the demand for providing real-time response to the dynamic changes of the e-logistic middleware services (e.g., being enabled by RFID technologies) continues to increase, the new opportunities provided by the architectural flexibility of *i*MSF will quickly lead to an urgent need of scalable computational means to perform the optimization and constraint satisfaction in an adaptive and distributed manner before the task of managing the highly dynamic distributed middleware services eventually becomes out of control for users. The computational means has to be *adaptive* due to the highly dynamic nature of the environment. And it has to be *distributed* as the assumption of having a metabroker with the accessibility to the complete set of status information of all the services in a networked service environment, as most centralized computational means require, can hardly lead to a solution which will be scalable to the size of the e-marketplaces that we are interested in.

Autonomy-Oriented Computing

With the advent of computing, we are fast entering a new era of discovery and opportunity. For example, in supply chain management and logistics, the supply chain managers will be able to best plan for their procurement order fulfillment activities on-the-fly by synthesizing up-to-minute inventory levels of different warehouses, goods locations along the supply chain, and large-scale simulations of consumer behavior at the downstream. The characteristic that the task of computing is seamlessly carried out in a variety of physical embodiments, in fact, has been observed in problems of many other domains, including life science, environmental science, and robotics, to name just a few. There is no single multipurpose or dedicated machine that can manage to accomplish a job of this nature. The key to success in such applications lies in a large-scale deployment of computational agents capable of autonomously making their localized decisions and achieving their collective goals.

Autonomy oriented computing is a new computing paradigm (Liu, Jin, & Tsui, 2005). AOC makes use of autonomous entities in solving computational problems and in modeling complex systems. As compared to other paradigms, such as centralized computation and top-down systems modeling, AOC has been found to be extremely appealing in the following aspects:

- To capture the essence of autonomy in natural and artificial systems.
- To solve computationally hard problems, for example, large-scale computation, distributed constraint satisfaction, and decentralized optimization, that are dynamically evolving and highly complex in terms of interaction and dimensionality.
- To characterize complex phenomena or emergent behavior in natural and artificial systems that involve a large number of self-organizing, interacting entities.
- To discover laws and mechanisms underlying complex phenomena or emergent behaviors.

Applying AOC to the service brokering in *i*MSF, the middleware services and the service brokers can be modeled as computational agents where the former ones take the role of service providers and the latter ones take the role of service brokers. In the subsequent subsections, how a set of autonomous services can self-organize in the service brokering process for e-logistics related applications is described. In particular, a bidding process coordinated by a service (discovery) broker together with an adaptive scoring scheme adopted by the service providers will first be introduced for achieving adaptive load balancing among the service providers. As the bidding process assumes that the decision making

of service brokers are independent of each other, it is equivalent to the study of the scenario with a single broker and multiple providers (later on named Single Broker Multiple Providers). This assumption sometimes could be too strong for applications like e-business as many e-business systems are developed to support business processes with inter-related steps. Our proposed way to relax the assumption and allow multiple service brokers to interact and coordinate (via local constraint refinement) for distributed decision making will be described (and will be named Multiple Brokers Multiple Providers).

Autonomous Service Brokering: Single Broker Multiple Providers

In general, to enable interaction among the agents to carry out AOC, two issues have to be figured out: (a) agent communication language and (b) agent interaction mechanism. In this subsection, all the explanation will be given in the context of load balancing for distributed computational tasks. However, the research issues discussed should be able to be generalized to a number of other application domains of similar nature.

A. Ontology for Resource and Performance Reporting

Services even with identical input/output interfaces can have different implementations, time-varying system load, time-varying cached data, and so forth. Thus, during service brokering process, there is a need of selecting from a set of functionally compatible services some suitable candidates for some forms of user-defined optimality. Specifying these performance related and state related information, say via ontology, is important to support autonomous service brokering in *i*MSF for overall computational performance optimization.

While the need of real-time system status imposes further requirements to the middleware framework so that the autonomy layer can be added, some existing middleware implementations have already been equipped with related "consciousness" regarding the real-time system information. In particular, the Globus Grid toolkit[5] is a distributed environment that enables flexible, secure, coordinated resource sharing among dynamic collections of individuals, institutions, and resources. The Open Grid Services Architecture (OGSA) of Globus, which embraced Web services starting with its GT3 version, has become the de-facto standard of Grid middleware and used in various Grid related projects (e.g., Geodise,[6] MyGrid[7]). For example, via the GT3 middleware, one can obtain state information of a Grid node from the service MasterForkManagedJobFactoryService. For network related states, they have not yet been available in GT3, and one could install the Ganglia system[8] for obtaining them in an XML format.

Based on the specific requirement of the application, different scoring schemes derived from that system state information can be adopted for service selection. Besides, some current/aggregated performance statistics obtained from the past history should also be an additional determining factor. Various proposals have been published in using RDF-Schema for ontology specification. For example, three related ontologies were proposed in Tangmunarunkit, Decker, and Kesselman (2003), namely resource ontology (e.g., OperatingSystem.TotalPhysicalMemory = 512MB), resource request ontology (e.g., MinPhysicalMemory = 1024MB), and policy ontology (that captures the resource authorization and usage policies). The actual resource matching based on these ontologies were also described in Tangmunarunkit et al. (2003). In addition, performance related ones (e.g., ResponseTime = 14min.) should also be considered and both types of information (resource related and performance related ones) should be considered for service selection, described in the following.

B. Brokering Via Bidding

With the assumption that service resource, performance, and status ontologies are ready and that the different participating middleware systems are able to provide related information for matchmaking, service brokering can then be performed. If the service provisioning environment is static (say, a cluster), the brokering process can just be a simple semantic matchmaking using ontologies. However, if the environment contains a huge number of services and is dynamic (say, the Web/Grid), adding an autonomy layer to each individual participating middleware system can allow the matchmaking to be more robust and adaptive to changes that are more stationary and thus hard to be modeled probabilistically. In this section, we describe how a bidding-like mechanism based on a dynamic scoring scheme can be incorporated in the middleware systems so that they as a whole can evolve dynamically so that an overall optimality criterion (e.g., load balancing) can be achieved, and yet there is no need to keep the performance and status information of the ever-growing set of participants at a bottleneck centralized broker for coordination.

Let I denote a particular service interface, $E_i(I)$ denote the estimated service time of the ith implementation for the service interface I, and $B_i(I)$ denote the value sent to the broker by the ith implementation for bidding the interface I to be performed. The bidding process is then described as:

The service discovery broker first notifies each of the service providers that host the required service implementations. Being notified, each service implementation will make use of the current estimated service time $E_i(I)$ as well as the current system load to compute a bid value as:

$$B_i(I) = (1 - L_i) \times \frac{1}{E_i(I)}$$

where L_i is the CPU usage of the node hosting the ith service implementation and then send the bid back to the broker. Note that L_i is a state information and $E_i(I)$ is a performance prediction. The broker then selects a service implementation according to the probability distribution:

$$P(i) = \frac{B_i(I)}{\sum_i B_i(I)}$$

After the selected implementation finished the assigned job, it will notify the broker the result. The broker will then return the actual service time A_i and the estimated service time of the ith service implementation will be updated as:

$$E_i^{t+1}(I) = (1-\alpha) \times E_i^t(I) + \alpha \times A_i$$

where α is the updating rate which defines the responsiveness of the system. Such an updating rule is able to capture smooth variation of the service performance. The sequence diagram of the bidding process is shown in Figure 4.

For cases where worst-case performance has to be controlled, the minimax strategy can be used for computing the bidding score. In particular, we can store the actual service times for a time window of size N and then we take the maximum instead of average to avoid selecting services with large service time fluctuation. The formula for the bidding score can then be reformulated as:

$$B_i(I) = (1-L_i) \times \frac{1}{\max_{t \in \{i,i-1,\ldots i-N+1\}} (A_t)}$$

In addition, one can also compute directly the standard deviation as uncertainty and incorporate it into the bidding score, for example,

$$B_i(I) = (1-L_i) \times \frac{1}{E_i(I) + \beta s_i}$$

where

$$s_i = \sqrt{\frac{1}{N-1} \sum_{t=i-N+1}^{i} (A_t - \overline{A_t})^2}$$

and β is used for controlling the degree of tolerance for performance fluctuation.

More detailed experimental results that validate the feasibility of the use of the bidding process and demonstrate its effectiveness can be found in Cheung et al. (2004).

Autonomous Service Brokering: Multiple Brokers Multiple Providers

What is being addressed in the previous section is essentially the simplest form of service provisioning that can be employed in middleware frameworks like *i*MSF. At one time, only one service selection decision would be made without considering the selection decisions of the other service (discovery) brokers, even though the decisions are sequential in nature and could affect each other.

A. Service composition in E-Logistics

Sequential decision making scenarios are common in e-business related problems like e-supply chain management and e-logistics as there is always ambiguity in determining the requirements and solutions of individual participants involved in a service chain, which may further result in uncertainties and dynamics in searching, coordinating, and integrating the solutions throughout the integration process. Mechanisms are sought for coordinating the flow of information among the participants (interfaced via their middleware systems) across the chain and then linking up business processes so as to fulfill various constraints and at the same time optimize the overall chain performance. To extend what was described in the previous section, the bidding mechanism between service brokers and service providers

Figure 4. The sequence diagram of the bidding process (Cheung et al., 2004)

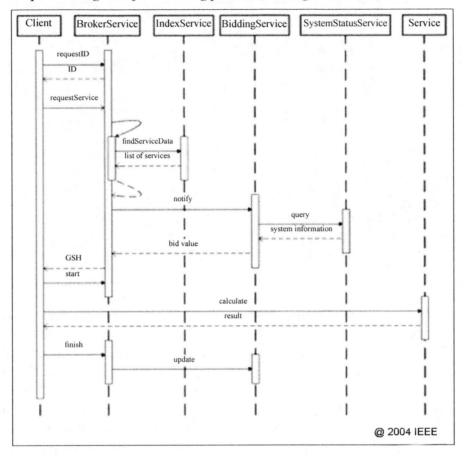

can be interleaved with an additional constraint coordination and refinement mechanism so that service brokers can coordinate and self-organize among themselves via *i*MSF while receiving bids from their individual groups of service providers until a solution emerges (Wang et al., 2006a, b; Wang, Wang, & Liu, 2007). Figure 5 gives an illustration of such an autonomous service composition for e-supply chain management.

As shown in Figure 5, a society of software agents, including a service dispatcher agent and a set of service broker agents and service provider agents is shown for characterizing the extended *i*MSF using an AOC approach. In particular, the *i*MSF is extended with the peer-to-peer coordina-tion among the service broker agents to further optimize their service provisioning quality. From the perspective of logistics, this corresponds to the scenario depicted in Figure 6, where a sup-ply chain is fulfilled through a set of services, including procuring components, preprocessing components, assembling components into prod-ucts, postprocessing products, and delivering components or products, where the customer and service providers are distributed in different locations. With regard to service requirements, the issues of time, cost, location, and so forth, are considered to be important attributes of quality of service (Menasce, 2004). Normally, a composite service that fulfills a supply chain is completed

Figure 5. An illustration of agent-mediated e-supply chain coordination

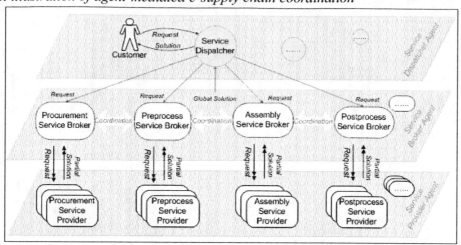

and delivered to the customer on or before the due date required by the customer and then a feasible solution with the lowest cost will be accepted. Within a composite service, a component service can only be scheduled to start after its preceding service is completed, and to end before its succeeding service starts. Moreover, when the customer and component service providers are distributed in different locations, one or more delivery services are embedded into the process.

A major challenge here is that available resources of the component services (including service providers and their solutions) are *undetermined* in advance in most situations. As only the requests of the composite service claimed by the customer are known, instead of the requirements of the component services that constitute the chain, the customer's request has to be trans-

formed into a set of estimated component service requests (or requirements) in the form of local constraints as aforementioned. Those estimated requirements could then be sent out to service providers and further refined based on real-time responses from service providers and real-time coordination among services.

To decrease the complexity of the integration process caused by adding delivery services on demand, we treat delivery as a type of standard service that could be provided by a certain global delivery company (e.g., DHL), and could be bound with any component service when necessary. After receiving service requests from the service dispatcher, service brokers will send them to service providers for collecting suitable solutions to each service. However, available solutions may be incompatible with each other to

Figure 6. A supply chain process

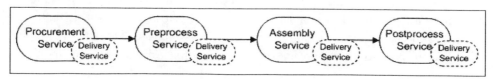

form a global solution. A service broker needs to coordinate with its *neighboring* (i.e., preceding and succeeding) brokers to refine the constraints for achieving new bids that could be involved in a global solution.

To illustrate the idea using an example, suppose the service dispatcher received a service request from a customer on July 1, described as "1200 products to XYZ Plaza by 20 July." This service is decomposed into four services, S_a, S_b, S_c, and S_d. Based on the request, the time constraint of the composite service is specified as start-time>=1 AND end-time <=20. According to average percentage of time spent on each service and average price of each service, the constraints of each service are estimated by the service dispatcher as shown in Table 1. Service requests based on these constraints are sent to service brokers to forward to service providers. After receiving these requests, service providers may send out bids. The bids satisfying the constraints are posted. Based on the bids received and posted after sending

out the initial request, each service broker may identify the bids that are compatible with bids of other services. Promising bids are also identified such as B201, B205, and so forth. Based on the information of promising bids of the preceding and succeeding service, each service broker may adjust its service constraint. As shown in the table, the time constraints of S_a and S_d are updated. After sending out updated service request based on the updated time constraints, new bids are received and posted, for example, B107 and B405.

In the middle of the composition process, the service broker of S_a may update its cost constraint by 5% up since there are too few bids available for this service. As a result, bid B106 is found from the ignored bids and then posted. Moreover, in the whole process, each service broker will report its posted bits that are compatible with the bids of the preceding and succeeding service to the service dispatcher for working out global solutions. Among four global solutions achieved so far, the one with the lowest cost is selected, which is

Table 1. An example illustrating service brokers coordination (Wang et al., 2006a)

	Service S_a	Service S_b	Service S_c	Service S_d
Average cost and duration according to history data	$1200 5 days	$2600 10 days	$800 3 days	$1200 2 days
Estimated constraint (cost, start-time, end-time)	($1320, 1, 6)	($2860, 5, 16)	($880, 15, 19)	($1320, 18, 20)
Bids received after sending out initial request (Bid_ID, cost, start-time, end-time)		(B201, $2650, 8,16) (B205, $2610, 7,15) (B210, $2650, 5, 14)	(B303, $800, 15, 16) (B304, $750, 17, 18)	(B402, $1160, 19, 20)
Updated time constraints (start-time, end-time)	(1, 7)			(17, 20)
Bids received after updating time constraints (Bid_ID, cost, start-time, end-time)	(B107, $1200, 1, 7)			(B405, $1150, 17, 18)
Updated cost constraint	$1386			
Bids posted after updating cost constraint (Bid_ID, cost, start-time, end-time)	(B106, $1370, 1, 6)			

composed of B107, B201, B304, and B402. Once the solution is accepted by the customer, the service dispatcher will ask service brokers to make commitments with related service providers.

Before we move on to the details of the coordination mechanism, note that the service composition considered in this chapter is a simplified situation where there is only one preceding or succeeding service of a component service. While this is not uncommon in logistics, it will also be interesting to see how the mechanism can further be extended to composite services that need multiple preceding and succeeding services.

B. Agent-Mediated Decision and Coordination Mechanism

Decision and coordination in supply chain integration, as described in the previous section, is usually complex and can be partitioned into a set of inter-related subproblems. Multi-agent theory has been known to be effective in addressing related distributed decision making. During service integration, decision and coordination among services by agents are modeled as a distributed constraint satisfaction problem in which solutions and constraints are distributed into a set of services to be solved by a group of agents. Finding a global solution to the composite service requires that all agents find the solutions that satisfy not only their own constraints, but also inter-agent constraints (Liu et al., 2002; Yokoo, 2001). The autonomous and interactive behavior of agents involved in decision and coordination for e-supply chain integration is elaborated as follows:

1. **Collect solutions:** After receiving the service request from service dispatcher, each service broker may forward the request to corresponding service providers for collecting solutions/bids of the service, where a bid is defined as:

    $$Bid_{ij} = [b\_id_{ij}, s\_t_{ij}, e\_t_{ij}, c_{ij}, loc_{ij}, des_{ij}]$$

    Bid_{ij}, the jth bid sent to service broker i for service i, contains five parts: b_id_{ij} denotes the ID number of the bid, which is associated with the private details of a bid; s_t_{ij} and e_t_{ij} denotes the start time and end time respectively scheduled for the service; c_{ij} denotes the cost claimed by the service provider; loc_{ij} denotes the location of the service; and des_{ij} denotes the destination of the service.

2. **Filter out dominated solutions:** For all bids received from service providers, the service broker will filter out dominated bids before posting them as candidate solutions. A newly received bid, $Bid_{i\beta}$ (bid β for service i), is identified as a dominated bid if it requires the same or higher cost as well as the same or more execution time compared with an existing solution Bid_{ij}, by satisfying the following condition:

    $$c_{i\beta} \geq c_{ij} \text{ AND } s\_t_{i\beta} \leq s\_t_{ij} \text{ AND } e\_t_{i\beta} \geq e\_t_{ij} \text{ AND } loc_{i\beta} = loc_{ij} \text{ AND } des_{i\beta} = des_{ij}$$

    Similarly, any existing solution Bid_{ij} will be filtered out if it is dominated by a new bid $Bid_{i\beta}$ by satisfying the following condition:

    $$c_{i\beta} < c_{ij} \text{ AND } s\_t_{i\beta} \geq s\_t_{ij} \text{ AND } e\_t_{i\beta} \leq e\_t_{ij} \text{ AND } loc_{i\beta} = loc_{ij} \text{ AND } des_{i\beta} = des_{ij}$$

    After a bid for a particular component service is removed as a dominated bid, its connections with other bids of its preceding and succeeding services are removed as well. By filtering out dominated solutions, the number of partial solutions is controlled in a reasonable scale.

3. **Identify mutually compatible solutions:** Each service broker will report its newly posted bids to its preceding and succeeding service brokers, so that each broker may identify its solutions that are compatible with

the solutions of its neighbors. We denote service u and service v as the preceding and succeeding service of service i, and $Bid_{i\beta}$, $Bid_{u\alpha}$, $Bid_{v\gamma}$ as a bid of service i, service u, service v, respectively. After posting $Bid_{i\beta}$, service broker i will connect it with $Bid_{u\alpha}$, an existing bid of its preceding service if the two bids are compatible by satisfying the following condition:

$$s\_t_{i\beta} > e\_t_{u\alpha} \text{ AND } loc_{i\beta} = des_{u\alpha}$$

The service broker will also link $Bid_{i\beta}$ with $Bid_{v\gamma}$, an existing bid of its preceding service if the two bids are compatible by satisfying the following condition:

$$e\_t_{i\beta} < s\_t_{v\gamma} \text{ AND } des_{i\beta} = loc_{v\gamma}$$

In the example shown in Figure 7, where each bid is posted with its start time and end time, the mutually compatible solutions are connected to each other.

4. **Figure out promising solutions towards a global solution:** In service composition, each service broker may utilize its own information and limited information from its neighbors for coordinating and achieving coherence among the decisions. To achieve this, a service broker needs to identify a promising solution of its preceding service (**Preceding Promising Solution, P_PS**), as well as a promising solution of its succeeding service (**Succeeding Promising Solution, S_PS**). Based on this premise, the broker may refine the constraints of its own service to seek new bids that could work with the P_PS and S_PS. This is because a new bid of the service will probably be more involved in a global solution if it is compatible with the P_PS as well as the S_PS. Then, the problem is how to identify a promising solution (P_PS and S_PS) to a service.

A more promising solution to a component service is here defined as those that are connected with more existing solutions to its preceding service and at the same time have more free time for its succeeding service to relax its start time constraint, or if it can connect with more existing solutions to its succeeding service and at the same time have more free time for its preceding service

Figure 7. Partial solution vs. global solution (Wang et al., 2006b)

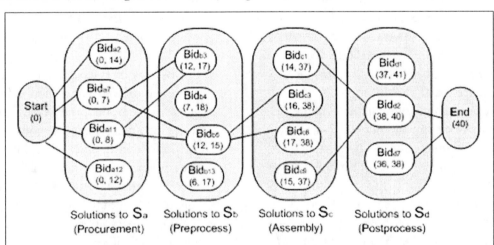

to relax its end time constraint. As shown in Figure 7, Bid_{b5} is identified as a P_PS from the viewpoint of S_c. This is because Bid_{b5} is connected with more bids of its preceding service S_a as well as leaving more time to its succeeding service S_c than any other bid of S_b. On the other hand, Bid_{b5} is also identified as a S_PS from the viewpoint of S_a since Bid_{b5} is connected with more bids of its succeeding service S_c as well as leaving more free time for its preceding service S_a than any other bid of S_b. In this way, each service broker may identify a P_PS among all the solutions of its preceding service, and a S_PS from all the solutions of its succeeding service. In Figure 7, the service broker of S_c will locate Bid_{b5} as the P_PS, and Bid_{d2} as the S_PS. Accordingly, a new bid of S_c would probably be more involved in a global solution if it could connect with Bid_{b5} as well as Bid_{d2}.

To quantify the degree of promising of a solution, the preceding promising value (*Pre_prom*) of Bid_{ij} can be defined as:

Pre_prom (Bid_{ij}) = *wp_conn* $*$ *Pre_conn* (Bid_{ij}) + *wp_tf* $*$ *Pre_tf* (Bid_{ij})

where *Pre_conn* (Bid_{ij}) measures the connectivity of Bid_{ij} with its preceding solutions; *Pre_tf* (Bid_{ij}) measures the free time Bid_{ij} leaves for its succeeding solutions; *wp_conn* and *wp_tf* denote the weight of *Pre_conn* and *Pre_tf* respectively. *Pre_conn* (Bid_{ij}) and *Pre_tf* (Bid_{ij}) are further detailed as:

Pre_conn (Bid_{ij}) = $(pre_{ij} - MINPRE_i)$ / $(MAXPRE_i - MINPRE_i)$

where pre_{ij} denotes the number of the preceding bids that connect with Bid_{ij}; $MAXPRE_i$ denotes the maximum value of pre_{ij} for $\forall j$; $MINPRE_i$ denotes the minimum value of pre_{ij} for $\forall j$, and

Pre_tf (Bid_{ij}) = $(MAXET_i - e\_t_{ij})$ / $(MAXET_i - MINET_i)$

where e_t_{ij} denotes the end time of Bid_{ij};

$MAXET_i$ is the maximum value of e_t_{ij} for $\forall j$; $MINET_i$ is the minimum value of e_t_{ij} for $\forall j$. The succeeding promising value (*Suc_prom*) of Bid_{ij} can be defined in a similar manner.

Based on the promising value, a promising bid can be selected by the service broker using different strategies, such as random selection strategy, elitist strategy, and tournament selection strategy. The random selection strategy chooses a bid at random. The elitist strategy selects the best bid, that is, the bid with the largest promising value. Tournament selection is one of many methods of selection in genetic algorithms which runs a "tournament" among a few individuals chosen at random from the population and selects the winner (the one with the best fitness) for crossover. Selection pressure can be easily adjusted by changing the tournament size.

5. **Refine constraints towards a global solution:** After selecting $Bid_{u\alpha}$ (a solution of the preceding service S_u) as the P_PS and $Bid_{v\gamma}$ (a solution of the succeeding service S_v) as the S_PS, the service broker of S_i may refine its service constraints Rq_i as:

    $Rq_i = [st_i, et_i, loca_i, dest_i]$

    where $st_i = e\_t_{u\alpha} + 1$; $et_i = s\_t_{v\gamma} - 1$; $loca_i = des_{u\alpha}$; $dest_i = loca_{v\gamma}$
    where 1 means one time unit, for example, one day.

    In this way, service brokers may achieve coordination and coherence among decisions of component services through a series of adjustments on constraints that are individually made but interact with each other. Furthermore, service brokers may communicate to figure out a global solution at regular intervals. One or more feasible global solutions could be generated, and the one with the lowest cost is reported as a bid to the customer. The sequence diagram illustrating the bidding as well as the coordination processes is shown in Figure 8.

To study the effectiveness of the proposed processes for autonomous service brokering, a simulation testbed has been developed for comparing experimentally how the quality of the solutions will be affected by variants of the bid picking strategies. More related details can be found in Wang et al. (2006b).

A Case Study

In this case study, a process for developing a system for collaborative warehouse management through

*i*MSF is illustrated. The following is a proof of concept implementation of such a collaborative management system based on *i*MSF.

In Figure 9, each shelf middleware system manages different shelf sensors for different goods. The shelf middleware systems/services connect with each other through available peer to peer protocols to enable warehouse management applications. The conceptual architecture of the shelf middleware system is shown in Figure 10. Implementation of the shelf middleware services has to rely on the support of the rules. In a warehouse, shelves are mounted with sensors in order to detect status of goods or products (e.g., their number of items on the shelf). With such smart shelves, it is possible to detect which goods or products are out of stock or misplaced. The rules

Figure 8. Agent-mediated coordination activities

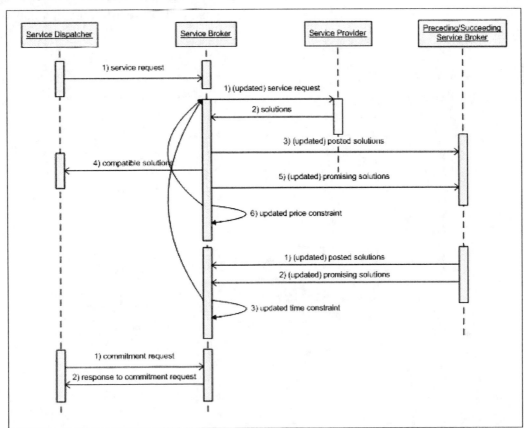

Figure 9. Proof of concept implementation for collaborative warehouse management

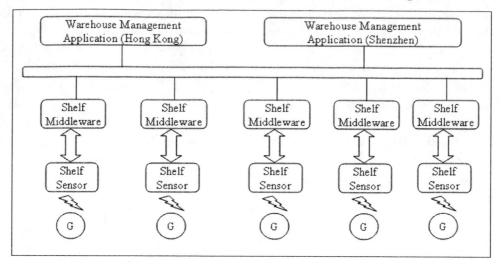

to check whether the product is out of stock are quite simple. To determine the time to restock is relatively harder since this is a decision to be made from a collection of sensor events. If every misplace event leads to a restock, the cost is usually not bearable. Thus, different restock rules could be defined, depending on the restock requirements. Further, even if a product is out of stock in one warehouse, it may be available in another warehouse. If the product is out of stock in both warehouses, then a reordering order may be sent out. After rules are defined for the middleware

Figure 10. Proof of concept implementation architecture for collaborative warehouse management (CWM)

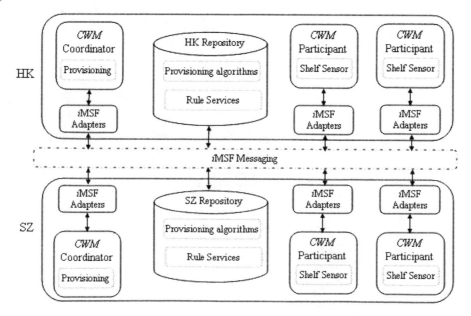

services, then the services will be made available through the publication broker. Event subscription information if any of these services will be conducted through the subscription brokers. The next step would be service provisioning. Usually a few provisioning algorithms would be provided, unless there are particular requirements to develop new algorithms. These new algorithms would be placed in the *i*MSF repository.

The proof of concept implementation architecture is shown in Figure 10. In this case study, we can assume that *i*MSF adapters are already available. Users then only need to define rule based middleware services and provisioning algorithms (see the repository modules in Figure 10). Although the general ideas about the process to define middleware services is given, the design methodology for JECA rules (Luo et al., 1998) is still an ongoing research topic, which is beyond the scope of this chapter.

CONCLUSION

In this chapter, a service oriented e-logistics middleware service framework for e-logistics infrastructure and network integration and development is introduced. It is designed to enable more responsive supply chains and better planning and management of complex inter-related systems, such as materials planning, inventory management, capacity planning, logistics, and production systems. The proposed adoption of SOA can support the evolution of logistics information system development and easy integration of new and legacy systems. To make the middleware framework not only responsive but also adaptive to the dynamic environment, the autonomy oriented computing approach can be applied to the *i*MSF so that the middleware systems and brokers involved in the *i*MSF can self-organize themselves in response to their individual time-varying properties and constraints via only local interaction, and yet some global performance of the *i*MSF can still be effectively optimized.

What being described in this chapter should further contribute to the understanding of how logistics middleware systems can seemingly be integrated and self-adapt for optimizing the global supply network. How the underlying autonomous mechanisms can further be optimized is what we are currently working on. For the future, the e-logistics service framework will deliver general guiding principles, and generic representation approaches of e-logistics middleware tools and technologies development, common modeling, simulation, optimization, integration, and evaluation tools and techniques. These will facilitate analysis and problem solving in any domain and pertain to any type of system. The e-logistics middleware service framework makes it possible to provide solutions based on the tools and technologies developed and offered through third parties, to offer a one-stop solution to the challenges of all aspects of e-logistics system integration and development. Rich functionalities on the integrating heterogeneous middleware systems enable fast-to-market e-logistics system integration and development.

In short, in this chapter, a service oriented intelligent middleware framework enabling integrated e-logistics infrastructure and networks is presented. The *i*MSF provides event processing capabilities for processing data streams from sensing devices, and provides intelligent and autonomous service brokering capability for services provisioning for middleware services. System integrators equipped with the SOA oriented middleware service framework development methodologies can provide enterprises with e-logistics solutions which are more flexible, interoperable, and cost effective. As such, both system integrators and enterprises will benefit from the methodologies and technologies resulting from this research to support dynamic interconnections among heterogeneous middleware systems existing today for e-logistics application development and integration.

LIMITATIONS AND FUTURE RESEARCH DIRECTIONS

For the architectural design part, our proposed architecture supports flexible logistics applications through *i*MSF enabled networks. The *i*MSF delimits a virtual boundary over contained set of IT networks. *i*MSF represents the building blocks of distributed architectures. It isolates the behaviors and resources of architectural elements behind clearly defined abstraction boundaries. While *i*MSF enables flexible architectural structures, challenges still remain. How to efficiently locate and configure an appropriate middleware service or service component becomes a critical challenge. Mapping service requirements to the service capability of available service providers is not a trivial task. The success of this *i*MSF needs a community of advocators. It demands further research for the design methodology for JECA rules as well as provisioning algorithms development to realize *i*MSF's potential of its responsiveness to user requirements.

For the autonomous service provisioning part, the objective of the proposed approach described in the chapter is to support e-business partners to achieve feasibility and coherence in a collaboration plan. It tries to find a way to automate the integration process as much as possible, particularly to obtain coherence and coordination among decisions proposed by difference partners. Due to their irregular and ill-structured nature, systems for such kind of collaborative planning problems may integrate the users into the problem-solving processes. This system can be viewed as an assistant where managers of business entities interact with the system to confirm or modify the requests and the solutions proposed by the system.

However, what we have focused on so far is the mechanism of automatic decision and coordination among software agents in service composition. The interaction of the system with the users involved in the development of composition plans and its impact on the performance of the approach will be investigated in further studies.

In addition, our current study on agent-based service composition is to investigate the decision and coordination among software agents in task allocation and integration through autonomous reasoning and constraint negotiation in the context of e-logistics. However, a generalized and structured methodology and framework for the combination of decision support and distributed artificial intelligence still remains open. Moreover, the objective of a collaborative problem-solving system proposed in our framework is to find a satisfactory or feasible solution. Strategic management of decision making for reaching optimized solutions in an effective and efficient way is also an important issue under further investigation.

ACKNOWLEDGMENT

This work was supported by a RGC Central Allocation Group Research Grant (HKBU 2/03/C) from the Hong Kong Government.

REFERENCES

Cai, D., Luo, Z., Qian, K., & Gao, Y. (2005, November 14-16). Towards efficient selection of Web services with reinforcement learning process. In *Proceedings of the 17th IEEE International Conference on Tools with Artificial Intelligence* (pp. 372-376). Los Alamitos: IEEE Computer Society Press.

Chappell, D.A. (2004). *Enterprise service bus.* O'Reilly.

Cheung, W.K., Liu, J., Tsang, K.H., & Wong, R.K. (2004, September 20-24) Dynamic resource selection for service composition in the Grid. In *Proceedings of the 2004 IEEE/WIC International Conference on Web Intelligence* (pp. 412-418). Los Alamitos: IEEE Computer Society Press.

Langendoerfer, P., Maye, O., Dyka, Z., Sorge, R., Winkler, R., & Kraemer, R. (2004). Middleware for location-based services, design and implementation issues. In Q. Mahmoud (Ed.), *Middleware for Communication*. Wiley.

Liu, J., Jin, X., & Tsui, K. (2005). *Autonomy oriented computing: From problem solving to complex systems modeling*. Kluwer Academic Publisher/Springer.

Liu, J., Jing, H., & Tang, Y.Y. (2002). Multi-agent oriented constraint satisfaction. *Artificial Intelligence, 136*(1), 101-144.

Luo, Z., & Li, J. (2005a, October 18-21). A Web services provisioning optimization model in a Web services community. In *Proceedings of 2005 IEEE International Conference on E-Business Engineering* (pp. 689-696). Los Alamitos: IEEE Computer Society Press.

Luo, Z., Li, J., Tan, C.J., Tong, F., Kwok, A., Wong, E., & Wang, H. (2006b, June 21-23). Intelligent service middleware framework. In *Proceedings of the 2006 IEEE Conference on Service Operation, Logistics and Informatics* (pp. 1113-1118). Los Alamitos: IEEE Computer Society Press.

Luo, Z., Sheth, A., Miller J., & Kochut, K. (1998, November 14). Defeasible workflow, its computation, and exception handling. In *Proceedings of 1998 ACM Conference on Computer-Supported Cooperative Work, Towards Adaptive Workflow Systems Workshop*.

Luo, Z., Zhang, J., & Badia, R.M. (2006a). Service Grid for business computing. In *Grid Technologies, Emerging from Distributed Architecture to Virtual Organizations* (pp. 441-468). WIT Press.

Luo, Z., Zhang, J., Cai, D., & Kun, Q. (2005b, November 15-17). An integrated services framework for location discovery to support location based services. In *Proceedings of the 2nd International Conference on Mobile Technology, Applications, and Systems* (p. 7).

Mahmoud, Q. (2004). *Middleware for communication*. Wiley.

Mahmoud, Q., & Zahreddine, W. (2005). A framework for adaptive and dynamic composition of web services. *Journal of Interconnection Networks, 6*(3), 209-228.

Menasce, D.A. (2004). Composing Web services: A QoS view. *IEEE Internet Computing, 8*(6), 88-90. IEEE Computer Society Press.

Ran, S. (2003, March). A model for web services discovery with QoS. *ACM SIGecom Exchange, 4*(1).

Shepard, S. (2005). *RFID: Radio frequency identification*. New York: McGraw-Hill.

Tangmunarunkit, H., Decker, S., & Kesselman, C. (2003, May 20). Ontology-based resource matching in the Grid: The Grid meets the Semantic Web. In *Proceedings of the 1st Workshop on Semantics in Peer-to-Peer and Grid Computing, in conjunction with the 12th International World Wide Web Conference*.

Vinoski, S. (2002). Where is middleware? *IEEE Internet Computing, 6*(2), 83-85. IEEE Computer Society Press.

Wang, M., Cheung, W.K., Liu, J., & Luo, Z. (2006a, June 26-29). Agent-based Web service composition for supply chain management. In *Proceedings of IEEE Joint Conference on E-Commerce Technology and Enterprise Computing, E-Commerce and E-Services* (pp. 328-332). Los Alamitos: IEEE Computer Society Press.

Wang, M., Cheung, W.K., Liu, J., Xie, X., & Luo, Z. (2006b, September 5-7). E-service/process composition through multi-agent constraint management. In *Proceedings of the 4th International Conference on Business Process Management* (LNCS 4102, pp. 274-289). Springer-Verlag.

Wang, M., Wang, H., & Liu, J. (2007, January 7-10). Dynamic supply chain integration through intelligent agents. In *Proceedings of 40th Hawaii International Conference on System Sciences* (p. 46). Los Alamitos: IEEE Computer Society Press.

Yokoo, M. (2001). *Distributed constraint satisfaction: Foundation of cooperation in multi-agent systems*. Berlin/New York: Springer.

ADDITIONAL READING

Balke, W., & Wagner, M. (2003, May 20-24). Towards personalized selection of Web services. In *Proceedings of the 12th International World Wide Web Conference (WWW 2003)* (pp. 20-24).

Zacharia, G., Moukas, A., & Maes, P. (2000). Collaborative reputation mechanisms in electronic marketplaces. *Decision Support Systems, 29*, 371-388.

Luo, Z., Sheth, A., Kochut, K., & Miller, J. (2000). Exception handling in workflow systems. *Applied Intelligence, 13*(2), 125-147.

Foster, I., Kesselman, C., Nick, J.N., & Tuecke, S. (2002, June). Grid services for distributed system integration. *IEEE Computer.* IEEE Computer Society Press.

IEEE Internet Computing Special Issue: Middleware for Web Services (2003). IEEE Computer Society Press.

Blythe, J., Deelman, E., Gil, Y., Kesselman, C., Agarwal, A., Mehta, G., & Vahi, K. (2003, June 9-13). The role of planning in grid computing. In *Proceedings of the 13th International Conference on Automated Planning and Scheduling* (pp. 153-163). AAAI Press.

Liu, J. (2003, August 9-15). Web intelligence (WI): What makes wisdom Web? In *Proceedings of 18th International Joint Conference on Artificial Intelligence* (pp. 1596-1601).

Aiello, M., Papazoglou, M., Yang, J., Carman, M., Pistore, M., Serafini, L., & Traverso, P. (2002, August 23-24). A request language for Web services based on planning and constraint satisfaction. In *Proceedings of the VLDB Workshop on Technologies for E-services* (pp. 76-85).

Berardi, D., Calvanese, D., Giuseppe, D.G., Lenzerini, M., & Mecella, M. (2004, June 3-7). Synthesis of composite e-services based on automated reasoning. In *Proceedings of the ICAPS 2004 Workshop on Planning and Scheduling for Web and Grid Services* (pp. 94-96).

Buhler, P., & Vidal, J.M. (2004, July 28-30). Integrating agent services into BPEL4WS defined workflows. In *Proceedings of the 4th International Workshop on Web Oriented Software Technologies*.

Casati, F., & Shan, M. (2001). Dynamic and adaptive composition of e-services. *Information Systems, 26*(3), 143-163.

Doshi, P., Goodwin, R., Akkiraju, R., & Verma K. (2004, June 6-9). Dynamic workflow composition using markov decision processes. In *Proceedings of the 2004 IEEE International Conference on Web Services* (pp. 576-582). Los Alamitos: IEEE Computer Society Press.

Kephart, J.O., & Chess, D.M. (2003). The vision of autonomic computing. *IEEE Computer, 36*(1), 41-50.

Norman, T.J., Preece, A., Chalmers, S., Jennings, N.R., Luck, M., Dang, V.D., Nguyen, T.D., Deora,

V., Shao, J., Gray, A., & Fiddian, N. (2004). Agent based formation of virtual organisations. *International Journal of Knowledge Based Systems, 17*(2-4), 103-111.

Raman, R., Livny, M., & Solomon, M. (2000, August 1-4). Matchmaking: Distributed resource management for high throughput computing. In *Proceedings of the 9th IEEE International Symposium on High Performance Distributed Computing* (pp. 290-291). Los Alamitos: IEEE Computer Society Press.

Tsui, K., Liu, J., & Kaiser, M. (2003). Self organized load balancing in proxy servers: Algorithms and performance. *Journal of Intelligent Information Systems, 20*(1), 31-50.

Cheung, W.K., & Liu, J. (2005). On knowledge grid and grid intelligence: A survey. *Computational Intelligence: An International Journal, 21*(2), 111-129. Blackwell Publishing.

Verma, K., Doshi, P., Gomadam, K., Miller, J., & Sheth, A. (2006, September). Optimal adaptation in web processes with coordination constraints. In *Proceedings of 2006 IEEE International Conference on Web Services* (pp. 257-264). Los Alamitos: IEEE Computer Society Press.

Luo, Z., Wong, E., Cheung, S., Ni, L., & Chan, W. (2006, October 26-28). RFID middleware benchmarking. In *Proceedings of the 3rd RFID Academic Convocation in conjunction with the China International RFID Technology Development Conference & Exposition*, Shanghai, China.

Luo, Z., Fei, Y., & Liang, J. (2006). On demand e-learning with service grid technologies. *Edutainment*, pp. 60-69.

Ng, H., Fok, W., Wong, E., & Luo, Z. (2006, October 24-26). Quality management using RFID and third generation mobile communications systems. In *Proceedings of 2006 International Conference on E-Business Engineering* (pp. 504-510). Los Alamitos: IEEE Computer Society Press.

Christopher, M. (2005). Logistics and supply chain management: Creating value-adding networks. *Financial Times* (3rd ed.). Prentice Hall.

Fawcett, S., Ellram, L., & Ogden, J. (2007). *Supply chain management: From vision to implementation.* Prentice Hall.

ENDNOTES

[1] http://www.cscmp.org/
[2] http://www.tdctrade.com/alert/cba-e0503ppl.htm
[3] http://www.epcglobalinc.org
[4] http://www.w3c.org/TR/WSDL
[5] http://www.globus.org
[6] http://www.geodise.org
[7] http://www.mygrid.org.uk
[8] http://ganglia.sourceforge.net/

Chapter XII

A Multi-Agent System Approach to Mobile Negotiation Support Mechanism by Integrating Case-Based Reasoning and Fuzzy Cognitive Map

Kun Chang Lee
Sungkyunkwan University, Korea

Namho Lee
Sungkyunkwan University, Korea

ABSTRACT

This chapter proposes a new type of multi-agent mobile negotiation support system named MAM-NSS in which both buyers and sellers are seeking the best deal given limited resources. Mobile commerce, or m-commerce, is now on the verge of explosion in many countries, triggering the need to develop more effective decision support systems capable of suggesting timely and relevant action strategies for both buyers and sellers. To fulfill a research purpose like this, two artificial intelligence (AI) methods such as CBR (case-based reasoning) and FCM (fuzzy cognitive map) are integrated and named MAM-NSS. The primary advantage of the proposed approach is that those decision makers involved in m-commerce regardless of buyers and sellers can benefit from the negotiation support functions that are derived from referring to past instances via CBR and investigating inter-related factors simultaneously through FCM. To prove the validity of the proposed approach, a hypothetical m-commerce problem is developed in which theaters (sellers) seek to maximize profit by selling their vacant seats to potential customers (buyers) walking around within reasonable distance. For experimental design and implementation, a multi-agent environment Netlogo is adopted. A simulation reveals that the proposed MAM-NSS could produce more robust and promising results that fit the characteristics of m-commerce.

INTRODUCTION

The modern mobile computing world is characterized by one of both ubiquitous connectivity and ubiquitous computational resources (Edwards, Newman, Sedivy, & Smith, 2004). Recent popular forms of mobile computing encompass omnipresent short-range communications (including both infrastructure-based technologies such as WiFi and peer-to-peer technologies such as Bluetooth), and also omnipresent long-range communications (such as cellular telephony networks). This maturing mobile environment justifies conservative estimates based on the 2000 Census report suggesting that by 2006 10% of U.S. workers will be completely mobile, with no permanent office location (Lucas, 2001). This trend will be fueling development of new mobile applications as advances in mobile technology increase coverage, data speeds, and usability (Barbash, 2001; Crowley, Coutaz, & Bérard, 2000; Parusha & Yuviler-Gavishb, 2004; Pham, Schneider, & Goose, 2000; Turisco, 2000).

In this sense, it is no wonder that **mobile commerce** (or **m-commerce**) replaces traditional forms of electronic commerce rapidly. Various types of m-commerce services include mobile shopping, location sensitive information service, traffic updates, and logistic tracking services, all of which utilize the concepts of customization, personalization, location sensitive, context awareness (Lee & Yang, 2003; Schilit, 1995; Schilit, Adams, & Want 1994; Wang & Shao, 2004; Want, Hopper, Falcao, & Gibbons, 1992; Want, Schilit, Adams, Gold, Petersen, Ellis, et al., 1995). M-commerce has been successfully activated in some industries, leading to competitive advantage (Rodgera & Pendharkarb, 2004; Varshney, 1999) and improved workflow as well as reduced costs and risk management (Miah & Bashir, 1997; Porn & Patrick, 2002; Turisco, 2000). However, such a success story is confined to specific applications where the decision support framework is not considered seriously. To reap better results from the users' view, decision makers engaged in a specific type of m-commerce should be supported more intelligently and robustly.

It cannot be overstated that decision makers under a specific m-commerce situation need more timely and robust decision support because they are in several types of contexts. For example, they cannot afford to receive detailed information from a decision support system because of the limited display capability of mobile devices they carry. Besides, they do not have enough time to consider all the related factors before making decisions because they are usually on the move. This kind of environmental limitations require that a **decision support framework** should be developed for enhancing decision making effectiveness for m-commerce users.

For this purpose, this chapter proposes a new kind of decision support framework named **MAM-NSS_**(multi-agent mobile negotiation support system) which can benefit both m-commerce buyers and sellers. MAM-NSS is based on a multi-agent mechanism in which buyers and sellers are respectively represented by agents. Each agent tries to coordinate with each other until reaching a compromised decision. Especially, the proposed MAM-NSS focuses on the fact that decision makers engaged in a specific m-commerce situation are often facing two kinds of needs: (1) to refer to past instances carefully and (2) mull over inter-related factors simultaneously. A literature survey shows that there exist few studies dealing with those research needs. To fill such a research void, this chapter proposes two important mechanisms like **case based reasoning** (CBR) and **fuzzy cognitive map** (FCM). The proposed MAM-NSS combining CBR and FCM is therefore expected to provide more robust decision support to m-commerce decision makers irrespective of buyers and sellers. To prove the validity of the proposed approach, a hypothetical m-commerce problem is developed in which theaters (sellers) seek to maximize profit by selling their vacant seats to potential customers (buyers) walking around within reasonable distance. For experimental design and implementation, a **multi-agent** environment *Netlogo*[1] is adopted.

BACKGROUND

Recent Trends in M-Commerce

Electronic commerce applications recently provided by mobile communication services include mobile information agents (Cabri, Leonardi, & Zambonelli, 2001, 2002; Mandry, Pernul, & Rohm, 2000-2001), online kiosks (Slack & Rowley, 2002), government applications (e.g., online selling by the postal service, Web-based electronic data interchange, or EDI, in trade applications), and direct online selling systems such as Internet-based (or Web-based) shopping mall systems and Internet-based stock trading systems. In the m-commerce and mobile communication services, their ease of use and multimedia approach to the presentation of information attract potential customers. Some countries and regions have put in tremendous efforts in pushing the development and deployment of m-commerce. These countries include the U.S., Japan, South Korea, Hong Kong, and the Scandinavian countries. There have been many different kinds of m-commerce applications deployed to businesses in these areas. In the U.S., the current rush to wireless communication methods was triggered by the U.S. Federal Communication Commission's auctioning of personal communication-service spectrum space (Senn, 2000). The collaboration of public and private sectors has facilitated the development of m-commerce businesses.

Recently, a large number of organizations have adopted m-commerce for business purposes in order to gain competitive advantages in the electronic market. To cite a few examples, NTT DoCoMo, Vodafone, Verizon, Sprint PCS, and AT&T Wireless have provided "cybermediation" for greater efficiency in supply and marketing

channels through m-commerce. M-commerce can benefit business transactions by providing more efficient payment systems, shortening time to markets for new products and services, realizing improved market reach, and customization of products and services (Barnes, 2002; Senn, 2000). Besides, innovative m-commerce applications have been constantly reshaping business practices in terms of enhancing customer service, improving product quality, and lowering cycle time in business processes (Seager, 2003).

As m-commerce supports online purchasing through an electronic channel, that is, the Internet, via electronic catalogs or other innovative formats, customers procure products, services, and information through m-commerce (Bailey & Lawrence, 2001). In m-commerce, potential customers can visit various "virtual malls" and "virtual shops," and browse through their catalogues to examine products in vast detail. New areas of business opportunities for retailers, producers, and consumers can be developed from these virtual markets on m-commerce. Mobile information agents provide an effective method to support the electronic marketplace by reducing the effort involved in conducting transactions (Wang, Tan, & Ren, 2002). Mobile agents can also help search other agents for contracting, service negotiation, auctioning, and bartering (Mandry et al., 2000-2001). Agents roam through Internet sites to locally access and elaborate information and resources (Omicini & Zambonelli, 1998). The introduction of mobile agents into the electronic market scenario reduces the load and the number of necessary connections to suppliers. In this way, the multi-agent approach is a feasible way to model and analyze complex m-commerce applications.

Among the three distinct identifiable classes of electronic commerce applications (i.e., **business-to-customer (B2C)**, **business-to-business (B2B)**, and intra-organization (Applegate, Holsapple, Kalakota, Radermacher, & Whinston, 1996), m-commerce generally falls into the B2C class. M-commerce provides Web presence with information about company products and services and facilities for both online and off-line purchasing. M-commerce also facilitates other business related activities, such as entertainment, real estate, financial investment, and coupon distribution. Usually, m-commerce sellers are required to make competitive offers in order to sell their products or services to the target customers within reasonable distance. In location-based m-commerce applications, sellers should compete with each other to appeal to potential buyers because there could be only a few buyers in a limited area. For sellers, making timely and attractive offers to buyers on the move is very challenging because the buyers continue to receive information and offers from competing sellers.

Intelligent Agents

Fundamentals

The proposed MAM-NSS is basically based on the multi-agents. Both sellers and buyers engaged in a certain m-commerce are represented by specific agents, and each agent is entitled to receiving proper decision support from the MAM-NSS. An intelligent agent (or agent) has various definitions because of the multiple roles it can perform (Applegate et al., 1996; Hogg & Jennings, 2001; Persson, Laaksolahti, & Lonnqvist, 2001; Wooldridge, 1997; Wooldridge & Jennings, 1995). An intelligent agent is simply a software program that simulates the way decision makers think and make decisions. It performs a given task based on the information gleaned from the environment to act in a suitable manner so as to complete the task successfully. It is able to adjust itself to the changes in the environment and circumstances, so that it can achieve the expected result (Paiva, Machado, & Prada, 2001).

The term "**intelligent agent**" can be disintegrated into two words: intelligence and agency. The degree of autonomy and authority vested in the agent is called its agency. It can be measured at least qualitatively by the nature of the interaction between the agent and other entities in the system in which it operates. An **agent** is an individual and it runs independently. The degree of agency will be enhanced if an agent represents a user in some way. Therefore, collaborative agents represent a higher level of agency because they cooperate with other agents or programs or entities, and so on. The agent intelligence can be interpreted as the degree of reasoning and learned behavior. It is the ability to understand the user's statement of goals and carry out the task delegated to it. Such intelligence can be easily found in the reasoning process of many decision or AI models. Intelligence enables agents to discover new relationships, connections, or concepts independently from the human user, and exploit these in anticipating and satisfying a user's needs (Bonarini & Trianni, 2001; Hu & Weliman, 2001; Schaeffer, Plaat, & Junghanns, 2001).

To retain the characteristics of "intelligence" and "agency," an intelligent agent should possess the abilities of mobility, benevolence, rationality, adaptability, and collaboration (Wooldridge, 1997). Mobility is the ability to move around an electronic network (Bohoris, Pavlou, & Cruickshank, 2000; Lai & Yang, 1998; Lai & Yang, 2000). Benevolence is the assumption that an intelligent agent does not have conflicting goals, and therefore it will always try to complete the assigned tasks (Hogg & Jennings, 2001; Jung & Jo, 2000). Rationality is the assumption that an agent will act in order to achieve its goals and will not act in such a way as to prevent its goals from being achieved, at least insofar as its beliefs permit (Hogg & Jennings, 2001; Persson et al., 2001). Adaptability indicates that an agent should be able to adjust itself to the habits, working methods, and preferences of its user (Jung & Jo, 2000). Collaboration is an ability to cooperate with other agents so that an agent can achieve what a goal decision maker wants to attain (Jung & Jo, 2000; Lee & Lee, 1997; Wu, Yuan, Tseng,

& Fuyan, 1999). This chapter places a strong emphasis on the collaboration ability. Although no single agent possesses all these abilities in a real situation, it is certain that these kinds of characteristics are those that distinguish agents from ordinary programs.

Multi-Agent System

To investigate a computational model that actually encodes and uses conflict resolution expertise, a focus can be placed on the **multi-agent framework**, which is adopted from the distributed AI problem (Bird, 1993; Chaib-Draa & Mandiau, 1992; Cooper & Taleb-Bendiab, 1998; Luo, Zhang, & Leung, 2001; Sillince, 1998; Sillince & Saeedi, 1999; Tung & Lee, 1999). Multi-agent systems have offered a new dimension for coordination in an enterprise (Bonarini & Trianni, 2001; Hu & Weliman, 2001; Kwon & Lee, 2002; Sikora & Shaw, 1998; Strader, Lim, & Shaw, 1998; Ulieru, Norrie, Kremer, & Shen, 2000; Wu, 2001). Incorporating autonomous agents into **problem-solving** processes allows improved coordination of different functional units to define tasks independently of both the user and the functional units under control (Cabri et al., 2002). Under a multi-agent system, the problem-solving tasks of each functional unit becomes populated by a number of heterogeneous intelligent agents with diverse goals and capabilities (Lottaz, Smith, Robert-Nicoud, & Faltings, 2000; Luo et al., 2001; McMullen, 2001; Ulieru et al., 2000; Wu, 2001).

The multi-agent system, in which multiple agents work collaboratively to solve specific problems, provides an effective platform for coordination and cooperation among disputing multiple entities in real world cases. For example, when a conflict occurs between buyers and sellers over a limited resource, it is difficult for a single authority or committee to reconcile it to the full satisfaction of all the entities concerned. Therefore, it is likely that the use of a multi-agent system for coordination will result in a more sys-tematic and organized method in reality without causing unnecessary emotional and behavioral side effects.

Decision Support Mechanisms

MAM-NSS is equipped by two decision support mechanisms such as CBR and FCM. First, **CBR** is a renowned artificial intelligence methodology that provides the technological foundations for intelligent systems (Kolodner, 1993). Given a case base where a number of **past instances** are stored, CBR consists of several phases: indexing cases, retrieving the appropriate candidate cases from the case base, approximating potential solutions from them, testing whether the proposed solutions are successful, and learning to upgrade the decision quality by updating the case base and retrieval mechanism. CBR is most applicable when there is (1) no decision model available; (2) a specific decision model is too hard to acquire; or (3) when past cases are available or easy to generate. With these CBR benefits, the CBR approach is extensively used for negotiation (Kowalezyk & Bui, 1999). Considering the advantages of CBR above, MAM-NSS adopts CBR to allow agents to refer to the past relevant instances that seem to explain a part of current decision making problem before making final decisions.

Second, **FCM** is utilized to provide agents with the capability of analyzing complicated **interrelationships of all the relevant factors** by viewing them simultaneously. FCM was introduced by Kosko (1986, 1987) in which fuzzy causality concept is introduced to represent uncertainty embedded in problem domain. In this way, FCM provides a more flexible and realistic representation of the domain knowledge. For example, Ray and Kim (2002) used it as a tool for understanding and controlling intelligent agents. Liu and Satur (1999) have used FCM as a decision support mechanism for interpreting geographic information as well as designing automatic context awareness function that is one of the important characteristics in m-commerce.

By integrating CBR and FCM, MAM-NSS is designed to provide decision makers on the move with more improved decision support functions. CBR is especially useful for m-commerce users who do not have sufficient time to consider all the constraints before making decisions. By retrieving appropriate past examples and suggesting them as a benchmarking point, CBR can help m-commerce users make fast decisions. There are many factors that are influencing m-commerce decisions either indirectly or directly. However, users cannot afford to consider all the causal relationships among those factors thoroughly in a situation when they need to move and there is not enough time. In that situation, FCM can provide an analytical and systematic way of investigating

causal relationships between all the factors related to the m-commerce situation.

MAM-NSS

Basics

Figure 1 depicts a hypothetical m-commerce situation where MAM-NSS is used to provide timely decision support to m-commerce users. Since the term "m-commerce" may cause different interpretations depending on situations, we need to define m-commerce conditions more clearly to make further discussions unambiguous. First, buyers are assumed to carry mobile devices such as PDAs or mobile phones. Second, buyers cannot have access to the telecommunication network

Figure 1. M-commerce situation

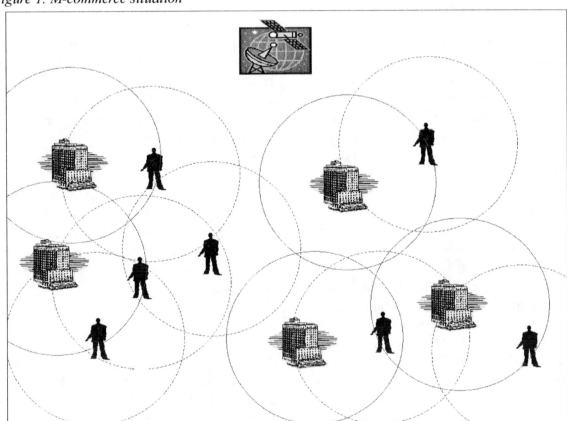

line because they are moving. Third, buyers make reservations with sellers through mobile devices. Fourth, sellers provide information about their services and goods through mobile devices.

The **negotiation process** of MAM-NSS is composed of four steps. Since MAM-NSS runs on a multi-agent framework, a buyer is represented by **B-agent** and a seller **S-agent** from now on. In other words, each agent possesses its own generic knowledge including either buyer's or seller's basic preference that has been predefined.

The first step is to identify buyers' and sellers' location. Such location identification is based on opt-in agreement with mobile telecommunication company.

The second step is for sellers to provide an offer to potential buyers within a fixed distance from where buyers can arrive after a reasonable time. For example, the seller's offer may include an appropriate product/service and its price quote, all of which are assumed to be obtained from CBR inference. For this CBR inference, sellers need various kinds of contextual information such as potential buyers' current location, weather, other events, and so forth.

The third step is for buyers to check the offer from sellers on their mobile devices in which buyers' personal preference is stored. Then the seller's offer is compared with the buyer's predefined preference. If the offer is not appealing, then the buyers modify the offer and send it to the seller. All this process is performed on a multi-agent basis.

The fourth step is for sellers to review the modified offer from buyers. At this time, FCM is used to induce an appropriate price level considering complicated causal relationships among qualitative and strategic factors simultaneously. Depending on marketing strategy, the sellers may change their price offer to lower or higher. Therefore, the sellers' decision can become more strategic and flexible in accordance with changing situations.

Architecture

MAM-NSS is composed of three entities like **B-agent**, **S-agent**, and **mobile telecommunication company**. B-agent is assumed to be downloaded into buyer's mobile device when she subscribes to the MAM-NSS service. The B-agent becomes personalized according to **buyer's personal preference** about specific products/services, and related prices, quality, brand, and other properties. Especially, B-agent specifies its own utility function following buyer's preference which is compiled from the online questionnaire when the buyer subscribes for the MAM-NSS service. Then the B-agent is stored in the memory of the buyer's carried **mobile device**. The B-agent uses this information to negotiate prices with sellers. S-agent is basically linked to the seller's back office system and negotiates with B-agent on the price of seller's products/services. S-agent's first price offer is composed referring to CBR inference, and relayed to potential buyers who are moving within a reasonable distance from the seller. When the buyer's modified price offer is entered, S-agent revises its price offer by using FCM and then feeds it back to the buyer. In this process, negotiations are going on until the final deal is struck between buyer and seller.

In the process of negotiations, the mobile telecommunication company acts as an intermediary between S-agents and B-agents. If the **sellers** and **buyers** subscribe to the MAM-NSS service provided by the mobile telecommunication company, then they can share the information needed in negotiation such as the location of sellers and buyers, price offer, and related product/service information.

S-Agent

The ultimate goal of S-agent is to maximize profit. For this purpose, S-agent seeks a potential buyer in the range acceptable to the buyer within a specific time limit. Then it calculates

Figure 2. MAM-NSS architecture

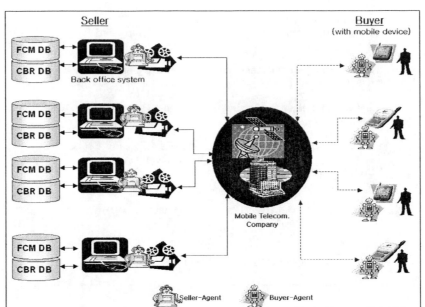

the bid price of the selling product/service based on CBR, and sends the offer including price and product/service to the potential B-agents through the mobile telecommunications company. A wide variety of past selling instances are stored in the case base, and CBR uses the **similarity index** (SI) below to select the candidate case that seems to fit most with the current selling situation. Once such case is chosen successfully, then the price offer can be made appropriately referring to the price information attached to the selected case.

$$SI_i = \sqrt{\sum_{j=1}^{n}(N_j - S_{ij})^2}$$

where N_j indicates jth attribute value of a new case (j=1,2,…,n), and S_{ij} denotes jth attribute value of ith case in case base of CBR (i=1,2,…, m). Netlogo source code for implementing the CBR mechanism using SI is listed in Table 1.

If the buyer accepts the price offered by S-agent, then the deal between seller and buyer will be completed and the buyer stops negotiating. But if the buyer is not satisfied with the price offered by S-agent, the price level is adjusted and then relayed to S-agent. Finally, S-agent decides whether or not to accept the newly-adjusted price, using the FCM inference. If another price offer is made, then negotiation proceeds to the next round. This process is repeated until all sellers and buyers find their appropriate partners that meet their respective goals. In Table 2, the FCM mechanism to be used for adjusting the **price offer** is represented in the Netlogo source code.

B-Agent

The buyer represented by B-agent seeks to maximize its own utility in the process of negotiating with the suppliers. Buyers can download B-agents from the subscribed telecommunication company's site and store them into their mobile devices. B-agents incorporate the following **util-**

Table 1. S-agent's CBR inference mechanism

```
to change-CBR-price
locals [temp_t temp_i temp_si temp_optimal_si temp_item]
set temp_t(1)
repeat seller_number [
ask seller with [reg_number = temp_t and mobile_service = 1 and
mdss_service = 1 ]
[
    set vacant_percent int (vacant_seat_number / seat_number * 100)
set temp_i (0)
    repeat length CBR_price [
set temp_si sqrt((F1_Current_Value - item (temp_i) CBR_F1_List) ^ 2
                    + ((F2_Current_Value - item (temp_i)
CBR_F2_List) ^ 2
                    + (F3_Current_Value - item (temp_i)
CBR_F3_List) ^ 2
                    + (F4_Current_Value - item (temp_i)
CBR_F4_List) ^ 2
                    )

    if (temp_i = 0)[set temp_optimal_si (temp_si) set temp_item
(temp_i)]
```

Table 2. S-agent's FCM mechanism to adjust price offer

```
to Buyer-decision-for-new-customer-offer
set temp (0)
set temp_i (1)
repeat customer_number [
set inference_list [ ]
set inference_list lput FCM_factor1 inference_list
set inference_list lput FCM_factor2 inference_list
set inference_list lput FCM_factor3 inference_list
set inference_list lput FCM_factor4 inference_list
set inference_list lput FCM_factor5 inference_list
set inference_list lput FCM_factor6 inference_list
set inference_list lput FCM_factor7 inference_list
set inference_list lput FCM_factor8 inference_list
set inference_list lput FCM_factor9 inference_list
set inference_list lput FCM_factor10 inference_list

show inference_list

set FCM_ACCEPT_REULT (FCM_inference inference_list )

ask buyer with [id_number = temp_i ] [
 if (reserve != 1 and mdss_service = 1)  [
 set temp_utility_adjustment (utility_adjustment)
 ask seller with [reg_number = temp2 ] [
  if (available_number_of_product > 0) [
    if (FCM_ACCEPT_REULT = 1 ) [
       show temp_new_price + "<---- accept"
       set vacant_seat_number (vacant_seat_number - 1)
       set temp1 (1) ask customer with [id_number = temp_i ]
                                   [set reserve (1)  set color black]
```

ity functions where $i=1,2,...,$m (number of sellers) and $j=1,2,...,$n (number of utility factors):

$$U_i = \sum_{j=1}^{n} W_{ij} \cdot F_{ij}$$

U_i denotes ith buyer's utility, W_{ij} buyer's preference for jth utility factor, and F_{ij} ith buyer's jth

utility factor. It is certain that $\sum_{j=1}^{n} W_{ij} = 1$. Examples of utility factors include not only price, product, and quality, but also **contextual information** such as the buyer's current location and environmental constraints. Table 3 shows the Netlogo source code for calculating the B-agent's utility function.

Table 3. B-agent's utility calculation

```
to search-buyer
set temp (1)
set temp_id (1)
repeat customer_number [
ask customer with [reserve != 1 and id_number = temp_id] [
set temp_distance (p_distance )
set temp_price (p_price )
set temp_time (p_time )
set temp_boxoffice_ranking (p_boxoffice)
set temp_genre (p_genre )
set temp_customer_x (current_x) set temp_customer_y (current_y)
set utility (0)
set temp_selected_theoter (0)

repeat seller_number [
 ask seller in-radius-nowrap (remaining_time / time_per_patch)
  with [available_product_number > 0 and reg_number = temp1][
set actual_distance (abs (sqrt((temp_customer_x - location_x) ^ 2
+ (temp_customer_y - location_y) ^ 2 ) ))

Convert_factor_point

set temp_util (  temp_P1 * temp_point_F1
          + temp_P2 * temp_point_F2
          + temp_P3 * temp_point_F3
          + temp_P4 * temp_point_F4
          + temp_P5 * temp_point_F5
          )
if (temp_util > utility) [ set utility (temp_util)
            set temp_selected_seller (reg_number)  ]
                ]
          set temp1 (temp1 + 1)] set temp1 (1)
```

Table 4. B-agent's price update process

```
ask buyer with [deal !=1 ] [
 set goal_utility (Current_utility + (utility_adjustment / 100) * Current_utility )
  set temp (selected_buyer)
     ask seller with [reg_number = temp ][
       if (available_number > 0) [
       if (p_temp > 0 ) [
 set temp_price_down_request int((goal_utility -
```

If B-agent gets the price offer from S-agent and this offer does not meet the buyer's goal utility, then the B-agent suggests a new price using the mechanism shown in Table 4. If the seller accepts the new price offered by the buyer, then the deal is complete. However, if no sellers accept this price, then the B-agent increases the price decreasing its goal utility. In this case, a new round of negotiation resumes.

EXPERIMENTS

Problem Description

Three Groups

The target problem here is that there are a number of movie theaters in an area, and customers want to go to the theater depending on their personal situations. Based on whether customers (i.e., buyers) and theaters (i.e., sellers) are using mobile devices or not, we categorize them into three groups for the sake of the experiment. Group 1, called "**non-mobile group**," is not using mobile devices. Therefore, customers either reserve tickets through non-mobile channels, such as telephone or cable Internet, or buy onsite from the box office. Theaters are assumed to contact customers through the non-mobile channels too. Group 2, called "**passive mobile group**," is using mobile devices but no negotiation functions. Therefore, customers in this group can get information about movies through mobile channels, but they cannot negotiate with theaters through agents. Theaters are also using mobile channels to send movie information to customers. Group 3, called "**active mobile group**," is assumed to use the proposed MAM-NSS for negotiation through mobile channels and agents. Customers and theaters are offering their own preferences such as price and vacant seats utilizing the negotiation mechanism based on MAM-NSS. The three groups will be compared with each other through Netlogo simulation experiments. In fact, group

1 is inappropriate for m-commerce situations because they are not reached by mobile devices. However, group 1 is included so that the other two groups can be compared.

Sellers

- Assumptions about theaters are as follows: they have 200 seats, cost $700 per show, and start to sell vacant seats an hour before the movie begins. List price for a ticket is $7, and movie genres a theater is showing have four types. All the theaters show the movie with box office ranking from 1 to 10. Every 20 minutes, theaters in group 2 are offering discriminated pricing strategies to buyers through mobile channels, depending on the vacancy rate: $6.5 if vacancy rate < 40%, $6.0 if 40% ≤ vacancy rate ≤ 50%, $5.0 if 50% ≤ vacancy rate ≤ 60%, $4.0 if 60% ≤ vacancy rate ≤ 70%, $3.0 if 70% ≤ vacancy rate ≤ 80%, and $2.0 if 80% ≤ vacancy rate. However, as noted previously, customers cannot negotiate with the theaters in group 2, indicating that they have no choice but to accept the price or not.

- Meanwhile, theaters in group 3 are offering different ranges of price using CBR inference where a case is composed of four input attributes (*current vacancy rate (%), remaining time before the show (minutes), box office ranking of the current show, approximate number of reachable customers*) and one output attribute (*ticket price*). Therefore, the price changes in accordance with the input attribute values theaters are currently facing. Theaters decide whether or not to accept the newly adjusted price offered by the buyers, using FCM as shown in Figure 3. If the FCM result is less than 1, then the theater rejects the buyer's price. Otherwise, the theater accepts the buyer's price, and the buyer's seat number and show time are specified accordingly. When conducting FCM analysis, input constructs should be transformed into an appropriate value considering the input conditions. Table 5 shows input constructs, its conditions, and transformed values.

Customers

The customer's utility function includes the following five factors: (1) D (distance from

Figure 3. Group 3 theaters'

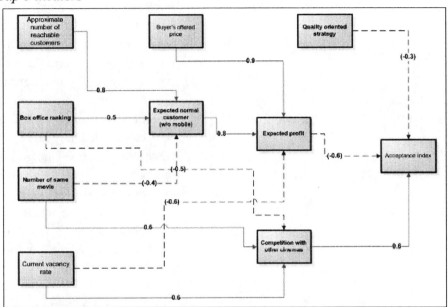

Table 5. Input constructs and conditions for theater's FCM

Input constructs	Condition	Transformed values
Approximate number of reachable customers	Many*	1
	Normal**	0
	Few***	-1
Box office ranking	Rank 1, 2	1
	Rank 3,4	5
	Rank 5,6	−0.5
	Rank 7 ~ 10	-1
Number of same movie	4 ~	1
	2 ~ 3	0.5
	1	-0.5
	0	-1
Current vacancy rate	40% ~	1
	30% ~ 40%	0.7
	20% ~ 30%	0.5
	10% ~ 20%	-0.5
	~ 10%	-1
Quality oriented strategy	Yes	1
	No	-1
Buyer's offered price	$7	1
	$5 ~ $7	0.7
	$4 ~ $5	0.5
	$3 ~ $4	0
	$2 ~ $3	-0.5
	$1 ~ $2	-0.7
	~ $1	-1

Many: # of reachable customers is greater than the total # of vacant seats of theaters
**Normal : # of reachable customers is equal to the total # of vacant seats of theaters*
***Few: # of reachable customers is less than the total # of vacant seats of theaters*

customer's current location to theater; for this experiment, it is adjusted between -18 and 18 on the Netlogo platform); (2) R (box office ranking of the movie); (3) G (movie genre); (4) P (newly adjusted ticket price that customers want); and (5) T (timeliness showing whether it is the exact time that customer wants). Using the five factors like this, ith customer's utility function is denoted as follows:

$$U_i = W_{D_i} \cdot D_i + W_{R_i} \cdot R_i + W_{G_i} \cdot G_i + W_{P_i} \cdot P_i + W_{T_i} \cdot T_i$$

- Table 6 addresses the various conditions and their converted values for the five utility factors.

MAM-NSS Simulation

Basics

The MAM-NSS simulation prototype was performed on the Netlogo platform which is a programmable multi-agent modeling environment for simulating natural and social phenomena, and particularly well-suited for modeling complex systems that develop over time. The target m-commerce problem described previously can be well represented by multi-agents composed of B-agents, S-agents, and interactions between

them for the effective negotiation. Therefore, MAM-NSS is capable of handling the problem very well.

As shown in Figure 4, MAM-NSS has six types of user interface components as follows:

1. *Behavior space* shows the customer's movement and the location of theaters. The human shape indicates a customer and the house shape represents a theater. Gray color means group 1, green color group 2, and pink color group 3. All customers are designed to move one unit of position over to the random direction at a time. The customers go out of simulation after they buy tickets.

2. *Graph* monitors the change of values such as number of customers, vacant seats, number of customers who have not bought a ticket, customers' average utilities, and theaters' average margins.

3. *Control button* prepares and prompts simulation.

4. *Slider* controls the initial conditions of simulation such as number of customers, number of theaters, and maximum number of vacant seats for each theater. Each theater has a number of vacant seats falling between 0 and the "vacant" number that is set by this slide.

Table 6. Customer's utility factors

Utility factor	Condition	Converted value
Distance from the theater (D)	In 20 minutes	50
	In 30 minutes	40
	In 40 minutes	30
	In 50 minutes	20
	More than 60 minutes	10
Box office ranking (R)	1,2	50
	3,4	40
	5,6	30
	7,8	20
	9,10	10
Movie genre (G)	Customer wanted	50
	Customer did not want	0
Ticket price (P)	For any new ticket price adjusted by customers	50 − (new ticket price / list price) * 50
Timeliness (T)	Exact time that customer wants	50
	Otherwise	0

5. *Monitor* shows the number such as rounds of simulation, remaining/elapsed before the next show starts, and simulation time.
6. *Command center* shows temporary data generated from the agent activities.

Results and Implications

Twenty six rounds of simulation were done on MAM-NSS with the initial conditions as follows. Total rounds of simulation is 35, number of theaters 12, number of customers ranging between 100 and 600 (evenly assigned to three groups), group 2 theaters sending discriminated price every 20 minutes, and theaters starting to offer discounted price 60 minutes before the show. The simulation results with MAM-NSS are summarized in Table 7 numerically, and in Figure 5 graphically. Under a 95% confidence level, statistical results in Table 8 reveal that in terms of average utilizes

and average profits, group 3 can yield the highest value compared with the other two groups. Its implication is as follows:

First, those users belonging to the passive mobile group can benefit from using the mobile devices. However, such mutual benefits increase much more when they use the negotiation support function provided by the proposed MAM-NSS.

Second, multi-agents are very convenient as well as effective for the m-commerce entities to handle them in their decision making process through the use of MAM-NSS. The reason is that agents are basically capable of autonomous operation once the entity's preference is predefined and stored into its memory. In the MAM-NSS environment, users do not have to bother themselves to interact with negotiation partners.

Third, both preference and conditions that users want their own agents to consider in the process of negotiations can be easily incorporated

Figure 4. MAM-NSS simulator

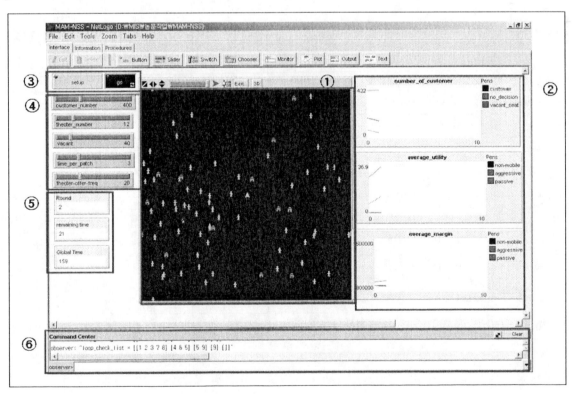

into agents. Since MAM-NSS is installed in the central server of the telecommunication company, it is very easy for users to use it.

Fourth, since m-commerce users are limited by narrow screen and specified functions of their mobile devices, and agents are capable of replacing users in the real negotiation process in an almost automatic manner, the use of negotiation support mechanism like MAM-NSS would greatly contribute to enhancing users' utilities and profits as well.

CONCLUDING REMARKS

To resolve the negotiation process between buyers and sellers in the context of m-commerce, we proposed a multi-agent mobile negotiation support system called MAM-NSS, in which all the buyers and sellers engaged in a m-commerce situation are represented by multi-agents embedded with each entity's preference and corresponding conditions. Its potentials were proved by the simulation experiments using the theaters' vacant seats

Table 7. Simulation result

Round	Average Utility			Average Profit		
	Non-mobile	Passive-mobile	Active-mobile	Non-mobile	Passive-mobile	Active-mobile
1	202	259	458	229	349	472
2	319	388	445	238	291	347
3	361	506	492	161	455	
4	299	365	482	134	295	506
5	311	441	410	66	442	468
6	289	406	587	189	208	496
7	285	431	559	423	253	353
8	309	432	497	178	439	377
9	259	428	477	166	390	355
10	402	439	568	197	220	477
11	326	355	436	332	204	480
12	332	405	498	215	266	410
13	327	485	506	190	356	502
14	252	398	412	152	451	403
15	331	463	533	220	386	473
16	445	393	448	243	400	371
17	305	438	508	101	321	470
18	265	354	534	274	318	362
19	245	370	480	99	395	484
20	387	464	565	52	475	498
21	381	449	488	204	429	401
22	339	444	551	176	347	493
23	245	411	512	143	328	501
24	371	444	562	178	330	471
25	334	431	597	159	344	481
26	225	413	547	103	236	502
27	338	416	492	192	283	498
28	225	392	457	124	288	451
29	319	434	487	169	397	449
30	213	322	401	255	305	457
31	321	491	492	190	409	401
32	257	381	474	220	348	287
33	350	469	509	157	411	483
34	247	426	492	147	431	388
35	349	350	406	264	337	441
Average	**215**	**290**	**347**	**131**	**243**	**308**

Figure 5. Utilities and margins by MAM-NSS simulation

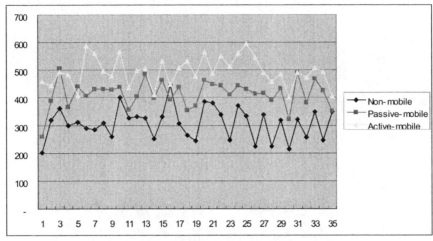

(a) Customers' average utilities

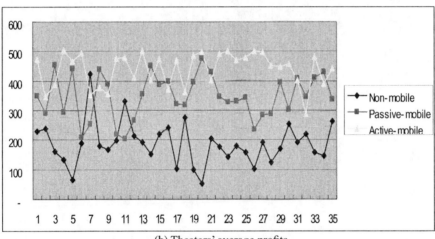

(b) Theaters' average profits

negotiation problems. Main contributions of the MAM-NSS are as follows.

First, CBR inference is incorporated to help S-agents decide an appropriate price for a vacant seat. Without using CBR, S-agents will have difficulty finding such appropriate price which is consistent with previous decision making results. Especially, such consistency in setting price for various situations is very important to customers who want be opportunistically exploited by sellers.

Second, FCM finds its great potential in the process of negotiation, due to its generalized inference capability in a presence of a number of inter-related factors. Without FCM, decision makers would feel very stressed to consider all the complicated causal relationships among the relevant factors and expect future inference results. In this chapter, FCM was used to help S-agents accept B-agent's price offer or not.

Table 8. Results of statistical test

Utility

(I) type	(J) type	Mean Difference (I-J)	Std. Error	Sig.	95% Confidence Interval	
					Lower Bound	Upper Bound
non-mobile	passive	-106.366	12.659	0.000	-137.812	-74.921
	active	-188.336	12.659	0.000	-219.782	-156.891
Passive-mobile	non-mobile	106.3665	12.659	0.000	74.921	137.812
	active	-81.9698	12.659	0.000	-113.416	-50.524
Active-mobile	non-mobile	188.3363	12.659	0.000	156.891	219.782
	passive	81.96982	12.659	0.000	50.524	113.416

Profit

(I) type	(J) type	Mean Difference (I-J)	Std. Error	Sig.	95% Confidence Interval	
					Lower Bound	Upper Bound
non-mobile	passive	-159.989	16.403	0.000	-200.735	-119.242
	active	-253.081	16.403	0.000	-293.827	-212.334
Passive-mobile	non-mobile	159.989	16.403	0.000	119.242	200.735
	active	-93.092	16.403	0.000	-133.838	-52.346
Active-mobile	non-mobile	253.081	16.403	0.000	212.334	293.827
	passive	93.092	16.403	0.000	52.346	133.838

Third, multi-agent schemes were found very meaningful in being used in the process of m-commerce negotiation. Such multi-agent approach has been proved useful and effective in a wide variety of problems in literature, but its potentials were not proved yet in m-commerce contexts. Therefore, this study adds meaning to literature in that sense.

This study has several positive implications for future m-commerce research. First, m-commerce is blooming as mobile devices are providing increased convenience in users' daily activities. However, there has been no important negotiation support system to leverage the potentials of m-commerce. In that meaning, this study will shed a positive light on using the generalized multi-agent framework for designing a mobile negotiation support system. Second, we proposed practical algorithms to be used in upgrading agents' capability in problem solving. Such algorithms would be used in the other business settings with minor adjustments.

But, this study has limitations in the point that (1) all data used in experiment are not real world data; (2) we do not suggest detailed mechanisms for extracting the buyer's preference; and (3) there is no comparing our negotiation algorithm with other algorithms. To compare the performance of negotiation algorithms is not simple. Measures need to be developed for comparing negotiation performance before comparing existing algorithms. Additionally, performance measures need to include not only quantitative factors, but also qualitative factors. These three limitations are left as future research topics.

ACKNOWLEDGMENT

This work was supported by grant No. B1210-0502-0037 from the University Fundamental Research Program of the Ministry of Information & Communication in the Republic of Korea, 2005.

REFERENCES

Applegate, L. M., Holsapple, C. W., Kalakota, R., Radermacher, F. J., & Whinston, A. B. (1996). Electronic commerce: Building blocks of new business opportunity. *Journal of Organizational Computing & Electronic Commerce, 6*(1), 1-10.

Bailey, M. N., & Lawrence, R. L. (2001). Do we have a new economy? *American Economic Review, 91*(2), 308-312.

Barbash, A. (2001). Mobile computing for ambulatory health care: Points of convergence. *Journal of Ambulatory Care Management, 24*(4), 54-60.

Barnes, S. J. (2002). The mobile commerce value chain: Analysis and future developments. *International Journal of Information Management, 22*(2), 91-108.

Bird, S. D. (1993). Towards a taxonomy of multi-agent systems. *International Journal of Man-Machine Studies, 39*, 689-704.

Bird, S. D., & Kasper, G. M. (1995). Problem formalization techniques for collaborative systems. *IEEE Transactions on Systems, Man, and Cybernetics, 25*(2), 231-242.

Bohoris, C., Pavlou, G., & Cruickshank, H. (2000). Using mobile agents for network performance management. In *Proceedings of Network Operations and Management Symposium "The Networked Planet: Management Beyond 2000"* (pp. 637-652).

Bonarini, A., & Trianni, V. (2001). Learning fuzzy classifier systems for multi-agent coordination. *Information Sciences, 136*(1-4), 215-239.

Cabri, G., Leonardi, L., & Zambonelli, F. (2001). Mobile agent coordination for distributed network management. *Journal of Network & Systems Management, 9*(4), 435-456.

Cabri, G., Leonardi, L., & Zambonelli, F. (2002). Engineering mobile agent applications via context-dependent coordination. *IEEE Transactions on Software Engineering, 28*(11), 1039-1055.

Chaib-Draa, B. (1995). Industrial applications of distributed artificial intelligence. *Communications of the ACM, 38*(11), 49-53.

Chaib-Draa, B., & Mandiau, R. (1992). Distributed artificial intelligence: An annotated bibliography. *SIGART Bulletin, 3*(3), 20-37.

Cooper, S., & Taleb-Bendiab, A. (1998). CON-CENSUS: Multi-party negotiation support for conflict resolution in concurrent engineering design. *Journal of Intelligent Manufacturing, 9*(2), 155-159.

Coursaris, C., & Hassanein, K. (2002). Understanding m-commerce. *Quarterly Journal of Electronic Commerce, 3*(3), 247-271.

Crowley, J. L., Coutaz, J., & Bérard, F. (2000). Perceptual user interfaces: Things that see. *Communications of the ACM, 43*(3), 54-64.

Edwards, W. K., Newman, M. W., Sedivy, J. Z., & Smith, T. F. (2004). Supporting serendipitous integration in mobile computing environments. *International Journal of Human-Computer Studies, 60*, 666-700.

Hogg, L. M. I., & Jennings, N. R. (2001). Socially intelligent reasoning for autonomous agents. *IEEE Transactions on Systems, Man, & Cybernetics Part A: Systems & Humans, 31*(5), 381-393.

Hu, J., & Weliman, M. P. (2001). Learning about other agents in a dynamic multiagent system. *Cognitive Systems Research, 2*(1), 67-79.

Jung, J. J., & Jo, G. S. (2000). Brokerage between buyer and seller agents using constraint satisfaction problem models. *Decision Support Systems, 28*(4), 293-304.

Kwon, O. B., & Lee, K. C. (2002). MACE: Multi-agents coordination engine to resolve conflicts among functional units in an enterprise. *Expert Systems with Applications, 23*(1), 9-21.

Lai, H., & Yang, T. C. (1988). A system architecture of intelligent-guided browsing on the Web. In *Proceedings of Thirty-First Hawaii International Conference on System Sciences* (pp. 423-432).

Lai, H., & Yang, T. C. (2000). A system architecture for intelligent browsing on the Web. *Decision Support Systems, 28*(3), 219-239.

Lee, W. J., & Lee, K. C. (1999). PROMISE: A distributed DSS approach to coordinating production and marketing decisions. *Computers and Operations Research, 26*(9), 901-920.

Lee, W. P., & Yang, T. H. (2003). Personalizing information appliances: A multi-agent framework for TV programme recommendations. *Expert Systems with Applications, 25*(3), 331-341.

Lottaz, C., Smith, I. F. C., Robert-Nicoud, Y., & Faltings, B. V. (2000). Constraint-based support for negotiation in collaborative design. *Artificial Intelligence in Engineering, 14*(3), 261-280.

Lucas, J. H. C. (2001). Information technology and physical space. *Communications of the ACM, 44*(11), 89-96.

Luo, X., Zhang, C., & Leung, H. F. (2001). Information sharing between heterogeneous uncertain reasoning models in a multi-agent environment: A case study. *International Journal of Approximate Reasoning, 27*(1), 27-59.

Mandry, T., Pernul, G., & Rohm, A. W. (2000-2001). Mobile agents in electronic markets: Opportunities, risks, agent protection. *International Journal of Electronic Commerce, 5*(2), 47-60.

McMullen, P. R. (2001). An ant colony optimization approach to addressing a JIT sequencing problem with multiple objectives. *Artificial Intelligence in Engineering, 15*(3), 309-317.

Miah, T., & Bashir, O. (1997). Mobile workers: Access to information on the move. *Computing and Control Engineering, 8*, 215-223.

Ngai, E. W. T., & Gunasekaran, A. (2005). A review for mobile commerce research and applications. *Decision Support Systems*. Retrieved August 20, 2007, from http://www.sciencedirect.com

Omicini, A., & Zambonelli, F. (1988). Co-ordination of mobile information agents in TuCSoN. *Internet Research: Electronic Networking Applications and Policy, 8*(5), 400-413.

Paiva, A., Machado, I., & Prada, R. (2001). The child behind the character. *IEEE Transactions on Systems, Man, and Cybernetics - Part A: Systems and Humans, 31*(5), 361-368.

Parusha, A., & Yuviler-Gavishb, N. (2004). Web navigation structures in cellular phones: The depth/beadth trade-off issue. *International Journal of Human-Computer Studies, 60*, 753-770.

Persson, P., Laaksolahti, J., & Lonnqvist, P. (2001). Understanding socially intelligent agents: A multilayered phenomenon. *IEEE Transactions on Systems, Man, and Cybernetics - Part A: Systems and Humans, 31*(5), 349-360.

Pham, T., Schneider, G., & Goose, S. (2000). A situated computing framework for mobile and ubiquitous multimedia access using small screen and composite devices. In *Proceedings of the Eighth ACM International Conference on Multimedia* (pp. 323-331).

Porn, L. M., & Patrick, K. (2002). Mobile computing acceptance grows as applications evolve. *Healthcare Financial Management, 56*(1), 66-70.

Rodgera, J. A., & Pendharkarb, P. C. (2004). A field study of the impact of gender and user's technical experience on the performance of voice-activated medical tracking application. *International Journal of Human-Computer Studies, 60*, 529-544.

Schaeffer, J., Plaat, A., & Junghanns, A. (2001). Unifying single-agent and two-player search. *Information Sciences, 135*(3-4), 151-175.

Schilit, W. N. (1995). *System architecture for context aware mobile computing*. Unpublished doctoral thesis, Columbia University.

Schilit, B. N., Adams, N. I., & Want, R. (1994). Context-aware computing applications. In *Proceedings of the First International Workshop on Mobile Computing Systems and Applications* (pp. 85-90).

Seager, A. (2003). M-commerce: An integrated approach. *Telecommunications International, 37*(2), 36.

Senn, J. A. (2000). The emergence of m-commerce. *Computer, 33*(12), 148-150.

Sikora, R., & Shaw, M. J. (1998). A multi-agent framework for the coordination and integration of information systems. *Management Science, 44*(11), 65-78.

Sillince, J. A. A. (1998). Extending electronic coordination mechanisms using argumentation: The case of task allocation. *Knowledge-Based Systems, 10*(6), 325-336.

Sillince, J. A. A., & Saeddi, M. H. (1999). Computer-mediated communication: Problems and potentials of argumentation support systems. *Decision Support Systems, 26*(4), 287-306.

Slack, F., & Rowley, J. (2002). Online kiosks: The alternative to mobile technologies for mobile users. *Internet Research: Electronic Networking Applications and Policy, 12*(3), 248-257.

Strader, T. J., Lim, F. R., & Shaw, M. J. (1998). Information infrastructure for electronic virtual organization management. *Decision Support Systems, 23*(1), 75-94.

Tung, B., & Lee, J. (1999). An agent-based framework for building decision support systems. *Decision Support Systems, 25*(3), 225-237.

Turisco, F. (2000). Mobile computing is next technology frontier for healthcare providers. *Health Care Financial Management, 54*(11), 78-80.

Ulieru, M., Norrie, D., Kremer, R., & Shen, W. (2000). A multi-resolution collaborative architecture for Web-centric global manufacturing. *Information Sciences, 127*(1-2), 3-21.

Varshney, U. (1999). Networking support for mobile computing. *Communications of AIS, 1*(1), 1-30.

Wang, F. H., & Shao, II. M. (2004). Effective personalized recommendation based on time-framed navigation clustering and association mining. *Expert Systems with Applications, 27*(3), 365-377.

Wang, Y., Tan, K. L., & Ren, J. (2002). A study of building Internet marketplaces on the basis of mobile agents for parallel processing. *World Wide Web, 5*(1), 41-66.

Want, R., Hopper, A., Falcao, V., & Gibbons, J. (1992). The active badge location system. *ACM Transactions on Information Systems, 10*(1), 91-102.

Want, R., Schilit, B., Adams, N., Gold, R., Petersen, K., Ellis, J., et al. (1995). *The PARCTAB ubiquitous computing experiment* (Tech. Rep. No. CSL-95-1). Xerox Palo Alto Research Center.

Wooldridge, M. (1997). Agent based software engineering. *IEEE Proceedings of Software Engineering, 144*(1), 26-37.

Wooldridge, M., & Jennings, N. (1995). Intelligent agents: Theory and practice. *The Knowledge Engineering Review, 10*(2), 115-152.

Wu, D. J. (2001). Software agents for knowledge management: Coordination in multi-agent supply chains and auctions. *Expert Systems with Applications, 20*(1), 51-64.

Wu, G., Yuan, H., Tseng, S. S., & Fuyan, Z. (1999). A knowledge sharing and collaboration system model based on Internet. In *Proceedings of IEEE International Conference on Systems, Man, and Cybernetics* (pp.148-152).

ENDNOTE

[1]   http://ccl.northwestern.edu/netlogo/

Chapter XIII
A Study of Malicious Agents in Open Multi–Agent Systems:
The Economic Perspective and Simulation

Pinata Winoto
Hong Kong Baptist University, Hong Kong

Tiffany Y. Tang
Hong Kong Polytechnic University, Hong Kong

ABSTRACT

This chapter focuses on the issue of malicious agents in open multi-agent systems (MAS) with discussions in relation to existing crime study, that is, supply-demand analysis and deterrence theory. Our work highlights the importance of mechanisms to make intervention to the MAS, in an attempt to deter malicious agents from maximizing their utilities through illegal actions. Indeed, in market-oriented MAS, human interventions sometimes would be necessary, given the condition that these interventions would not destroy the ecological stability of the agent society. However, an automatic intervention mechanism seems to be the winning card in the end, among them, the reputation mechanism, agent coalition formation, and so forth. It is our hope that our work can shed light on these issues as well as their deployment in MAS.

INTRODUCTION

In the context of human criminal studies, crimes can broadly be grouped as economically driven crimes and non-economically driven crimes. Economically driven crimes (or *economic crime*, for short) are primarily driven by financial gains and presumably follow the utilitarian concept; that is, it is controlled by manipulating its pains (punishments) and gains (rewards). Generally, if there are victims left by a crime, it is called a predatory crime. In human society, crime is a complex phenomenon. In the agent society, crime is less complex due to a specific agent's intention/purpose, for instance, violating committed contract for bidding agent, sending misleading information for advertising agents, entering restricted area for search agents, and so forth.

The context of this chapter is on the study of malicious agent society, which is characterized by economic and predatory crimes. However, the model used is based on the economic model of 'human' crime, which is still a controversial issue. For example, it is commonly assumed in the model that all criminals follow rational choice behavior, while in the real world many 'real' criminals are addicted to alcohol/drugs. Yet, rational choice model may fit better in agent society, since all agents are preprogrammed to make rational decisions to maximize rewards. Therefore, one of the potential applications of this study is to seek optimal multi-agent system (MAS) policy, especially when heterogeneous agents may behave maliciously.

Controlling malicious agents in MAS is not new. Through punishment and probability of arrest/conviction (Winoto, 2003a, b), we make an effective way of governing these malicious agents. Specifically, agents with common goals are deployed in a malevolent game: compete with other agents by means of malevolent actions. In this chapter, we will discuss several general models that may be used in open MAS. Open MAS are characterized by heterogeneous agents who may use various strategies (including malevolent behaviors), enter and exit the system freely, and compete to maximize individual utility. An example of open MAS is an open electronic marketplace, where agents (including human agents) make transactions for goods or services using specific negotiation protocols.

Motivational Examples

The following examples illustrate the potential application of our work.

- **Scenario 1.** Suppose there is an electronic marketplace where agents may sell/buy a used laptop using a bargaining protocol. Assume that all negotiations are initiated by the sellers. In order to facilitate the bargaining, the authority (electronic market) allows an agent to negotiate with multiple agents simultaneously. However, a nasty buyer may replicate her buying agent into several different agents who negotiate with the same seller. As the seller rejects one of them, the buyer may replicate an agent to join the negotiation again. Since the seller cannot detect the 'real' identity of the buying agents, the seller will be exploited until the buyer makes his final decision. More seriously, if the seller uses a learning algorithm in its decision, such as predicting the market price, then the buying agent will manipulate the seller's belief. What puts a seller at a disadvantage is that it reveals the same information (e.g., item being sold); thus replicating itself into multiple sellers cannot help much. To deal with this issue, the authority may prohibit any buyer/seller from using more than one agent in the same negotiation, for instance, by checking the identity of all agents' owners or by asking for some deposit for each agent registered in the market; however, the former will burden the server while the latter will burden the users. Alternatively, the authority may check the identity of agents randomly and impose punishment for an offense.

- **Scenario 2.** Suppose 1,000 deep-sea autonomous voyagers from a joint mission of various enterprises are sent to explore the bottom of the Atlantic Ocean. In order to facilitate their mission, 10 mobile power stations are sent to accompany them such that they can recharge their batteries without going to the surface. The service is on a first-come-first-serve and limited-charging-time basis. However, some voyagers may collude to occupy a slot by blocking others from reaching it, especially during rush time. To deal with it, the power stations may track the collusion by asking the voyagers to register their identity before charging. However, not all collusion can be tracked down with certainty, which makes the system vulnerable. In order to deter it from happening, a punishment will be imposed on colluded voyagers.

- **Scenario 3.** Suppose in a Web-based auction site, after each transaction, a buying agent can provide a numerical rating to reflect the trustworthiness of the selling agent it dealt with, and vice versa. The overall trustworthiness of each agent will then be measured by the average feedback scores it receives. This type of the reputation system is very popular in the e-business world, such as on eBay, Yahoo! Auction, and so forth, and works effectively in fostering trust among e-strangers. However, an agent can go dirty when it finds out that it had become 'untrustworthy' by replicating itself, that is, create another virtual identity; and the fresh new comer is clean again. This exploitation can greatly undermine the reputation system, and agents involved should be punished.

Many other types of offenses in MAS can be found easily. Hence, it is vital to find effective policies before the deployment of open MAS. The objectives of this chapter are to:

1. Bring up the issue of malicious agents and discuss them in relation to deterrence theory.
2. Compare some policies through simulation, which is the first step towards the understanding of the governance strategy by means of deterrence theory.
3. Propose some governance schemes.

Organization of This Chapter

In the next section, we will provide a detailed discussion on the two main issues: the economic theory of crime and controlling the malicious agents with macropolicy. We will then discuss our simulation studies in two ways:

- An agent-based simulation on the market for offenses where we will show through deterrence theory the behaviors of malicious agents and their tendency to commit crimes.
- Through some modified macropolicies, we will show how to deter malicious agents from committing a crime in open-agent systems.

We will then discuss another way of preventing other benevolent agents from being cheated/attacked by malicious agents through a free-market-like policy: reputation scheme. A comparison will be made to compare the pros and cons of applying these strategies in open MAS.

MALICIOUS AGENTS AND DETERRENCE THEORY

We will first describe the theoretical framework of crime from an economists' perspective.

The Economic Theory of Crime: Background Information

The first study of crime through modern economic analysis is the seminal work by Gary Becker (1968). Today, many works still follow genuine Beckerian theory or mix it with other methods such as game theory and information processing (e.g., Marjit & Shi, 1998; Sah, 1991). However, all of them aim to minimize the social cost of crime based on economic principles.

Basic Theoretical Framework

Some basic theoretical frameworks in criminal studies are:

- **Microlevel:** Agent decision to participate in an illegitimate activity (crime). At this level, we are only interested in the decisions made by agents in committing crimes. In economics, these decisions represent human cognitive processes, while in our discussion they represent decisions made by autonomous software and their designers. Basically, the decisions depend on:

    1. *The expected gain from that illegitimate activity.* There are three major factors affecting the expected gain:
        - Net return from an illegitimate activity, U_1. This value equals the return from the illegitimate activity minus direct costs incurred from that activity. For instance, suppose it costs $10 for an agent to intrude into a system (e.g., cost incurred from hacking the system), and the data inside the system are worth $25. Then the agent's net return from intruding the system is $25 − $10 = $15.
        - Perceived probability of conviction, p_c. This value depends on many aspects such as the technology used by the agent and other agents/systems. Following our previous example, if an agent has an advanced technology in evading common intrusion detectors, then the perceived probability of conviction from intruding a system is close to zero. However, this value may increase if the intruded system has an advanced intrusion detector.
        - Net return if convicted, U_2, which is a negative value. Depending on when the agent consumes the net return from the crime, this value may equal U_1 minus punishment or just the punishment. Following our previous example, suppose the penalty for an intrusion is $1000 and an intruder agent is caught several days after intruding into a system. Then, its net return from conviction is −($1000 − $15) = −$985. However, if the agent is caught during the intrusion, the net return of conviction is −$1000.

    In economics, a common assumption is that a human offender behaves as if to maximize his expected utility (e.g., Becker, 1968; Ehrlich, 1996; Fender, 1999; Sah, 1991). Formally, the combination of those factors could be represented by the von Neumann-Morgenstern expected utility:

    $$\text{EU}_{\text{crime}} = (1 - p_c)\, U_1 + p_c U_2$$

    2. *Certain gain(s) from legitimate activities, U_{legal}.* In our context, this gain represents what an agent made from doing other legitimate actions. For example, rather than allocating computational power to intrude into a system, the agent in our previous example can search for other systems which provide similar data for free. Or, it may perform other tasks which generate revenue and legitimately buy the data for $25.

    3. *Taste (or distaste) and preference for crimes, U_{taste}.* In human crime, this value represents 'a combination of moral values, proclivity for violence, and preference for risk' (Ehrlich, 1996). In MAS, this value may be assigned by agent designers, which may fall within interval $(-\infty, \infty)$. A completely malevolent agent can be seen as that

with $U_{taste} = -\infty$, because it always commits crime (very low moral value). And a completely benevolent agent can be seen as that with $U_{taste} = +\infty$, because it never commits crime.

Generally, a person/agent will commit crime if $EU_{crime} > U_{legal} + U_{taste}$. The righthand side of the formula constitutes the minimum value (threshold) for an agent to enter the illegitimate market. If the value is big, then there might a group of agents who never commit crime regardless of the penalty or conviction rate. For instance, assume that the crime is riskless (penalty = 0 or conviction rate = 0) and the highest EU_{crime} = Constant > U_{legal}. If there is a fraction of agents whose $U_{taste} > EU_{crime} - U_{legal}$, then they will not commit crime regardless of how light the penalty or how low the conviction rate is. Therefore, given any value of EU_{crime} < riskless EU_{crime}, there is a fraction of agents who will not commit crime when their $U_{taste} > EU_{crime} - U_{legal}$. The number of them will increase when EU_{crime} decreases.

- **Macrolevel (Ehrlich, 1996):** The market for offenses. At this level, the criminal activities within a society are represented as a market, where we have both supply and demand for crimes.

1. The **supply side** is determined by the distribution of 'taste of crime,' U_{taste}, or 'legal income,' U_{legal}, in the population. As described before, different 'taste of crime' represents different thresholds for those agents (or people in Ehrlich model) to commit crime. Therefore, higher expected return from crime causes higher participations in crime (the upward sloping of supply curve, see Figure 1).

2. The **demand side** is determined by the tolerance of crime that is inversely related to the demand for self-protection and public enforcement. Higher self-protection induces lower expected return from crime, therefore reduces the crime rate. And a higher level of public enforcement causes lower crime rates too (downward sloping of demand curve, see Figure 1).

Figure 1. The market for offenses (from Ehrlich, 1996)

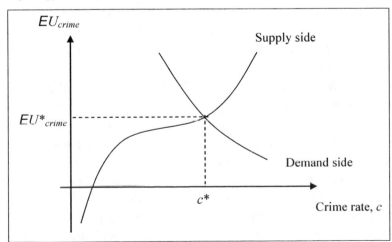

- **Other innovations made on:** Many innovations of the classical economic model of crime have been made especially during the past decades. Among them are:

  1. **Dynamic model:** Some recent work has begun to explore dynamic deterrence models (e.g., Davis, 1988; Leung, 1995). The reason is that static models cannot accommodate many phenomena, including recidivism, discount factor of future punishment, accumulation of criminal skills (learning), and so forth. Some modifications of the classical model include:

     ° Using a multiple-period rather than one-period framework. In the latter model, each agent has only one opportunity to choose whether or not to commit crime; while in the former model, each agent has many opportunities to choose from. This model can accommodate the study of recidivism (Leung, 1995).

     ° Adding the discount factor for future consumption and future punishment (Leung, 1995). The underlying assumption is that some agents/people are not concerned with what will happen in the future, while some are. Hence, future consumption/punishment may affect the agent current decision.

  2. **Information process and social Interactions:** Sah (1991) added the Bayesian inference techniques into his model. The inference process is used to model how a potential offender predicts the probability of conviction from the information given by others (cohorts, etc.). Under this model, Sah is able to show how different crime rates might occur under the same economic fundamentals. Generally, a potential offender is a social agent, equipped with the capabilities to recognize its environment, and therefore produces the dynamics of its society.

  3. **Experimental economics:** One of the studies conducted focuses on nonpredatory crime, that is, bribery (Abbink, Irlenbusch, & Renner, 1999). Another equation-based simulation was conducted by İmrohoroğlu, Merlo, and Rupert (1996). These approaches use empirical data to justify the existing model. Our previous studies (Winoto, 2003a, b) and that reported in this chapter also use this approach.

     While much literature in economics has shown the existence of (theoretical) multiple equilibria in the crime market (e.g., Fender, 1999; İmrohoroğlu et al., 1996; Sah, 1991), no experiments have been done to justify it. In this chapter, we will discuss the existence of multiple equilibria based on a model proposed by Fender (1999).

Some Notations and Terminology

Some terms used in the multi-agent simulations discussed in the next section include

- **Law enforcement** consists of agents who act as police officers to fight crimes. Their actions are restricted by resource constraints.

- **The level of law enforcement** represents the effectiveness of law enforcement to fight crimes. A higher level of law enforcement means higher spending to fight crimes and a higher chance to catch criminals. Up to now, there has been no consensus in the literature on what is the functional form to describe the apprehension rate (İmrohoroğlu et al., 1996; Pyle, 1983). Fender (1999) uses an increasing and concave function (diminishing marginal productivity of law enforcement) to represent the probability of punishment, that is, $p = min[1, G(E)/C]$, where p denotes the probability of punished, E denotes the total expenditure in law enforcement, and G is the number of criminals punished.

- **Market for offenses (crime market)** is an abstract market where the demand and supply of crimes are coordinated. It is adopted directly from (Ehrlich, 1996).
- **Honest agent** is an agent who will not commit crime under any conditions (programmed not to harm other).
- **Potential offender** is an agent who will commit crime if he thinks it is worthy to do so.

Fender's Equilibrium Theory (1999)

Through conventional mathematical derivations, Fender (1999) has shown that in the long run, multiple equilbria of the market for offenses may exist (either stable or unstable equilibria). His model is solely based on Beckerian. According to his analysis, the underlying intuition for the existence of the multiple equilbria is:

1. If the law enforcement is constant and the crime rate is high and increases, then the conviction rate decreases (due to the diminishing marginal productivity of law enforcement). Thus, an illegitimate activity becomes more attractive and the number of criminals increases.
2. If the law enforcement is constant and the crime rate is very low, then any marginal crime could be detected easily.

The basic assumptions in Fender (1999) are:

1. The economy consists of a population of n heterogeneous agents.
2. $(n-m)$ agents never commit crime (honest citizens).

3. The remainder, m, are potential offenders.
4. Honest citizens always work and receive w_h.
5. Potential offenders receive w_p from legitimate work; w_p is generated from a uniform distribution $[w_h - \alpha, w_h + \alpha]$.
6. If a potential offender succeeds in his crime, his payoff is u_s.
7. But if he fails, he will be punished so that he will obtain u_p, which equals u_s – penalty.
8. Denote the number of criminals as C, then the number of noncriminals (workers) as $n-C$, and the number of crime per non-criminal as $C/(n-C)$.
9. Only law-abiding agents are potential victims. If the average loss from crime is l, then expected loss of each law-abiding agent is $lC/(n-C)$.
10. The government collects tax for enforcement expenditure E from all workers; the tax (in \$) is equally distributed to all agents no matter how much they earn from work, which equals $E/(n-C)$.
11. Every potential offender follows the von Neumann–Morgenstern Expected Utility, so that he will commit crime if,

$$pu_2 + (1-p)u_1 - w_h + (lC+E)/(n-C) > 0 \tag{1}$$

From those assumptions, we can complete the mathematical analysis as follows (see Fender, 1999, for details):

1. There is a critical value w^* that satisfies $pu_2 + (1-p)u_1 - w^* + (lC+E)/(n-C)=0$; which means that the agent is indifferent between committing crime and work. Those agents whose $w_h > w^*$ would not commit crime, but those whose $w_h < w^*$ would.
2. Under the uniform distribution, the proportion of agents whose $w_h < w^*$ is $[w^*-w_h+\alpha]/2\alpha$

3. Therefore, the number of criminals is $C = [w^* - w_h + \alpha]m/2\alpha$ or $w^* = w_h + \alpha + 2\alpha\, C/m$. Plug this equation into equation (1) and we obtain:

$$p = \left[\frac{1}{u_s - u_f}\right]\left[u_s - 2\alpha\frac{C}{m} - w_h + \alpha + \frac{lC + E}{n - C}\right]$$

(2)

Equation (2) represents the relationships between punishment rate p and the number of criminals C (EC locus). Moreover, p is bounded to [0, 1].

Another relationship between p and C (PP locus) can be derived from the relationship between the expenditure of law enforcement and the punishment rate as follows:

$$p = \min\left[1, \frac{G(E)}{C}\right]$$

(3)

Fender (1999) does not provide any numerical values in his paper. Figure 2 is based on parameters: $n = 2000$, $m = 1000$, $E = \$100000$, $u_s = \$2000$, $u_f = \$500$, $\alpha = \$1000$, $w_h = \$2000$, and $l = \$2000$.

Using equations (2) and (3), Fender (1999) shows the locus and equilibria, and he also makes some conjectures about it. Basically, he believes

that the stable equilibria shown in Figure 2 are points A (0% crime) and D (100% crime). He believes that point B is an unstable equilibrium. Our simulation will test his conjectures and show that the point-wise equilibrium may be violated when the society consists of only a small number of overlapping generations of agents.

Applying Deterrence Theory in MAS

As state before, when we talk about malicious agent behavior, we only focus on economic-driven behavior following a utilitarian concept: that is, it is controllable by manipulating its pains (punishments) and gains (rewards). Some malicious agents may be generated for destructive purposes without making any profit, for example, spreading a computer virus. Those agents are not controllable through punishment. In this chapter, we will not discuss such behavior except for economic-driven offenses. The most common assumptions in the deterrence theory are *normative rationality* in terms of the von Neumann–Morgenstern expected utility (EU) theory, risk neutrality of offender, and utilitarianism of authorities (law enforcement) who are able to minimize social cost subject to budget constraints (Becker, 1968; Fender, 1999; Hohl, 1998).

Figure 2. Theoretical multiple equilibria in EC and PP locus

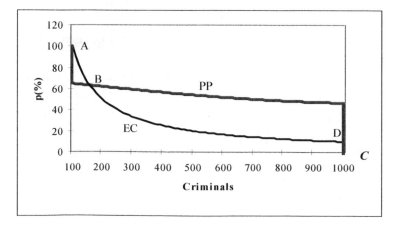

However, in human society, those assumptions are weakly supported by empirical studies. For instance, many empirical studies show the violation of the EU principle in some crimes (Carroll, 1978; Carroll & Weaver, 1986). And most of the people do not show risk neutrality. Many crimes are committed under the influence of alcohol/drugs. Furthermore, despite the fact that capital punishment could deter murder effectively (Ehrlich, 1975, 1996), many countries prohibit it for reasons other than economic concerns.

Nevertheless, in MAS, those assumptions may hold due to the following reasons:

1. The decisions made by artificial agents may follow a preprogrammed computation that is less likely to deviate from its economic interest.
2. Designers are more rational than law enforcement authority in human society, because they are more profit-oriented and most of the offenses made by agents are less serious than those made by criminals in human society.

Back to the Ehrlich model (c.f. Figure 1), the intersection point of supply and demand curves is the equilibrium point of the market. A shift of the supply curve to the right, for example, due to the decrease of EU_{legal}, will cause higher offense rate and lower EU_{crime}. Conversely, a shift of the demand curve to the left, for example, due to the expansion of authority's effort against offenses, will reduce both the offense rate and EU_{crime}.

However, the demand curve may not be a steep downward curve. If the authority and victims only put a little effort into fighting an increasing number of offenses, which may happen due to a simple MAS design, then the demand curve will be a flat downward curve (or a horizontal line). For instance, the authority in Scenario 1 may use a simple policy such as randomly checking a fraction of buying agents, whose number is proportional to the number of offenses. In this case, any shift of the supply curve will not change EU_{crime} but the offense rate. If the policy gets worse, for example, the accuracy of detecting an offense decreases as the number of workload/offenses increases, we may have an upward demand curve (e.g., D^{up} in

Figure 3. The effect of upward demand curve

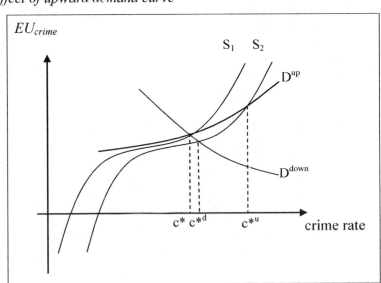

Figure 3). Thus, a shift of the supply curve to the right (S_1 to S_2) may bring the system into chaos (much higher offense rate, c^{*u}). Conversely, if the supply curve shifts to the left, for example, due to the increasing of EU_{legal}, then the offense rate falls faster in D^{up}. Hence, it is important for MAS designers to prevent offenses with appropriate tools. In Scenario 1, it can be accomplished by introducing a reputation mechanism, for example, providing incremental bonus (incentive) for buyers who have never been detected in any offenses.

Recall that the contour of a supply curve depends on EU_{crime}, EU_{legal}, and U_{taste}. If some malicious agents do not follow the EU principle, then the supply curve may show a kinked curve (S1 in Figure 4). The steep supply curve S1 represents a large number of 'irrational' agents who tend to offend regardless of the value of EU_{crime}. Conversely, the flat supply curve S2 represents a large number of highly 'rational' agents, whose $U_{taste} = 0$. When EU_{crime} is higher than the broken point in S2, the offense rate becomes sensitive to EU_{crime}. A shift of demand curve to the left (D to D') has little effect on the offense rate in S1, but a huge effect on the offense rate in S2 (Figure 4). Consequently, raising EU_{legal} will not help much in S1, as most agents are 'irrational.'

Thus, preventing them from entering the system is more effective.

Despite the appropriateness of applying deterrence theory to control malicious agents in MAS, difficulties still remain in implementing it, among them:

1.   Difficulty in convicting an offense. In Scenario 1, it may happen that a competitor of a buyer intentionally replicates the buyer's agent such that the authority falsely disqualifies the buyer (framing).

2.   Difficulty in deciding the severity of punishment. Killing an artificial agent may not really attain the deterrence goal, because they can be replicated with different identities. Imposing fines or seizing money deposits might be more effective.

3.   Difficulty in punishing an offender. Since the users of open MAS may spread in various physical locations, suing the user may not be that simple.

4.   Low economic benefit. The expenditure against offenses may be higher than the economic benefit from it.

5.   Agents/users might have different interpretations on which actions belong to offenses.

Figure 4. The effect of various supply curves

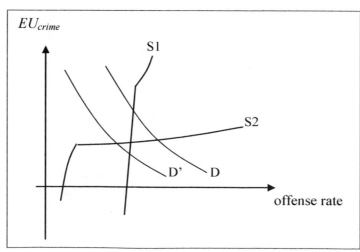

These are the obstacles in the implementation of open MAS, and more seriously are the incentives of the deployment of malicious agents.

Controlling Malicious Agents Through Preventive Schemes

Macropolicy Governing Agents in Open MAS: Previous Study

Although as shown in previous sections, it is vital to study the macropolicy in open MAS, very few researchers have addressed it. Most of the related work can be divided into three main categories:

- Preventive mechanism. Studying either detection mechanisms to avoid malicious host/agents or enhanced security system to prevent intrusion by malicious host/agents. For instance, security safe ambient (Bugliesi & Castagna, 2001), partially-typed semantics (Hennesy & Riely, 1999), and execution tracing (Vigna, 1998) can be used to check a malicious host/agent. Sandbox security (e.g., cryptographic and authorizing technique) and blackbox security (Hohl, 1998) can be used to prevent the attack by a malicious host/agent.
- Curative mechanism. Dealing with after-attack situation or fault-tolerance. For instance, replicating agents to replace a 'killed' agent (Fodoruk & Deters, 2002; Schneider, 1997).
- Agent data security. Recently proposed in Bonatti, Kraus, and Subrahmanian (2003) is a safe data and information management mechanism to prevent agents from releasing information, especially sensitive and private information to third parties. They called this type of agents 'secure agents.'

Our work is the complement of all the above work, which mainly considers the macropolicy in deterring malicious agents. For instance, intrusion detection and sandbox security can be considered the investments made by MAS authority. Replicating mechanism, blackbox security, and data security can be considered the level of self-protection by benevolent agents.

Our Work in Controlling Malicious Agents: Preventive Schemes

Controlling malicious behavior can be accomplished in various ways. For example, a seller may reject all buying agents who show the same behavior, record their identities, and so forth. The authority may impose a severe punishment or increase the security level of the system, for example, detecting all agents' identities, checking all past transactions, testing their behavior, and so forth. In order to control malicious agents, the authority may adopt one of the following preventive schemes: *active-users scheme, levy scheme, third-party scheme, or hybrid scheme*, which follows some macrodeterrence theories.

Active-Users Scheme (AU)

In an active-users scheme, the effort against offenses is left to all users (self-protection). The authority may help users by maintaining past records of offenses such as their *modus operandi*, prevention/detection methods, and so forth. Depending on the nature of the offense, the authority may or may not punish the offender. The characteristics of this scheme are:

1. Agents adaptively adjust their security level based on owners' risk preference, capability, budget, and so forth. Thus, there is a self-adjustment process in the MAS (a downward demand curve in Figure 3).
2. Increasing agents' productivity will only shift the demand curve.
3. Some risks for potential offenders, because they may not know the security level of each agent.

This scheme will be chosen if:

1. The designer cannot predict/prevent most offenses or the prevention technology is beyond the capability of the system designer.
2. The centralized prevention is more costly, or less effective, than distributing or delegating it to users.
3. Usage of the system is not sensitive to the offense occurred in the system (e.g., in Scenario 2).

Levy Scheme (LV)

In a levy scheme, the effort concentrates on the MAS authority. But the expense will be distributed evenly to all users in terms of levy, collected on a per usage or per agent basis. For instance, the authority collects a certain fee from each transaction in Scenario 1 in order to maintain the extra cost to detect offenses. The characteristics of this scheme include:

1. The authority adjusts the security level based on its goal and capability.
2. The authority controls the punishment level.
3. It may generate lower risks for potential offenders because the security level may not change frequently.
4. The levy can be used for two purposes, that is, detecting an offense and increasing the expected gain from not committing an offense (shift of demand curve and supply curve, respectively).

This scheme will be chosen when:

1. The designer knows most types of offenses and he has an effective prevention technology.

2. The centralized prevention is either cheaper, more effective, or both than distributing it to all users.
3. The usage of the system is very sensitive to the offenses occurred in the system.

Third-Party Scheme (TP)

In a third-party scheme, the effort against offenses is left to third parties, such as law-enforcement authority or relevant department/company. Under this scheme, the designers may not be responsible for preventing the offense, but they will pay a certain fee to third parties. The characteristics of this scheme include:

1. It produces higher risks for potential offenders, since the third party may have the best technology.
2. The security level of the MAS depends on exogenous factors.
3. Less burdens for both agents and system designers.

Hybrid Scheme (HB)

In a hybrid scheme, the effort against offenses is distributed to more than one party. The characteristics of this scheme are the mixture of the characteristics of all other schemes, which lead to a higher risks for potential offenders because they may not know the exact information of the security level. This scheme will be chosen if:

1. The designer cannot predict part of the possible offenses in the system, thus cannot prevent all the offenses.
2. The centralized prevention is more costly, or less effective, than distributing or delegating it to users at a certain level (diminishing marginal productivity of levy scheme).
3. There are third parties with a lower marginal prevention cost at some points.

Scheme Selection

Several factors must be considered in selecting an optimal scheme. For instance, what is the most effective scheme from users' perspective? What is the most profitable scheme from an authority's perspective? What is the most flexible one to anticipate future offenses? Centralized scheme (levy and third-party) may be more effective when the flow of users in/out of the open MAS is very fast, because most users have little experiences in facing an offense. However, if the benevolent-agents are smart and aware of the offenses, the active-user scheme (AU) may be better. A novice-benevolent-risk-averse agent will prefer levy scheme more than active-users scheme. The reason is that a novice-risk-averse agent may not know the information of the offense rate and have no idea of how to prevent any offense. Since it is a risk-averse agent, it will prefer to pay a levy and let the system handle it. On the contrary, an expert-benevolent-risk-averse agent or benevolent-risk-seeking agent may choose AU or LV. In addition, depending on the level of its trust toward system capability, the orientation of a benevolent-agent may change. If an agent believes that the system can better prevent the offense than it does, then the agent may prefer LV over AU. Also, based upon the agent's trust toward the third parties, it may also prefer TP over AU or LV.

From the authority's point of view, the best scheme must be the one that can generate the highest expected profit. Depending on the nature of the system and agents' characteristics, the sensitivity of the authority's profit with respect to the offense rate can be positive, zero, or negative. Generally, an increase in offense rate causes a lower profit, due to higher prevention cost and lower users' usage rate. However, the risk of being victimized may induce users' willingness to pay levy, leading to an increase of authority's profit. Given this situation, an equilibrium point may be reached, on which the maximum profit is generated, that is, when the marginal cost of deterring offenses

equals the marginal gain of doing it. However, analysis of it may not be tractable unless we make several assumptions. For instance, the distribution function of users' valuation (willingness to pay levy), the distribution function of usage rate, the estimated number of malicious agents (with their characteristics), the effectiveness of the technology for offense and how to prevent it, the loss caused by offenses and their punishment, and so forth are all known.

SIMULATION STUDIES

In order to gain a clear picture of controlling and managing malicious agents in open MAS, we simulate a market for offenses (simulation part 1), followed by showing how to govern the society through various preventative schemes discussed earlier in the chapter (simulation part 2). The objective of the first simulation is to test the stability of theoretical equilibria of the offensive behaviors in open MAS. The objective of the second simulation is to test the effectiveness of various preventive schemes.

Part 1: The Agent Market for Offenses

Overlapping-Generation Model

Most of the assumptions used in our simulation are to those in Fender (1999). However, the following more realistic assumptions are added into the simulation:

A1. *The society follows a 10-generation overlapping model.*

In a *10*-generation overlapping model, all agents live for *10* periods of time. During each period, a new generation will be born while the oldest dies. We choose 10 periods for simulation purposes only. A very long period (e.g., greater than 50) should be avoided because it will lead to a low heterogeneity of our agent society (note that we are interested in open MAS). Similarly, a very short period (e.g., 1 or 2 periods) is too short for agents to have a chance to observe their society.

A2. *The society consists of two types of agents: honest citizens (potential victims) and potential offenders, consisting of half of the population (see Table 1).*

This half-half fraction is chosen arbitrarily for simulation purposes only. Intuitively, a higher fraction of honest citizens may reduce the crime rate, and vice versa.

A3. *The society runs for 300 periods.*

We have tested other periods and found that 300 periods are considerably enough for observing simulation results.

A4. *Honest citizens always work legally and earn $2000 each period.*

A5. *During each period, potential offenders can choose either to commit crime or to work legally, but not both.*

If an agent chooses to work legally, his return is derived from the uniform discrete distribution {$1000, $1001, $1002, ..., $3000}. If he succeeds in committing crime, then his return is $2000; if he fails, he only earns $500. Again, the value here is chosen for simulation purposes only.

A6. *Honest citizens are the only potential victims.*

A7. *There is only one type of economic crime.*

A8. *Each offender will be punished by probability p that depends on the level of public expenditures devoted for law enforcement.*

We use an increasing and concave function $p = E^{0.4}/C$, as shown in Figure 6 (PP locus), where E is the expenditure devoted to law enforcement.

A9. *All workers shall pay tax to the government for law enforcement; the amount is determined by dividing the expenditure equally to all workers.*

A10. *Each agent knows exactly the past punishment rate, the expected gains from work, crime, and the expected punishment.*

A11. *Each agent consumes his income; no saving exists in the economy.*

Interactions

Figure 5 shows the interaction among three types of agents: potential offenders, honest citizens, and the government. The simulation process follows the algorithm:

- **Step 1:** Initialize the agents' wage; generate it randomly from uniform distribution.
- **Step 2:** For period = 1 to 300, repeat step3 until step4.
- **Step 3:** All potential offenders make decisions whether or not to commit crime. If a potential offender succeeds, he will get some money from his victim ($2000). If he fails, he will be punished (receive $500).
- **Step 4:** The oldest generation dies, and a new generation is born. Update the social parameters, for example, crime rate, punishment rate, number of criminals, and so forth.

Each potential offender could only commit one crime during each period. In the next period, a new generation will be born and the old one dies, and the simulation continues. All agents are interacted to produce a time series data, for example, crime rate, punishment rate, and so forth.

Table 1. Simulation setup

Generation:	1	2	.....	10
Number of potential offenders / generation	0.5N	0.5N	.....	0.5N
Number of honest citizens / generation	0.5N	0.5N	.....	0.5N
Total number of agents /generation	N	N	.....	N

Figure 5. The process of the simulation with three type of agent: potential offender, honest citizen, and government

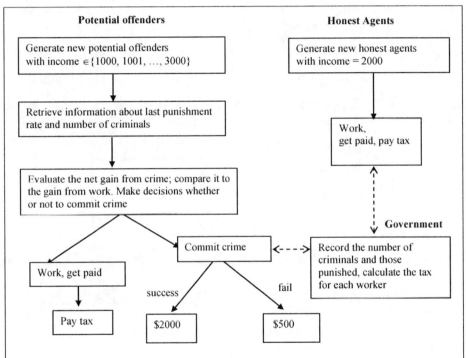

Experiments and Results

In this experiment, two groups of treatment are conducted, each with 12 different initial punishment rates. The difference between those two groups is the agents' population, that is, 4,000 agents in the first experiment and 1,000 agents in the second experiment. Moreover, every combination runs 5 times, hence totally 120 trials are

conducted. The first two treatments of experiments are as follows:

T1. Benchmark data. Agent's life-span L equals 10, the number of population in each generation N equals 200, therefore totally 4,000 agents are used in each round. The uniform distribution of wage is {1001, 1002, …, 3000}. The theoretical unstable equilibrium is attained when the initial punishment rate equals 63%.

T2. Reduce the population in each generation such that *N* equals 50, or 1000 agents are used in each round. The theoretical unstable equilibrium is attained when the initial punishment rate equals 55% (see the intersection between PP and EC in Figure 6).

Figures 7 to 9 shows some of the results of the experiments. The main result is that the theoretical unstable equilibrium is not an unstable equilibrium in the 10-generation overlapping model. The stability of this equilibrium is maintained if the range of the initial punishment rate is between 45% until 69%. If the initial punishment rate is greater than 69% or less than 45%, the equilibrium point will move to 0% or 100% crime rate, respectively.

We conjecture that the "residual" stable crime rate appears for two reasons:

- The multiple-period framework allows agents to make current decisions based on the information collected from the previous period. This mechanism generates the dynamic process of agents' decisions and

Figure 6. Theoretical multiple equilibria in EC and PP locus when the population is reduced to 1000

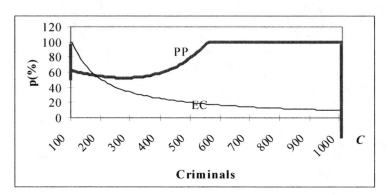

Figure 7. The high crime rate (100%) stable equilibrium is attained when the initial punishment rate equals to 45% (population = 4000).

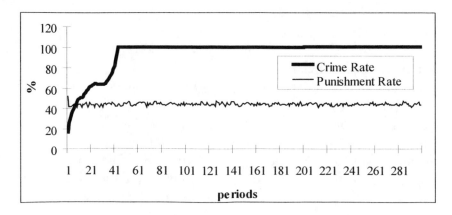

254

Figure 8. The low crime rate (12.5%) stable equilibrium is attained when the initial punishment rate equals to 60% (population = 4000).

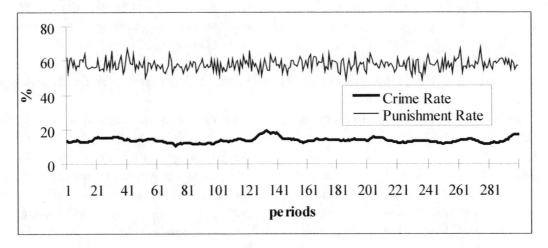

Figure 9. The zero crime rate (0%) stable equilibrium is attained when the initial punishment rate equals to 69% (population = 4000).

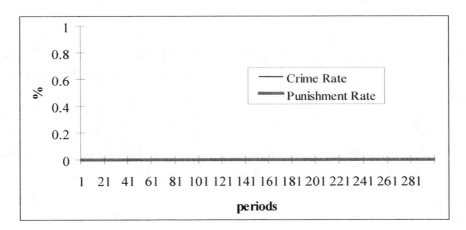

the outcomes. A propagation of previous information (punishment rate, and so forth) has helped the society to stabilize its overall crime rate.

• The finite "small" number of agents causes a nonsmooth distribution. Consequently, the statistical value (especially mean value of crime rate) is not as what is predicted in the theoretical analysis. The second experiment below (Figure 10) justifies it.

Figure 10 shows the stability of a 'theoretically' unstable equilibrium point. All the data are taken from the last 50 periods. In Figure 10, we compare the stable band between population 4,000 and population 1,000. An unexplained phenomenon is shown here. In low population society (population size = 1000), the theoretical unstable initial punishment rate (55%) is on the right side of the stable band. Hence, if the society's initial punishment rate is higher than 55% and lower than 73%, then the crime rate will erratically increase or decrease within a small range as shown in Figure 8. However, in a high population society

(population size = 4000), the theoretically unstable initial punishment rate is not at the edge of stable band, but within it slightly to the left. There is no explanation of how the population size affects the band of stable equilibria here.

Part 2: Governing Malicious Agents

Since there are too many inter-related variables in the system, it is hard to say which scheme is good for specific open MAS. Rather than using formal analysis, again we use a simulation to evaluate a hybrid scheme (combination of AU and LV). Although this simulation provides a sample of the governance policy in a specific domain, similar simulation can be conducted to generalize it to analyze other situations. We suggest more simulations should be conducted in the future.

Dynamic Model

We assume agents are heterogeneous and some are more patient for their future consumption than others; some are risk-neutral while others are risk-averse. Agent's risk aversion is simulated here

Figure 10. The "stable" range of the "theoretically unstable" equilibrium shown as the relation between initial punishment rate and the long run crime rate

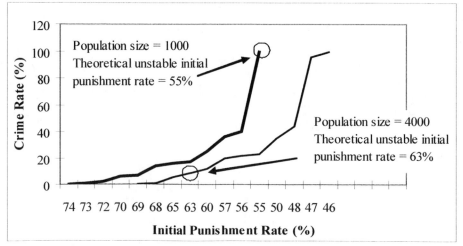

because more studies in agent's decision under uncertainty in MAS have considered it as one of the agent's properties. And based on their learning abilities (intelligence), agents need different time to acquire the same amount of skills, predict, or detect an offender, and so forth.

Suppose that EU_{legal} is not constant, but can be derived from a function $\Pi(\omega, class, C)$, where $\omega \in [0, \tau]$ is agent's experience in legal activity, $class \in \{benevolent, malevolent\ not\ detected, malevolent\ detected\}$, and C is the number of offenders during that period. Here, $\Pi_\omega > 0$, which means that as more legitimate activities are performed by an agent, the more income it make. Also assume that $\Pi(detected) < \Pi(not\ detected) < \Pi(benevolent)$, and $\Pi_C > 0$. In other words, different classes mean different expected gains; and a high offense rate increases EU_{legal}.

Let the expected gain from an offense, R, depend on the victim's self-protection, ρ_v, and offender's productivity $\Theta(\tau, s)$, where τ is its length of stay in the system and s is its offending technology, where $\Theta_\tau < 0$ and $\Theta_s > 0$. Obviously, a high level of victim's self-protection may reduce the loss from the offense and increase the chance to convict the offender. In each period, offender i may be convicted by probability $p_c(\Theta, o, E)$, where o is its offense records, and E is the authority's effort in that period. If an offender is convicted, it will be punished with imprisonment (killed or banned from using the service for certain periods) $\theta(r)$ and/or fine $\lambda(r)$, where r is its crime records, $\theta_r > 0, \lambda_r > 0$. Therefore, a potential offender may commit an offense if its perceived expectation satisfies $EU_{crime} > EU_{legal}$. And,

$$EU_{crime} = (1 - p_c)\,U(R) - p_c\,(U(\theta, \lambda) + DU)$$

where $U(\theta, \lambda)$ is the punishment in terms of money, consisting of the perceived fine λ and the opportunity cost during imprisonment, and DU represents other loss, which are randomly determined. Generally, (see equation (A)), where L is the expected length of time that the agent will use the system, τ is agent's present time (age) in the system, θ represents the length of imprisonment, L-τ-θ represents the predicted rest of time after the agent can use the system again, $\sum_t \beta^t \Delta\Pi(\tau+t)\varphi(\tau+t)$ represents the opportunity cost of imprisonment converted to the present valuc. β is the discount factor of future utility, $\Delta\Pi(\tau+t)$ is the expected gain at time $\tau+t$, $\varphi(\tau+t)$, which is is the indicator value that equals to 1 if the offender expects that it still uses the system at time $\tau+t$ or 0 otherwise. $\sum_\iota \Delta\Pi(\tau+\theta+\iota)\varphi(\tau+\theta+\iota)$ represents the loss due to the change of agent's status from *not detected* to *detected*, where $\Delta\Pi(\tau+\theta+\iota)$ is the difference between $\Pi(detected)$ and $\Pi(not\ detected)$, and κ is the indicator valuc cqual to 1 if the offender has no conviction record and otherwise equals 0. A smart agent may predict the parameters in the above equations more accurately.

Benevolent agents tend to be more passive in the simulation, since they only control their self-protection. They predict the probability of being victimized, loss from being victimized, life span in using the system, and the level of authority's effort. Then, they decide how much should be invested on the self-protection scheme. We assume that most of the benevolent agents will be cautious and invest as much as possible. The authority receives the total lump sum tax ΣT in every period, that is, $\Sigma T = T(N\text{-}C)$, where $(N\text{-}C)$ is the number of participants in the system. We also assume that the authority follows "the rule

Equation A.

$$U(\theta, \lambda) = U\left(\lambda + \sum_{t=1}^{\theta} \beta^t \Delta\Pi(\tau+t)\varphi(\tau+t) + \kappa \sum_{\iota=1}^{L-\tau-\theta} \Delta\Pi(\tau+\theta+\iota)\varphi(\tau+\theta+\iota) \right)$$

of thumb," that is, always maintains current effort if the offense rate decreases, but raises it if the offense rate does not decrease. If the authority faces budget deficit, they will reduce their effort. The simulation follows the following steps:

- **Step 1:** Initialize the agents, authority, and society.
- **Step 2:** For period = 1 to 100, repeat Step 3 until Step7.
- **Step 3:** Randomly match potential offenders and potential victims. All potential offenders make decisions. If a potential offender succeeds in its action, then it will get something from its victim. If it fails, it will be convicted.
- **Step 4:** Authority starts to arrest the offenders who have not been convicted.
- **Step 5:** Authority decides on a new policy to fight offenses.
- **Step 6:** Each agent decides its new self-protection scheme.
- **Step 7:** Some agents exit and some enter the system.

Experiment Results and Discussions

Altogether seven groups of treatments were conducted, each with four combinations of control variables. Moreover, every combination runs for three times, hence totally 84 trials were conducted. Since each combination was repeated three times, the result is not statistically significant. The seven treatments are:

- **T1. Benchmark:** Agent's life span L equals 10 and the number of population in each generation N equals 100. Therefore, 1,000 agents are used in each round. There is no capital punishment, except for discrimination toward convicted offenders (negative reputation). Bias from any predicted value is set to 50% for most stupid agents, and roughly 10% for most intelligent agents.

- **T2.** Reduce discrimination towards convicted offender (equality treatment).
- **T3.** Increase random disutility from punishment, *DU*.
- **T4.** Increase the fine (fine can be used to financially support the authority).
- **T5.** Lessen the contact between potential offenders and potential victims by half (coalition formation among benevolent agents).
- **T6.** Impose a strict policy; that is, authority always increases the security level, plus a higher subsidy for supporting authority expenses in the initial period.
- **T7.** Increase the life span to 50% longer.

Table 2 shows one of the results of the experiment. Column T2 shows the effect of equality (in terms of gain) for both potential offenders previously convicted and not convicted. If both of them are treated equally, then those ex-convicted will be more likely to behave benevolently, especially after they realize the risk of offense. The results show a strong reduction of recidivism to approximately 40% and the offense rate to approximately 17%. The results of increasing punishment (pain and fine) do not change the offense rate much (column T3 and T4). Therefore, we can conclude that most offenders act less rationally or they would not overestimate the punishment. Column T5 shows the effect of isolation between potential victims and offenders (coalition formation). Under perfect isolation, all potential offenders cannot find potential victims. In current simulation, we reduce the opportunity of which potential offender will meet potential victims from 10 in benchmark to 5. Obviously, the results show that the isolation reduces the offense rate. And column T6 shows the effect of tough authority in deterring offense. The results show that the policy is not effective in reducing the offense rate because the authority may eventually go bankrupt (due to limited income from system usage and a high cost of maintaining system security). In T7, we use longer generation and life span, that is, 15 periods instead of 10 periods. By increasing the length of agent's life span, the length of sentences becomes relatively shorter. Intuitively, a relative short of sentence would undermine the punishment, thus increase the crime rate. Results from the simulation confirmed our hypothesis.

It can be seen from the results that some policies may be more effective than others. For example, building a coalition among benevolent agents is more effective than increasing the punishment. And equality treatment to convicted offenders may reduce the recidivism in the system. However, the result discussed in this experiment should not be generalized into other domains. The assumptions in the system may implicitly redirect the result. For instance, we do not attach the usage rate to the offense rate. In reality, agents' usage rate also depends on the reputation of the system. If the usage rate increases as the offense rate decreases, then the bankruptcy in T6 may be avoided. The recidivism may also be eliminated if all convicted offenders are killed and banned from the system forever. Certainly, all the policies must be based on real situations. For instance, capital punishment can be carried out in Scenario 1 but not in Scenario 2. Note that in this simulation we do not simulate agents with social network, trusteeship in coalition formation, or reputation mechanism. Finally, agents in an open MAS may reside in the system for a certain period of time, and they will learn whether to use the system or not, and malicious agents will learn whether to commit offense or not, or to recognize certain groups of vulnerable victims.

LESSONS LEARNED AND DISCUSSION

Controlling Malicious Agents Through Punishment

Two conclusions could be drawn from the simulation study on the market for offense (experiment part 1):

- When the population is 1,000, the theoretical unstable initial punishment rate is at the lowest 'critical value.' Any reduction of the initial punishment rate would cause it to be unstable (consistent with the theoretical result). But an increase of the value will not cause the crime to reduce automatically (stabilized). Only if we further raise the punishment rate until 74%, we can get unstable equilibrium (the crime rate moved to zero). Therefore, the theoretical result is only right in half part (unstable in one side).

- When the population equals 4,000, both movements of the initial punishment rate towards higher or lower value would not destruct stability. But the magnitude needed to destroy the stability is different for different directions. As shown in Figure 10, it needs more reduction of the punishment rate before the system becomes unstable (63% to 46%). But it only needs a few additions (63% to 68%).

Table 2. The effect of various policies on several main variables in the system

	Average change in % relative to benchmark					
	T2	T3	T4	T5	T6	T7
Offense Rate	**-16.64**	9.22	6.55	**-42.93**	0.95	**38.36**
Recidivism	**-40.21**	5.26	4.11	-9.20	2.38	**21.97**
Self Protection	-0.33	0.00	0.19	-0.38	-0.14	**14.04**
Conviction Rate	7.86	-4.07	4.95	2.34	2.94	**-17.13**

The results on applying some macropolicies to govern agent society (experiment part 2) show that some policies may be more effective than others. For example, building a coalition among benevolent agents is more effective than increasing the punishment. And equality treatment to convicted offenders may reduce the recidivism in the system. Certainly, all the policies must be based on real situations. For instance, capital punishment can be carried out in Scenario 1 but not in Scenario 2. Note that in the simulations we discussed, we do not simulate agents with social network, trusteeship in coalition formation, or reputation mechanism.

Controlling Malicious Agents Through Reputation Mechanism

Nevertheless, a reputation-based mechanism, like those on e-Bay™, can help deter malicious agents from cheating other agents and manipulating associated transactions through close to market-oriented norms. The most important indicator is the *trust* towards which other agents show willingness to interact, including cooperation and coordination. It has been studied extensively in MAS (Elfoson, 1998; Jonker & Treur, 1999; Marsh, 1992, 1994; Schillo, Funk, & Rovatsos, 2000; Tang, Winoto, & Niu, 2003), where it is the *'attitude an agent has with respect to the dependability/capabilities of some other agent (maybe itself)'* (Jonker & Treur, 1999).

In our previous work (Tang et al., 2003), we employ a trust-mechanism into a personalized agent-based investment system called I-TRUST, where broker agents can win human users' trust through their 'hard-work': making money for their clients. Clients can punish or appreciate their broker agents by reducing/increasing the amount of money delegated to agents to invest in the next investment period.

From the agents' perspective, they will try to avoid being punished (losing their clients' trust) by not only investing to the best of their expertise, but also considering their clients' risk preferences. For example, in one specific investment period, agents might consider buying $5000 worth of stock A after a careful estimation of market information; however, his client might disagree and suggests buying only $2000. In this case, broker agents must sacrifice their expertise and follow, to some degree, their clients' advice, and thus, invest, say, $2500 worth of stock A. Therefore, broker agents might try to balance their investing expertise and feedback from their clients. Results of the third experiment show that the broker agent struggles to locate an optimal trade-off to retain clients' trust. We found out that the controllability of a user/client toward the autonomous system plays a significant role for trust building. In one of the experiments, users could add/limit the amount of money they are willing to delegate to their agents. This simple implementation of control has already been observed to make a difference for trust building, compared with more complex forms of control when a user can further specify the type of stocks she allows her agents to buy/sell, while at the same time, retains the authority to buy/sell other types of portfolios.

It is convenient to draw a conclusion that this type of *automatic deterrence policy*, to some degree, could achieve a relatively better governing outcome since agents (both benevolent and malicious agents) *have to* observe some norms in the society in order to survive. Nevertheless, there are still rooms for malicious agents to exploit in order to maximize their profit. Hence, further studies could be conducted to investigate these policies as well as the effectiveness of them. In addition, from the deployment's perspective, it is not trivial to apply the mechanism (such as those on e-Bay™) in MAS. As such, further studies are necessary.

Fighting with Malicious Agents Through Coalition Formation

Our simulation results reported that some policies may be more effective than others. For example, allowing benevolent agents to form coalitions is more effective than increasing the punishment. It is not difficult to agree with this claim due to the ways malicious agents attack. When benevolent agents can build a community and increase their total degree of protecting themselves, the chance for the malicious agents to launch the attack will be lower, even if the authority will maintain current deterrence efforts. In our current simulations, however, we did not go further to investigate the strategies of forming coalitions, for instance, how the benevolent agents can strategically identify other agents in the neighborhood to quickly form a coalition through some protocols.

When we are allowing benevolent agents to form coalitions, would we also allow malicious agents to form coalitions? This is a tough question, which could add much more instability and unpredictability into the open MAS.

As far as the simulation variables can be injected, the results would be interesting for us to see a broader picture of controlling malicious agents in open MAS.

CONCLUSION

In this chapter, we bring up the issue of malicious agents in open MAS with several examples and discussions in relation to the existing crime study, that is, supply-demand analysis and deterrence theory. In addition, we also propose various governance schemes from the perspective of both the system designers and users, in four preventive schemes, with detailed descriptions of their characteristics. Finally, by comparing with some policies through a simulation, we have made a first step towards an empirical analysis of the governance strategies by means of deterrence theory.

Our work highlights the importance of mechanisms to make intervention to the MAS, in an attempt to deter malicious agents from maximizing their utilities through the illegal actions. Indeed, in market-oriented MAS, human interventions sometimes would be necessary, given the condition that these interventions would not destroy the ecological stability of the agent society. However, an automatic intervention mechanism seems to be the winning card in the end, among them, the reputation mechanism, agent coalition formation, and so forth. It is our hope that our work can shed light on these issues as well as their deployment in MAS.

REFERENCES

Abbink, K., Irlenbusch, B., & Renner, E. (1999). An experimental bribery game. *The Journal of Law, Economics and Organization, 18*(2), 428-454.

Becker, G. S. (1968). Crime and punishment: An economic approach. *Journal of Political Economy, 76*, 169-217.

Bonatti, P., Kraus, S., & Subrahmanian, V. S. (2003). Secure agents. *Annals of Mathematics and Artificial Intelligence, 37*(1-2), 169-235.

Bugliesi, M., & Castagna, G. (2001). Secure safe ambients. In *Proceedings of the 28th ACM Symposium on Principles of Programming Languages*, London (pp. 222-235).

Carroll, J. S. (1978). A Psychological approach to deterrence: The evaluation of crime opportunities. *Journal of Personality and Social Psychology, 36*(12), 1512-1520.

Carroll, J. S., & Weaver, F. (1986). Shoplifters' perceptions of crime opportunities: A process-tracing study. In D. B. Cornish & R. V. Clarke (Eds.), *The*

Reasoning Criminal: Rational Choice Perspectives on Offending (pp. 19-38). Springer.

Davis, M. (1988). Time and punishment: An intertemporal model of crime. *Journal of Political Economy, 96*, 383-390.

Ehrlich, I. (1975). The deterrent effect of capital punishment: A question of life and death. *American Economic Review, 65*, 397-417.

Ehrlich, I. (1996). Crime, punishment, and the market for offenses. *Journal of Economic Perspectives, 10*(1), 43-67.

Elfoson, G. (1998). Developing trust with intelligent agents: An exploratory study. In *Proceedings of the First International Workshops on Trust* (pp. 125-138).

Fedoruk, A., & Deters, R. (2002). Improving fault-tolerance by replicating agents. In *Proceedings of the First International Joint Conference on Autonomous Agents and Multi-Agent Systems*, Bologna, Italy (pp. 737-744).

Fender, J. (1999). A general equilibrium model of crime and punishment. *Journal of Economic Behavior & Organization, 39*, 437-453.

Hennesy, M., & Riely, J (1999). Trust and partial typing in open systems of mobile agents. In *Proceedings of the 26th ACM Symposium on Principles of Programming Languages (POPL '99)*, San Antonio, TX (pp. 93-104).

Hohl, F. (1998). Time limited blackbox security: Protecting mobile agents from malicious hosts. In G. Vigna (Ed.), *Mobile Agents and Security* (LNCS 1419, pp. 92-113). Springer.

İmrohoroğlu, A., Merlo, A., & Rupert, P. (1996). *On the political economy of income redistribution and crime* (Staff Rep. No. 216). Federal Reserve Bank of Minneapolis Research Department.

Jonker, C. M., & Treur, J. (1999). Formal analysis of models for the dynamics of trust based on experiences. In *Proceedings of MAAMAW'99* (LNAI 1647, pp. 221-232). Springer.

Leung, S. F. (1995). Dynamic deterrence theory. *Economica, 62*, 65-87.

Marjit, S., & Shi, H. (1998). On controlling crime with corrupt officials. *Journal of Economic Behavior & Organization, 34*, 163-172.

Marsh, S. (1992). Trust and reliance in multi-agent systems: A preliminary report. In *Proceedings of the 4th European Workshop on Modeling Autonomous Agents in a Multi-Agent World (MAAMAW'92)*, Rome.

Marsh, S. (1994). *Formalizing trust as a computational concept.* Unpublished doctoral thesis, University of Stirling, Department of Mathematics and Computer Science, UK.

Marsh, S. (1994). Trust in distributed artificial intelligence. In C. Castelfranchi & E. Werner (Eds.), *Artificial Social Systems* (LNAI 830, pp. 94-112). Berlin: Springer-Verlag.

Pyle, D. J. (1983). *The Economics of Crime and Law Enforcement.* New York: St. Martin Press.

Sah, R. K. (1991). Social Osmosis and Patterns of Crime. *The Journal of Political Economy, 99*(6), 1272-1295.

Schneider, F. B. (1997). Towards Fault-Tolerant and Secure Agentry. In M. Mavronicolas & P. Tsigas (Eds.), *Proceedings of the 11th International Workshop on Distributed Algorithms* (LNCS 1320, pp. 1-14). Berlin: Springer.

Schillo, M., Funk, P., & Rovatsos, M. (2000). Using trust for detecting deceitful agents in artificial societies [Special issue: Deception, Fraud, and Trust in Agent Societies]. *Applied Artificial Intelligence Journal, 14*, 825-848.

Tang, T. Y., Winoto, P., & Niu, X. (2003). I-TRUST: Investigating Trust between Users and Agents in a Multi Agent Portfolio Management System. *Electronic Commerce Research and Applications, 2*(4), 302-314. Elsevier Science B.V.

Vigna, G. (1998). Cryptographic Traces for Mobile Agents. In G. Vigna (Ed.), *Mobile Agents and Security* (LNCS 1419, pp. 137-153). Springer.

Winoto, P. (2003a). A Simulation of the Market for Offenses in Multiagent Systems: Is Zero Crime Rates Attainable? In J. S. Sichman, F. Bousquet, & P. Davidsson (Eds.), *Multi-Agent-Based Simulation II* (LNCS 2581, pp. 181-193). Springer.

Winoto, P. (2003b). Controlling Malevolent Behavior in Open Multi-Agent Systems by Means of Deterrence Theory. In *Proceedings of 2003 IEEE/WIC International Conference on Intelligent Agent Technology*, Halifax (pp. 268-274).

Section V
Cross–Fertilized Techniques in Business Antomation

Chapter XIV
Features for Killer Apps from a Semantic Web Perspective

Kieron O'Hara
University of Southampton, UK

Harith Alani
University of Southampton, UK

Yannis Kalfoglou
University of Southampton, UK

Nigel Shadbolt
University of Southampton, UK

ABSTRACT

There are certain features that distinguish killer apps from other ordinary applications. This chapter examines those features in the context of the Semantic Web, in the hope that a better understanding of the characteristics of killer apps might encourage their consideration when developing Semantic Web applications. Killer apps are highly transformative technologies that create new e-commerce venues and widespread patterns of behaviour. Information technology, generally, and the Web, in particular, have benefited from killer apps to create new networks of users and increase its value. The Semantic Web community on the other hand is still awaiting a killer app that proves the superiority of its technologies. The authors hope that this chapter will help to highlight some of the common ingredients of killer apps in e-commerce, and discuss how such applications might emerge in the Semantic Web.

INTRODUCTION

The **Semantic Web** (SW) is gaining momentum; as more researchers gravitate towards it, more of its technologies are being used, as more standards emerge and are accepted. There are various visions of where the technology might go, what tasks it might help with, and how information should be structured and stored for maximum applicability (Berners-Lee, Hall, Hendler, O'Hara, Shadbolt, & Weitzner, 2006; Berners-Lee, Hendler, & Lassila, 2001; Marshall & Shipman, 2003; Uschold, 2003). What is certainly clear is that no one who wishes seriously to address the problems of knowledge management in the 21st century can ignore the SW.

In many respects, the growth of the SW mirrors the growth of the World Wide Web (**WWW**) in its early stages, as the manifest advantages of its expressivity became clear to academic users. However, once the original phase of academically led growth of the WWW was over, to the surprise of many commentators, the Web began its exponential growth and its integration with many aspects of ordinary life. Technologies emerged to enable users to, for example, transfer funds securely from a credit card to a vendor's account, download large files with real time video or audio, or find arbitrary Web sites on the basis of their content.

It is important to note that the growth of the WWW had three separate but linked components, which led to three distinct sets of incentives. In the first place, there were quick wins from putting documents on the Web. Prior to the WWW, there was a culture of privacy about documents, and the proselytisers of the Web had to convince a lot of people that documents should be published and made available to all. This was perceived as a risk by document owners, and involved breaking down preconceptions about ownership, privacy, confidentiality, and commercial advantage. To do this, when a person or an organisation posted documents, there had to be immediate and tangible gains from the individual act of publication, an increase in one's social circle, an expansion of business, or a wider set of business opportunities. Without such immediate and individual gains, independent of any future network effects, fewer documents would have appeared on the Web.

However, those network effects also had to come into play. Because network effects are part of the context, and largely independent of individual decisions, they can sometimes be overlooked. But the creation of large business markets online happened because more and more people started an online existence. The evolution of social networks or multiplayer game scenarios depends on the critical mass of people spending a certain amount of time indoors next to the computer. But the network effects really do kick in, and an online presence for a business or a person is now so much more rewarding both financially and socially because so much of so many lives takes place online. *In extremis* the failure to engage with the WWW can now mean the serious loss of business.

Further, the tools had to be available for the WWW to take off. If creating HTML pages was at all difficult, involving steep learning curves, flaky or expensive software, and advanced design skills, then the WWW could not have integrated so easily with the rest of the environment. And if publishing documents online led to a backward step in information processing, if the documents, by being published, were somehow removed from an organisation's standard information management practices, then such an organisation might end up worse off than before, which would have strangled the global Web at birth, and it would have remained an academic tool. For instance, an organisation's practice in posting documents on the WWW might well compromise its version management control; one could imagine a situation that a draft document, posted on the Web, might be edited by two different people in parallel, and then two incompatible versions would circulate in parallel. The right tools, and the right management practices, needed to be in place in advance, to prevent such hiccups occurring; certainly not all of the potential pitfalls of posting pages would be easily predictable in advance.

The SW aims to do for data what the WWW did for documents. Most realistic visions of the future of a successful SW include a version of the WWW's exponential growth. The SW infrastructure should be put in place to enable such growth. With a clean, scalable, and unconstraining infrastructure, it should be possible for users to undertake all those tasks that seem to be required for the SW to follow the WWW into the stratosphere, such as publishing their RDF, converting legacy content, annotating, writing ontologies, and so forth. There must be immediate gains, network effects, and the tools and management practices to prevent backward movement. Without this, the culture of privacy and data protection that now surrounds the storage of data will prevent the SW from taking off. This culture needs to be adapted to the new technologies without compromising the important aspects of privacy (Berners-Lee et al., 2006).

However, that something is possible does not entail that it is inevitable. So the question arises of how developers and users might be persuaded to come to the SW. This type of growth of a network has often been observed in the business literature. Many technologies depend for their usability on a large number of fellow users; in this context, **Metcalfe's Law** (Gilder, 1993) states that the utility of a network is proportional to the square of the number of users.

Technologies which have this effect are called *killer apps*. Exactly what a killer app is, to a large extent, is in the eye of the beholder; in the WWW context, killer apps might include the Mosaic browser, Amazon, Google, eBay, or Hotmail. Hotmail attracted over 30 million members in less than three years after its 1996 launch; eBay went from nothing to generating 20% of all person-person package deliveries in the U.S. in less than two years. Of course, the WWW was a useful enough technology to make its own way in the world, but without the killer apps it might not have broken out of the academic/nerdy ghetto. By extension, it is a hope of the SW community that the SW might take off on the back of a killer app of its own.

The dramatic development of the WWW brought with it a lot of interest from the business community, and the phenomenon of killer apps has come under much scrutiny (Downes & Mui, 2000; Kelly, 1998). Attempts have been made to observe the spread of killer apps and to generalise from such observations; the tight development cycles of WWW technology have helped such observations.

In this chapter, our aim is to consider the potential for development of the SW in the light of the killer app literature from the business community. Of course, it is impossible to forecast where the killer app for the SW will come from. But examination of the literature might provide some pointers as to what properties such an application might have, and what types of behaviour it might need to encourage.

KILLER APPS AND THE SEMANTIC WEB

Killer apps emerge in the intersection between technology, society, and business. They are technological in the broad sense of being artificial solutions to perceived problems, or artificial methods to exploit perceived opportunities (which is not to say that they need to have been developed specifically with such problems or opportunities in mind). Mere innovation is not enough. Indeed, a killer app need not be at the cutting edge of technological development at all. The killer app must meet a need and be usable in some context such as work, leisure, or commerce. It must open up some kind of opportunity to bring together a critical mass of users.

To do this, killer apps have a number of features which have been catalogued by commentators. In this section, we will examine and reinterpret such features in the context of the SW. We reiterate that these features may not all be necessary, and they certainly are not sufficient; however, they can act as an interesting framework for our thoughts on this topic.

The main point, of course, about a killer app is that it enables a superior level of service to be provided. And equally clearly, the SW provides an important opportunity to do this, as has been argued from the beginning (Berners-Lee et al., 2001). There are obvious opportunities for any knowledge-based task or enterprise to improve its performance once knowledge sources are integrated and more intelligent information processing is automated.

The Bottom Line: Cost vs. Benefit

Merely providing the opportunity is not enough. Cost-benefit analysis is essential (Downes & Mui, 2000). There are several aspects to costs. Obviously, there are financial costs; will people have to pay for the killer apps on the SW? Maybe not; there are many examples of totally free **Internet applications**, such as Web browsers, search engines, and chat messengers. Such applications often generate large revenues through online advertising. According to the Interactive Advertising Bureau in the UK[1] and PriceWaterHouseCoopers,[2] the market size of online advertising in the UK for 2004 was £653.3 million, growing more than 60% in one year. Free products may be very important in this context (Kelly, 1998), and indeed killer apps are often cheaper than comparable alternative products (Christensen, 1997). Such costs are not the only ones incurred. There are also important resource issues raised by any plan to embrace the SW.

Conversion cost: As well as investing in technologies of certain kinds, organisations and people will have to convert much of their legacy data, and structure newly-acquired data, in particular ways. This immediately requires resources to support the development of ontologies, the formatting of data in **RDF**, the annotation of legacy data, and so forth, not to mention potential costs of exposing data in RDF to the wider world (particularly where

market structures reward secrecy). Furthermore, the costs of developing smart formalisms that are representationally adequate (the fun bit) are dwarfed by the costs of populating of informational structures with sufficient knowledge of enough depth to provide utility in a real-world application (Ellman, 2004). Note also that such a process will require ascent of some very steep learning curves.

Much of the legacy data will be sitting around in relational databases, and this very large quantity of relational data is a major target of the SW initiative (Berners-Lee et al., 2006). The mappings from relational data to RDF are fairly straightforward, and so the hope is that these conversions can lead to quick wins, particularly when an organisation has a lot of data distributed across its various divisions. However, the danger here (an important potential cost) is that an organisation takes data out of its usual information management streams. Seduced by the vision of the SW, it puts all its RDF in one place and sends it out into the world, without the tools to deal with it. Even though the older tools and structures were not perfect, they will of course have provided a useful service, and a firm which takes its data out of its old-fashioned environment can find itself subtracting value from the data, not adding it. Joining the SW community without the proper information management regime in place is a recipe for disappointment and disillusionment.

Maintenance cost: In a very dynamic domain, it may be that ontologies have to be updated rapidly (Buckingham-Shum, 2004; Klein & Fensel, 2001). The properties of **ontologies** are not as well-understood as they might be; areas such as mapping ontologies onto others, merging ontologies, and updating ontologies are the focus of major research efforts. It is currently unknown as to how much such maintenance effort would cost over time.

Organisational restructuring costs: Information processing is integrated into an organisation in subtle ways, and organisations often subconsciously structure themselves around their information processing models (Eischen, 2002), a fact implicitly accepted by the knowledge engineering community (Schreiber, Akkermans, Anjewierden, de Hoog, Shadbolt, de Velde, & Wielinga, 1999). Surveys of organisations, for example, reveal that ontologies are used in relatively primitive ways; indeed, in the corporate context, the term 'ontology' is a generic, rarely defined catch-all term. Some are no more than strict hierarchies, some are more complex structures allowing loops and multiple instantiations, still others are in effect (sometimes multilingual) corporate vocabularies, while others are complex structures holding metadata (McGuinness, 2002). Whatever their level of sophistication, corporate ontologies support the systematisation of large quantities of knowledge, far from the traditional artificial intelligence (AI) view of their being highly detailed specifications of well-ordered domains. Ontologies may refer to an internal view of the organisation such as marketing, research and development (R&D), human resources, and so forth, or an external one such as types of supplier and supplies, product types, and so forth. A recent survey showed that only a relatively small number (under a quarter) of corporate ontologies were derived from industry standards. The big issue for many firms is not representational adequacy but rather the mechanics of integration with existing systems (Ellman, 2004).

With a complex organisation, working in a dynamic environment, it may appear a daunting task to develop an ontology to cover it. However, ontologies can be fairly **lightweight**, and their main aim is to facilitate communication and translation, not to produce a philosophically accurate model of the world. A series of small, overlapping ontologies, with *ad hoc* translators between them, will allow SW technologies to add value to data without too large of an immediate cost. For instance, an application to fill in one's tax return will use data from one's bank about one's bank account. That data will be represented against the background of the bank's ontology, but the tax application need not commit to the whole of the bank's ontology, nor even translate all the concepts therein. It will suffice to translate the few concepts relevant to the problem where the two ontologies overlap. And this, in such technology, is what happens.

Tim Berners-Lee has argued, informally, that worries about ontology development are overblown. As the context of use and maintenance grows, and the number of users increases, the committees required to develop and maintain ontologies do not grow as quickly. The effort of developing large ontologies is obviously greater than the effort of developing smaller ones, but the effort *per user* will decrease. As long as ontology builders do not over-elaborate, it should be cost-effective for most organisations to devote resources to ontologies (Berners-Lee, 1998-2005; Berners-Lee et al., 2006).

Transaction costs: On the other hand, it is also true that if the SW does alter information gathering and processing costs, then the result will inevitably be some alteration of firms' management structures. The result will be leaner firms with fewer management layers, and possibly different ways of processing, storing, and maintaining information. Such firms may provide opportunities for new SW technologies to exploit, and a gap in the market from which a killer app may emerge.

It has long been argued that the size and structure of firms cannot be explained simply by the price mechanism in open competitive markets (Coase, 1991). The allocation of resources is made using two mechanisms: first (between firms and consumers) by distributed markets and coordinated by price, but also (within firms) by the use of authority within a hierarchy (i.e., people get ordered to do things). The question then is how this relates to a firm's structure: when a firm needs some service, does it procure it from outside and pay a market price, or does it get it done in-house, using workers under some contractual obligation, and why?

It is generally thought that such organisational questions are determined by the *transaction costs* within a firm (Williamson, 1975, 1991), in other words, the costs of actually making a transaction, costs such as information processing and acquisition (for example, the costs of researching the market to find the best price), information asymmetries (outside firms know more about the nature of their product than purchasers of those products), uncertainty, incomplete contracting, and so on. The promise of the SW is that many of the information gathering costs will be ameliorated, and at least some of the asymmetries will be evened up. The general result of this is likely to be a continuation of trends that we have seen in economies since the widespread introduction of IT, which is the removal of middle management ('downsizing'), and the outsourcing of many functions to independent suppliers. In the SW context, of course, many of those independent suppliers could well be automatic agents, or providers of Web services. If the SW contains enough information about a market, then we might well expect to see quite transformative conditions and several market opportunities. The killer app for the business aspects of the SW may well be something that replaces the coordinative function of middle management.

But we should add a caveat here: the marginal costs of information gathering will be ameliorated, but equally there will, as noted above, be possibly hefty sunk costs up front, as firms buy or develop ontologies, convert legacy data to RDF, lose trade secrets as they publish material, and so on. These initial costs may prove an extensive barrier to change.

Reducing costs: Here we see the importance of the increase in size of the user base. For example, the costs of developing and maintaining ontologies are high but can be shared. Lightweight ontologies are likely to become more important (Uschold & Gruninger, 2004); not only are they cheaper to build and maintain, but they are more likely to be available off the shelf (McGuinness, 2002). Furthermore, they are more likely to be easily understandable, mappable, maintainable, and so forth. The development of such lightweight multipurpose ontologies will be promoted as the market for them gets bigger.

Similar points can be made about ontology development tools. Better tools to search for, build, or adapt ontologies will spur their use or reuse, and again such tools will appear with the demand for them. And in such an environment, once an ontology has been developed, the sunk costs can be offset by licensing the use of that ontology by other organisations working in that domain (if, that is, the value of retaining the ontology as a trade secret does not exceed the potential licensing income). The costs, in such a networked environment, will come down for everybody too over the period of use; if a single firm took on the costs of developing an ontology for a domain (and licensed that ontology to its competitors, thereby avoiding the reduplication of effort over the marketplace as a whole), that firm could also take on, individually, the maintenance costs. Organisations that specialised in ontology maintenance and training for users could spring up, given sufficient demand for their services.

Increasing benefit: Similar to data restructuring, there has to be some discernible benefit for organisations putting their data in RDF, and these benefits will become more apparent with more published data in RDF. The issue here that a killer app might help with is that there seems to be little or no advantage for an individual firm in moving first. A firm that publishes its data in RDF early incurs costs early and takes a risk but gets little benefit; if it removes its data from its traditional information management stream, it might even reduce the value of that data. A firm that publishes RDF late incurs the costs later, takes on little risk, and gets more benefits. Nevertheless, being first in a new market is a distinct advantage (Christensen, 1997; Downes & Mui, 2000), but late entrants can also succeed if they outperform existing services (Evans & Wurster, 2000) (e.g., Google). So there is a prisoner's dilemma to be sorted out.

Berners-Lee (2003) argues that the killer app for the SW is the *integration*. Once distributed data sources are integrated, the sky becomes the limit. This of course could be true, but it will be hard to convince data providers to publish in RDF and join the SW movement without concrete examples of how they could benefit from doing so. This is probably supported by Berners-Lee's (2004) suggestion that we need to 'Justify on short/medium term gain, not network effect.' The network effect is a long-term goal, and to achieve this goal we need to show short-term gains. Integration alone might not be seen as a gain on its own, especially when considering costs and privacy issues. On the other hand, one target application area for the SW is in adding value to the large amount of relational data sitting around in isolation in databases across organisations or sectors. Being able to manipulate data from heterogeneous sources, in effect cross-referencing data from different databases and being able to infer information about objects represented in different ways (or even to determine when apparently two objects are identical), are very powerful capacities.

In a survey for business use cases for the SW, researchers of the EU *KnowledgeWeb*[3] emphasised the importance of proper targeting for SW tasks (Nixon, 2004) to avoid applying SW technologies to where they do not offer any clear benefit, which may discourage industry-wide adoption.

The survey concluded that the areas which seem to benefit more from this sort of technology are *data integration* and *semantic search*. It was argued that these areas could be accommodated with technologies for *knowledge extraction, ontology mapping,* and *ontology development.* Similarly, Uschold and Gruninger (2004) argue that ontologies are useful for better *information access, knowledge reuse, semantic-search,* and *inter-operability.* They also list a number of assumptions to be made to progress towards a fully automated semantic integration of independent data sources. Fensel, Bussler, Ding, Kartseva, Klein, Korotkiy, et al. (2002) describe the beneficial role of ontologies in general *knowledge management* and *e-commerce* applications. They also list a number of obstacles that need to be overcome to achieve those benefits such as *scalable ontology mapping, instantiation,* and *version control.* Other obstacles, such as *trust, agent coordination, referential integrity,* and *robust reasoning* have also been discussed (Kalfoglou et al., 2004).

Leveraging Metcalfe's Law

The relevance of Metcalfe's Law, that the utility of a network is proportional to the square of the number of its members (Gilder, 1993), is clear in the context of this examination of the nature of the costs; it is often cited in other contexts as an explanatory variable for killer apps (Downes & Mui, 2000; Evans & Wurster, 2000; Kelly, 1998). There are two stages to the process of growing a network: first, get the network's growth accelerating and, second, preserve the network once it is in place in order to create a *community of practice.* The economics of network effects show that a network of users of some technology or other good has three equilibrium positions. The first is where the number of users is zero. The second is a point where the network is satiated. Any further growth of the network will merely drag in users whose use of it is marginal, while any diminution will be felt as a loss of opportunity by the defecting network members. These are stable equilibria. But between these points is an unstable equilibrium, where users and costs, demand and supply, are balanced. However an increase in the size of the network, even a small one, will increase the benefits to users dramatically and bring new users in, while a diminution in the size of the network, even a small one, will decrease the benefits to network members dramatically and drive existing users away. It is essential for any network to get beyond the unstable equilibrium in the middle to promote growth (Varian, 2004).

Communities of practice: A community of practice (Wenger, 1998) is a group of people sharing some work or leisure-related practice; they need not all belong to the same organisation, but they do need some congruence of role. The community of practice that springs up around such a practice acts as a kind of support network for practitioners. It provides a language (or informal ontology) for people to communicate with, a corporate memory, and a means of spreading best practice. Such informal communities are self-selecting.

This self-selection, and informality, makes a community very hard to develop, because the community is a second-order development. So, we might take the example of Friend of a Friend (***FOAF***[4]). FOAF is a basic ontology that allows a user to express simple personal information (e-mail, address, name, etc.) as well as information about people they know. Many SW enthusiasts considered FOAF to be *cool* and *fun* and started publishing their FOAF ontologies. Currently, there are millions of FOAF RDF triples scattered over the Web, perhaps more than any other type of SW annotations.

Social network applications: There exist many Web applications that allow users to represent networks of friendships such as Friendster,[5] Okrut,[6] LinkedIn,[7] TheFacebook,[8] SongBuddy,[9]

to name just a few. However, FOAF has simply become a more convenient format for representing, publishing, and sharing this sort of information. Even though none of the applications above are entirely based on FOAF, some of them have already begun reading and exporting their data in FOAF format. FOAF is certainly helping spread RDF, albeit in a way limited to part of the SW community, and could therefore be regarded as a facilitator or a medium for possible killer apps that could make use of available FOAF files and provide some useful service.

Sustaining network growth: However, one interesting obstacle in the way of FOAF creating the nexus of users that will launch the SW is that a network is generally self-selecting and second order. A network is a network *for* something. There is absolutely nothing wrong with the FOAF method of providing a format for describing oneself. One obvious benefit of FOAF is that, as a pretty simple and flat ontology, it provides a relatively painless way of ascending the learning curve for non-users of SW technology. However, to sustain the network growth, there will still need for something underlying such networks, some practice, shared goal, or other practical purpose. Furthermore, it is also important to ascertain the invariants of the community or network experience, and ensure that the technologies and practices that produce growth preserve those invariants (Berners-Lee et al., 2006).

It may well be that a potentially more fruitful approach would be to support existing communities and try to expand SW use within them, so that little Semantic Webs emerge from them, as SW technologies and techniques reach saturation point within them (Kalfoglou et al., 2004). And because communities of practice overlap, and converge on various boundary objects and other linking practices and artefacts (Wenger, 1998), it may be that SW practice might spread beyond these islands of best SW practice. There are many obvious aids to such a development strategy; for instance, good-quality ontologies could be hand-crafted for particular domains. But also, it turns out that a number of the best SW tools at the moment also support this 'filling out' technique. For instance, **CS AKTive Space** (Shadbolt, schrae-fel, Gibbins, & Harris, 2003) specifically enables people to find out about the state of the discipline of computer science in the UK context, a limited but useful domain. Flink (Mika, 2005) generates FOAF networks for SW researchers. CS AKTive Space and Flink are winners of the 2003 and 2004 Semantic Web Challenges, respectively.

Open systems and social aspects: One other useful aspect of Flink is that it integrates FOAF profiles with ordinary HTML pages, and therefore sets up an explicit link between the SW and the WWW. Direct interaction with other existing systems increases the value of a system by acquiring additional value from those systems (Kelly, 1998). One good example is Protégé, an ontology editor from Stanford (Noy, Sintek, Decker, Crubezy, Fergerson, & Musen, 2001). By being open source and extendable, Protégé allowed many existing systems and tools to be linked or integrated with it, thus increasing its use and value. For this reason, and for being free, Protégé has quickly become one of the most popular ontology editing tools available today.

This openness is of course built into the very conception of the SW; the integration of large quantities of data, and the possibility of inference across them, is where much of the power stems from. As with the WWW, this does require a major programme of voluntary publication; for example, the possibility of e-commerce, and of the use of the WWW as a marketplace or as a version of the High Street stems from being able simply and conveniently to compare prices across retailers. The SW would add value (or reduce information processing costs) by allowing agents to do that and more. (Hendler, 2001).

And as with the WWW, if this process takes off, then more and more vendors would have to publish their data in RDF, even if they are initially reluctant. The argument in favour of such coercion is that everyone benefits eventually, and that early movers (those who make data such as price lists or stock lists available) not only gain, but force laggards to follow suit.

Privacy and trust: However, transparency and the removal of restrictions to publication are not undiluted goods. It may be that certain pieces of information benefit some organisations only as long as they withhold them from public view (trade secrets), that issues such as privacy and anonymity will rear their heads here, or even that differing intellectual property regimes and practices will lead to competitive advantage being lost in some economies.

In particular, integrating large quantities of information across the Internet and reasoning across them raises potential problems from a number of directions (Berners-Lee et al., 2006). First, it is the integration of information that threatens to allow harmful inference; information is quite often only harmful when seen in the right (or rather, wrong) context. The increasing numbers of mashups are potentially very dangerous here, as there is no ownership or chain of responsibility for information and its use. And the SW is the tool *par excellence* for putting previously harmless data in sensitive contexts. Second, publication of information (for instance, FOAF information) in a friendly and local context can quickly get out of one's control. There have been many warnings already about the placing of information on personal yet public spaces such as **MySpace**; there is anecdotal evidence that interviewees are being Googled to discover aspects of their private lives, to supplement the information about their public faces as presented on their CVs. Again, the SW is an important technology in this context, (O'Hara & Shadbolt, 2008).

It is often argued that standard data protection legislation is adequate for the new online contexts, but that policing is the problem. As it stands, traditional restrictions on the gathering of information are becoming decreasingly relevant as information crosses borders so easily; more plausible is policing restrictions on how information can be used once collected. The SW will require the development of formalisms that promote transparency of information use and accountability of information users. It should be possible to state one's preferences about the use of data in which one features (who, for example, can see the data, who can use it, and who can pass it on and to whom). Sometimes it will be necessary to override privacy preferences (for example, for the purposes of criminal investigation). It should be possible to receive metadata about how one's data have been used. And if there is a problem, and one's policies have been transgressed, it should be possible to hold someone to account. These requirements demand both social institutions and technical innovation. Various initiatives of the World Wide Web Consortium (**W3C**) address some aspects of these problems. The platform for privacy preferences (**P3P**) allows different agents a common view of privacy preferences, and is aimed at enhancing user control of data by simplifying the expression of privacy preferences. However, P3P contains no enforcement methods if preferences are violated (Cranor, Langheinrich, Marchiori, Presler-Marshall, & Reagle, 2002). The **Policy-Aware Web** initiative tries to provide rule-based formalisms to express preferences and increase transparency and accountability in the open and distributed environment of the WWW; SW technology can be important here by allowing rules and even proofs based on those rules to be exchanged between agents (Weitzner, Hendler, Berners-Lee, & Connolly, 2005).

Furthermore, formalising or externalising knowledge, for example, in the creation of ontologies, can have a number of effects. First of all, knowledge that is codified can become more 'leaky'; that is, it is more likely to leave an organisation. Second, it will tend to reduce the competitive advantage and therefore income of certain experts. Third, much depends on whether a consensus exists about the knowledge in the first place.

As Buckingham Shum (2004) argues, in a complex world where politics renders certain ideas untenable, incomplete knowledge requires whole perspectives to be understood, expertise must be combined in the light of discussion and argument, and information must be interpreted on the fly, and the consensual nature of ontologies (agreed conceptualisations of the world) can be unrealistic to insist on. The more lightweight the ontology, the easier it will be to adapt different viewpoints to it. Wilks (2004) argues that there are strong links between formal ontologies and the natural language terms which they borrow, and as with natural languages, insisting on a formal and verifiable common understanding of shared vocabulary (or isomorphic structures in formal ontologies) is too strong a constraint. Ontologies have to be flexible enough to allow interaction to take place.

Such flexibility, which will promote ease of use and therefore also help foster a user community, does mean that certain issues, such as referential integrity, will need to be faced (Bilenko, Mooney, Cohen, Ravikumar, & Fienberg, 2003), with some sufficiently lightweight methods for dealing with such problems of ambiguity as they arise.

Creativity and Risk

Killer app development cannot follow from careful planning alone. As we have noted already, there is no algorithm for creating a killer app. They tend to emerge from simple and inventive ideas; they get much of their transformative power by destroying hitherto reliable income streams for established firms. Christensen (1997) points out that most killer apps are developed by small teams and start-ups. Examples include **Google**,

eBay, and **Amazon**, which were all created by a few dedicated individuals. Giant industrial firms are normally reluctant to support risky projects because they are generally the ones profiting from the very income streams that are at risk (Downes & Mui, 2000).

However, even though most Semantic Web applications have so far been built in research labs and small groups and companies, there is clear interest expressed by the big players as well. So, for example, Hewlett Packard has produced **Jena** (Caroll, Dickinson, Dollin, Reynolds, Seaborne, & Wilkinson, 2003), a **Java** library for building SW applications, and IBM has developed **WebFountain** (Gruhl, Chavet, Gibson, Meyer, Pattanayak, Tomkins, & Zien, 2004), a heavy platform for large scale analysis of textual Web documents using natural language analysis and taxonomies for annotations. Adobe has perhaps gone further than many; Acrobat v5 now allows users to embed RDF metadata within their PDF documents and to add annotations from within Web browsers which can be stored and shared via document servers.

The SW provides a context for killer app development, a context based on the ability to integrate information from a wide variety of sources and interrogate it. This creates a number of aspects for the potential for killer apps. First of all, SW technologies might essentially be expected to enable the retrieval of data in a more efficient way than possible with the current WWW which is often seen as a large chaotic library. In such a world, the killer app might be some kind of retrieval technique to supplant Google, and therefore must do it without extra cost, without painful effort, and it should be the case that enough major players realise that Google is failing in some respect.

On the other hand, it may be that the SW might take off in an original and unpredictable direction. The clean infrastructure that the W3C ensures is in place could act as a platform for imaginative methods of collating and sifting through the giant quantities of information that is becoming avail-

able. This might result in a move away from the Web page paradigm, away from the distinction between content providers and consumers, as for example with efforts like CS AKTive Space (Shadbolt et al., 2003), or a move towards a giant, relatively uniform knowledge base (of the CYC variety) that could cope with all those complexities of context that foiled traditional AI approaches (Berners-Lee, 1998). The ultimate vision of the SW that prevails should affect not only the standards developed for it (Marshall & Shipman, 2003), but also where we might look for killer apps.

Personalisation

Personalisation has been a common thread in the development of killer apps. Customers tend to become more loyal to services they can customise to their liking (Downes & Mui, 2000). Many of today's killer apps have some level of personalisation; Amazon for example makes recommendations based on what the customer buys or looks at; **Auto Trader**[10] and **Rightmove**[11] save customers' searches and notify them via e-mails when a new result to their query is available; personalised Web services attract more customers (if done properly) and provide better tailored services (Evans & Wurster, 2000).

Personalisation is often the key to providing the higher service quality than the opposition; the service itself need not be provided in any better ways, but the personalised aspect gives it the extra that is needed to see off the alternatives. Such a connection could be indirect; for example, an Amazon-style recommender system, linked with an advertising platform, could help find alternative revenue streams and therefore drive down the cost to the consumer. The appearance of consumer choice in particular has two effects. First of all, the consumer may actually do a lot of the work for the service provider, for example, by negotiating a series of yes/no questions to an automated call centre, or by diligently scrolling through a set of

FAQs, the consumer may well diagnose his own problems. Second, the consumer's preferences get met so much more quickly (Markillie, 2005).

It goes without saying that personalisation is a hot topic on the SW; as well, annotated knowledge sources can be matched against RDF statements about individual consumers to create recommendations or targeted products. Indeed, the individual need not even do this voluntarily; as more information becomes available, for example, through the FOAF network or other RDF from third parties, users can become very known quantities indeed (raising all the privacy concerns we have noted elsewhere). There may well be major advantages to be had in systems that can feed information discreetly into recommender systems (Cox, Alani, Glaser, & Harris, 2004; schraefel, Preece, Gibbins, Harris, & Millard, 2004). Nevertheless, it is the personalisation aspect that has much potential for the SW, as long as the provision of enough information for the system to work interestingly is not too painful.

Semantic Web Applications

There have been few sustained attempts to try to promote SW applications. For instance, the important work of the **W3C** naturally is focused on the standards that will create the clean platform that is a necessary but sadly not sufficient condition for the SW to take off. But one of the most interesting and inspired is the series of Semantic Web challenges, which we will discuss briefly in the next section.

THE SEMANTIC WEB CHALLENGE

The annual International Semantic Web Challenge[12] (**SWC**) has been a deserved success, sparking interest and not a little excitement. It has also served to focus the community. Applications should 'illustrate the possibilities of the Semantic Web. The applications should integrate, combine, and deduce information from various sources to assist users in performing specific tasks.' Of course, to the extent that it does focus the community, the SWC will naturally influence the development of the SW.

Submissions to the SWC have to meet a number of minimum requirements:

- First, the information sources used:
    - ° Should be geographically distributed.
    - ° Should have diverse ownerships (i.e., there is no control of evolution).
    - ° Should be heterogeneous (syntactically, structurally, and semantically).
    - ° Should contain real world data, that is, are more than toy examples.
- Second, it is required that all applications assume an open world, that is, assume that the information is never complete.
- Finally, the applications should use some formal description of the meaning of the data.

Second, there are desiderata that act as tie-breakers:

- The application uses data sources for other purposes or in another way than originally intended.
- Using the contents of multimedia documents.
- Accessibility in multiple languages
- Accessibility via devices other than the PC.
- Other applications than pure information retrieval.
- Combination of static and dynamic knowledge (e.g., combination of static ontologies and dynamic workflows).
- The results should be as accurate as possible (e.g., use a ranking of results according to validity).

• The application should be scalable (in terms of the amount of data used and in terms of distributed components working together).

In the light of our discussion, these are interesting criteria. Many of them are straightforwardly aimed at ensuring that the characteristic possibilities of the SW are realised in the applications.

Semantic Web Challenge Winners

Let us have a close look at the winners of the SWC over the past three years: CS AKTive Space (CAS), the 2003 winner; **Flink**, the 2004 winner; and **ConFoto**, the 2005 winner. These applications were selected among 35 submissions over the past three years[13] and represent the best applications of Semantic Web applications in terms of SWC participant applications.

2003 winner, CS AKTive Space (CAS): In 2003, the first SWC took place, and a new style of exploring domain information by means of collecting and processing Semantic Web data won the first prize: CS AKTive Space (CAS).[14] CAS is a smart browser interface for a Semantic Web application that provides ontologically motivated information about the UK computer science research community. The scenario used is based on a real-world community request: the desire for a funding council to be able to get a fast overview of the council's domain from multiple perspectives. This requires bringing together data from heterogeneous sources and constructing methods to present the possible relations in the data to a user quickly and effectively. CAS developers report that this scenario applies equally to any stakeholder of the domain. While a funding body may wish to know, for instance, how funding in an area is distributed geographically, a researcher may also wish to know who is working on which topic, where they are located, and who is in the community of practice for the top researchers in their area (Glaser, Alani, Carr, Chapman, Ciravegna, Dingli, et al., 2004).

Figure 1. CS AKTive Space, 2003 SWC winner

CAS (Figure 1) provides multiple ways to look at and discover simple information or rich relations within the computer science domain of the UK. It facilitates querying, exploring, and organising information in ways that are meaningful to the users: where one user or group may be interested in seeing the relationship between funding, research area, and geographical region, another may be interested in who the top researchers are in AI and their address data. With CAS, people can formulate and see at a glance rich results like these, without having to string together large, complex queries.

CAS explores a wide range of semantically heterogeneous and distributed content relating to computer science research in the UK. There are almost 2,000 research active computer science faculty, 24,000 research projects represented, many thousands of papers, and hundreds of distinct research groups. These entities are described by a number of existing sources, such as institutional information systems (university Web sites, research council databases), bibliographic services, and other third party data sets (geographical gazetteers, UK Research Assessment Exercise submissions) (Shadbolt et al., 2003). This content is gathered on a continuous basis using a variety of methods, including harvesting and scraping of publicly available data from institutional Web sites, bulk translation from existing databases, and direct submissions by partner organisations, as well as other models for content acquisition. CAS supports both regularly scheduled harvesting to identify and deal with changes to existing data sources, and on-demand harvesting in response to changing user requirements (Ciravegna, Dingli, Guthrie, & Wilks, 2003) or update notifications from component sources.

The content is mediated through an **OWL** ontology[15] which was constructed for the application domain and which incorporates components from other published ontologies. The content currently comprises around 25 million RDF triples, and CAS developers provide scalable storage and retrieval technologies and maintenance methods to support its management (Harris & Gibbins, 2003). As a user interacts with the system, the interface generates queries (expressed in **RDQL**) which are evaluated by the triplestore server. The strength of CAS 'lies in the separation of users from the activity "under the hood", which enables users to answer questions that they might not be able to phrase' (Glaser et al., 2004). CAS also uses the **OntoCoPI** service (Alani et al., 2003) to get information about an individual's community of practice. OntoCoPI performs a spreading activation search, with weights assigned to the relationships in the ontology, in order to determine the ranked list of people with whom an individual associates. For example, the notion of a community of practice is calculated from a given researcher's co-authors, the projects that they are involved with, the institutions with which they are affiliated, and the topics in which they conduct research.

2004 winner, Flink: In 2004, the Flink[16] system won the SWC. Flink extracts, aggregates, and visualises online social networks (Mika, 2005). It employs semantic technology for reasoning with personal information extracted from a number of electronic information sources including Web pages, e-mails, publication archives, and FOAF profiles. The acquired knowledge is used for the purposes of social network analysis and for generating a Web-based presentation of the community. Flink is thus intended as a portal for anyone who is interested to learn about the work of the Semantic Web community, as represented by the profiles, e-mails, publications, and statistics. The data collected by Flink can also be used for the purpose of social network analysis, in particular, learning about the nature of power and innovativeness in scientific communities.

Flink is presented as a "who is who" of the Semantic Web by means of presenting professional work and social connectivity of Semantic

Web researchers. Flink defines the community of Semantic Web researchers as those researchers who have submitted publications or held an organizing role at four international conferences related to the Semantic Web. That data set represents a community of 608 researchers from both academia and industry, covering much of the United States, Europe, and, to a lesser degree, Japan and Australia.

Flink (Figure 2) takes a network perspective on the Semantic Web community, which means that the navigation of the Web site is organized around the social network of researchers. Once the user has selected a starting point for the navigation, the system returns a summary page of the selected researcher, which includes profile information as well as links to other researchers that the given person might know. The immediate neighbourhood of the social network (the ego-network of the researcher) is also visualised in a graphical form. The profile information and the social network is based on the analysis of Web pages, e-mails, and publications. The displayed profile

information includes the name, e-mail, homepage, image, affiliation, and geographic location of the researcher, as well as interests, participation at Semantic Web related conferences, e-mails sent to public mailing lists, and publications written on the topic of the Semantic Web. The full text of e-mails and publications can be accessed by following external links. At the time of the SWC contest, Flink contained information for about 5,147 publications authored by members of the community and 8,185 messages sent via five Semantic Web-related mailing lists.

The navigation from a profile can also proceed by clicking on the names of co-authors, addressees, or others listed as known by this researcher. In this case, a separate page shows a summary of the relationship between the two researchers, in particular, the evidence that the system has collected about the existence of this relationship. This includes the weight of the link, the physical distance, friends, interests, and depictions in common as well as e-mails sent between the researchers and publications written together.

Figure 2. Flink, 2004 SWC winner

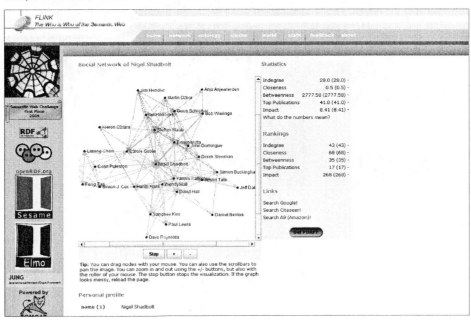

The information about the interests of researchers is also used to generate an ontology of the Semantic Web community. The concepts of this ontology are research topics, while the associations between the topics are based on the number of researchers who have an interest in the given pair of topics. An interesting feature of this ontology is that the associations created are specific to the community of researchers whose names are used in the experiment. This means that unlike similar lightweight ontologies created from a statistical analysis of generic Web content, this ontology reflects the specific conceptualisations of the community that were used in the extraction process. Also, the ontology naturally evolves as the relationships between research topics change.

2005 winner, CONFOTO: CONFOTO[17] won the 2005 SWC. CONFOTO is a browsing and annotation service for conference photos. It combines recent Web trends (tag-based categorisation, interactive user interfaces, syndication) with the advantages of Semantic Web platforms (machine-understandable information, an extensible data model, the possibility to mix arbitrary RDF vocabularies). CONFOTO offers a variety of tools to annotate and browse pictures. Simple forms can be used to create multilingual titles, tags, or descriptions, while more advanced forms allow the relation of pictures to events, people, ratings, or copyright information. CONFOTO provides a tailored photo browser and gallery generator for pictures, and a generic RDF browser for other resource types. Although a central repository is used to store resource descriptions, it is not necessary to copy photo files to the server: The application supports uploaded pictures as well as pictures linked via a URL or described in external RDF/XML documents. RSS export functions facilitate photo sharing, and a SPARQL interface enables and encourages extended data reuse.

Figure 3. Confoto, 0 2005 SWC winner

SWC and Killer Apps

Of particular interest to our work is the relation of the SWC criteria and the literature on killer apps. What the SWC is intended to uncover are new ways of exploiting information, particularly distributed information, and demonstrating the power of interrogation. In this respect, the challenge can only raise the profile of the SW, and help extend its community to more people and organisations. The SWC is an excellent vehicle for demonstrating where added value may come from. And as we have seen, increasing the size of the network will bring with it exponentially increasing benefits.

But the SWC looks unlikely to furnish us with a (prototype of a) killer app, because the criteria focus on interesting results, rather than on usability, superiority, or the alteration of old habits. Some of the criteria are slightly double-edged. For example, it is essential that an application uses a formal description of the meaning of the data. This, of course, is a deliberate attempt to ensure that one of the most contentious aspects of the SW (one of the most commonly-cited causes of scepticism about the SW's prospects) is incorporated, that is, the use of ontologies to provide understanding of terms in knowledge from heterogeneous sources. However, the way the challenge is constructed means that what is bound to happen in many if not most applications is that the developers will create their system with a possibly very painstakingly constructed ontology in mind, rather than taking the more difficult option of employing a very lightweight system that could work with arbitrary ontologies. This is particularly true for 2003 and 2004 SWC winner applications whereas the 2005 winner relies less on a rich ontological substratum but the application does not make use of rich ontological terms. The situation is somewhat similar to the knowledge engineering Sisyphus challenges, where knowledge engineering methodologies were tested and compared by being applied to the same problem. However, as the methodologies were applied by their developers, the results were less than enlightening; a later attempt to try to measure how difficult methodologies were to use by nonspecialists suffered from an unwillingness of most developers to discover this key fact about their methods (Shadbolt, O'Hara, & Crow, 1999).

There is little here to create the genuine community (as opposed to a large network) to promote the idea that users have something of a responsibility not to free ride and to publish RDF data. Neither is there much to promote personalisation within that community. There is little to protect privacy, little to reduce the pain of annotating legacy data or building ontologies, and, although the focus of the SWC is the results of information-processing, little to ensure that such processing can integrate into the organisational workflow. Surprisingly few of the traditional requirements, from a business perspective, appear in the SWC criteria.

Table 1 summarises how the SWC winners fit into some of the general features of killer apps that we discussed earlier. It is interesting to observe the similarity between the first two applications which both targeted the same domain and community. CS AKTiveSpace however focused on gathering information about where researchers are located and what areas they research, with some social network analysis based on addresses, research areas, projects, as well as co-authorship relations. Flink on the other hand is mainly for social network analysis based almost entirely of co-authorship information. The third winner, Confoto, allows for user networks to be formed and hence, in theory, it has the potential of generating a network value. Having said that, Confoto currently stands hopeless against giant rivals that proved extremely popular in the general WWW community, such as **Flickr**, although the latter is probably based on less advanced technology. Generally speaking, all three winners are specifically designed and built for a relatively small community (i.e., computer scientists or academics in general) which inherently restricts their

Table 1. Some general characteristics of Semantic Web Challenge winners

Application	Cost	Added Value	Community	Open Standard	Personalisation	Bootstrapping
CSAktiveSpace	Main cost is in data gathering, cleaning, and merging Simple browsing skills required for using existing service Use is free	Provides knowledge about researchers that is not easily obtainable from elsewhere But current system is only limited to Computer Science domain	Service targets a specific community No support for user interactions Metcalfe's network value can not be generated from the system	Based on RDF and knowledge base technology No API for external access, import/ export, or for adding extensions Not open source	Users can not change or add to the data held in the system No user specific features are provided	System is setup with data from various sources Various techniques are used to collect the data No dependent on number of users
Flink	Main cost is in gathering and preparing the data A learning curve to understand the system output Use is free	Produces graph network profiles for researchers based on co-authorship information Limited to specific bibliographic sources for Semantic Web researchers	Service targets a specific community No support for user interactions Metcalfe's network value can not be generated from the system	Based on RDF and knowledge base technology No API for external access, import, export, or for adding extensions Some is open source	Users can not change or add to the data held in the system No user specific features are provided	System is setup with data from various sources Various techniques are used to collect the data No dependent on number of users
Confoto	Main cost is in developing the system Data gathering cost is minimal Use is free	Provides conference attendees with a semantically enriched photo annotation and sharing tool General field is currently dominated by Flickr	Users can interact by annotating each other's photos, profiles, and events Can generate a network value	Based on RDF and XML Users can import /export data No API for external access Not open source	Some non-persistent personalisation is provided Users can added to the data held in the system	System needs to be bootstrapped with user data (photos) Entirely dependent on user input

number of potential beneficiaries and limits their potential spread.

None of this, let us hastily add, is intended as a criticism of the SWC, which has publicised the SW and drawn a lot of attention to the extra power that it can provide. Our point is merely that there is a lot more to finding a killer app than producing an application that does brilliant things.

DISCUSSION

Killer apps must provide a higher service quality and evolve (pretty quickly) into something perceived as indispensable, conferring benefits on their users without extra costs or steep learning curves. Individual users should coalesce into a community of practice, and their old habits should change in accordance with the new possibilities provided by the app. This is particularly important as the SW is likely to impose new costs on users in the short term, for example, through having to annotate legacy content, develop ontologies, and so forth. As the SW is in a relatively early stage of development, it is not currently clear exactly what threats and opportunities it provides (and, of course, the future form of the SW will conversely depend on what applications for it are successful). There has been some speculation about how the SW will develop and what extensions of the WWW will be appropriate or desirable. For instance, consider a recent attempt by Marshall and Shipman (2003) to understand potential development routes for the SW, which sets out three distinct but related visions of the SW:

1. *SW technology could bring an order and consistent structure to the chaotic Web.* Information access would be assisted by semantic metadata. This vision envisages that humans will continue to be the chief agents on the SW, but that information could now be represented and stored in ways to allow its use in situations far beyond those

foreseen by its original authors. In other words, the SW will extend the existing Web, but exactly how is hard to predict. In order for potential SW killer apps to respond to this vision should, ideally, have the following properties in particular:

°   They should help foster communities of users (that is, at some level, users should want to interact and share experiences with others). SW technology is expected to facilitate knowledge sharing and bring more people together.

°   Users should not feel submerged in a mass, but should retain their individuality with personalised products. With more machine readable information becoming available (e.g., FOAF), better personalisation should be feasible.

°   Users should be able to bring as much of their legacy content up to date with relatively painless maintenance techniques. So, for example, applications should be able to leverage comparatively simple ontologies; ontology construction, merging, and selection should be made easier, and we should be able to move away from handcrafting; annotation methods and interfaces have to be easy. In all these cases, the existence of a community of interested users will provide the initial impetus for the user to ascend the learning curve.

2. *The Web will be turned, in effect, into a globally distributed knowledge base, allowing software agents to collect and reason with information and assist people with common tasks, such as planning holidays or organising diaries.* This vision seems close to Berners-Lee et al.'s (2001) SW grand vision. In many ways, it is the composition of the other visions, assuming machine processing and global representation of knowledge. It is also a vision that will require more from a potential killer app:

    ° They should exploit integrated information systems to make inferences that could not be made before, or to bring sources of information (for example, data held in relational databases) together in the context of more powerful information processing tools. Showing added value is key to encourage businesses and content providers to participate in the SW.

    ° They should help remove the rather painful need to annotate, build ontologies, and so forth.

    ° The new application should fit relatively smoothly into current work or leisure experiences. Little change in habits is acceptable, assuming some returned benefit, but too much change is a problem.

3. *The SW will be an infrastructure made up of representation languages, communication protocols, access controls, and authentication services for coordinated sharing of knowledge across particular domain-oriented applications.* Information is used largely for the original purposes of its author, but much more machine processing will take place. If this vision prevails, then a prospective SW killer app should pay special attention to:

    ° It should not compromise other important aspects of users' lives, for instance by threatening privacy to a dangerous degree, either by making inappropriate surveillance possible, or by facilitating torts such as identity theft.

    ° Furthermore, if such a vision comes to pass, then the opportunities for killer apps are all the greater, in that any such standards-driven platform approach should make it possible for as many applications to flourish on top of it as possible. Whether such applications will ever be acknowledged as SW apps

rather than general WWW applications is open to debate.

These conditions for each vision of the SW are, of course, necessary yet not sufficient. Furthermore, all of the conditions apply, to some degree, to each of the three visions.

We have seen that killer apps appear when there are opportunities to make progress on costs, communities, creativity, and personalisation. All new technologies begin with a handicap in these areas. They impose costs of retooling, learning curves, and business process rescheduling. There is always a chicken and egg problem with the development of a community of users: the technology of necessity precedes the community. The risks of creative thought become clearer at the outset; the benefits only appear later. And the dialectic between personalisation and creating economies of scale often means that the latter are pursued long before the former. As an added handicap, it is often the case that the costs are borne disproportionately by early adopters. The opportunities of the SW are also therefore counterbalanced by the risks. We note that we cannot predict where new killers will come from. The transformations that such applications wreak make the future very different from the present. Hence we cannot be concretely prescriptive. But the general requirements for killer apps that emerge from our review of the business/management literature suggest certain routes for development, in addition to sensible lists of characteristics such as the criteria for the SW challenges, or the conditions listed at the beginning of this section.

Therefore, it is probably uncontroversial to assume that any SW killers will have to provide: (1) a service that is not possible or practical under more traditional technologies, (2) some clear benefit to developers, data providers, and end users with minimum extra costs, and (3) an application that becomes indispensable to a user-base much wider than the SW researchers community. Additionally, research should be focusing on four important areas. First of all, perhaps most important, the cost issue should be addressed. Either the potentially large costs of annotating, ontology development, and so forth should be mitigated or side-stepped by thinking of types of applications that can work with minimal non-automatic annotation, low cognitive overhead, or ontologies that sacrifice expressivity for simplicity. Second, another way of improving the cost/benefit ratio is to increase benefits, in which case, the fostering of user communities looks like a sensible way forward. This means that applications in real-world domains (preferably in areas where the Internet was already important, such as media/leisure or e-science) look more beneficial than generic approaches. Third, creativity is important, so radical business models are more interesting than simply redoing what the WWW already does. And fourth, personalisation needs to be addressed, which means that extended user models are required.

When we look at the three visions outlined by Marshall and Shipman, vision III appears to be the one most amenable to the development of killer apps, in that it envisages a platform upon which applications sit; the form of such applications is left relatively open. In contrast, vision I, for example, does not see too much of a change for the WWW and the way it is used, and so there are fewer opportunities opening up as a result. Indeed, when we look at vision I, assuming that the SW does improve the navigation of the chaotic Web, it might even be appropriate to say, not that there is a killer app for the SW, but rather that the SW is the killer app for the WWW. Whereas with Marshall and Shipman's vision III, the vision is of a garden in which a thousand flowers bloom. Applications can be tailored to context; ontologies can remain as lightweight as possible, and data can be exploited as required. Publication of data can be tempered, thanks to privacy-enhancing tools, by supporting restrictions on its use. On this vision, it is the painstaking, pioneering, and often tedious negotiations of standards that will be key; such standards need to support the right kind of research.

CONCLUSION

Killer apps are very difficult things to monitor. They are hard to describe, yet you know one when you see one. If you are finding difficulty persuading someone that something is a killer app, it probably is not. A lot depends on the boot-strapping problem for the SW; that is, if the SW community is small, then the chances of someone coming up with a use of SW technology that creates a genuinely new use for or way of producing information are correspondingly small. For it is finding the novelty that is half the battle. There is unlikely to be much mileage in simply reproducing the ability to do something that is already possible without the SW. Furthermore, it is likely that a killer app for the SW will exploit SW technology integrally; merely using RDF will not quite do the trick (Uschold, 2003). The willingness of the producers of already-existing killers to use SW technology, like Adobe, is encouraging, but again will not necessarily provide the killer app for the SW. The checklist of criteria for the SWC gives a good list of the essentials for a genuine SW application.

This is not simply a matter of terminology. The SW is more than likely to thrive in certain restricted domains where information processing is important and expensive. But the ambitions of its pioneers rightly go beyond that. For that to happen, killer apps need to happen. We hope we have given some indication, via our examination of the business literature, of where we should be looking.

ACKNOWLEDGMENT

This work is supported under the Advanced Knowledge Technologies (AKT) Interdisciplinary Research Collaboration (IRC), which is sponsored by the UK Engineering and Physical Sciences Research Council under grant number GR/N15764/01. The AKT IRC comprises the Universities of Aberdeen, Edinburgh, Sheffield, Southampton, and the Open University. The views and conclusions contained herein are those of the authors and should not be interpreted as necessarily representing official policies or endorsements, either expressed or implied, of the EPSRC or any other member of the AKT IRC.

REFERENCES

Alani, H., Dasmahapatra, S., O'Hara, K., & Shadbolt, N. (2003). Identifying communities of practice through ontology network analysis. *IEEE Intelligent Systems, 18*(2), 18-25.

Berners-Lee, T. (1998). *The Semantic Web road map.* Retrieved August 20, 2007, from http://www. w3.org/DesignIssues/Semantic.html

Berners-Lee, T. (1998-2005). *The fractal nature of the Web.* Retrieved August 20, 2007, from http://www.w3.org/DesignIssues/Fractal.html

Berners-Lee, T. (2003). Semantic Web: Where to direct our energy? Keynote speech at the *2nd International Semantic Web Conference (ISWC2003),* Florida. Retrieved August 20, 2007, from http://www.w3.org/2003/Talks/1023-iswc-tbl/

Berners-Lee, T. (2004). WWW 2004 keynote. In *Proceedings of the 13th International World Wide Web Conference,* New York. Retrieved August 20, 2007, from http://www.w3.org/2004/Talks/0519-tbl-keynote/

Berners-Lee, T., Hall, W., Hendler, J.A., O'Hara, K., Shadbolt, N., & Weitzner, D.J. (2006). A framework for Web science. *Foundations and Trends in Web Science, 1*(1), 1-129.

Berners-Lee, T., Hendler, J., & Lassila, O. (2001, May). The Semantic Web. *Scientific American.*

Bilenko, M., Mooney, R., Cohen, W., Ravikumar, P., & Fienberg, S. (2003, September/October). Adaptive name matching in information integration. *IEEE Intelligent Systems,* pp. 2-9.

Buckingham Shum, S. (2004, January/February). Contentious, dynamic, information-sparse domains ... and ontologies? *IEEE Intelligent Systems,* pp. 80-81.

Carroll, J.J., Dickinson, I., Dollin, C., Reynolds, D., Seaborne, A., & Wilkinson, K. (2003). *Jena: Implementing the Semantic Web recommendations* (Tech. Rep. No. HPL-2003-146). HP Laboratories Bristol.

Christensen, C.M. (1997). *The innovator's dilemma.* Harvard Business School Press.

Ciravegna, F., Dingli, A., Guthrie, D., & Wilks, Y. (2003). Integrating information to bootstrap information extraction from Web sites. In *Proceedings of Workshop on Information Integration, IJCAI'03,* Acapulco, Mexico.

Coase, R. (1991). The Nature of the Firm. In E.O. Williamson & S.G. Winter (Eds.), *The nature of the firm: Origins, evolution and development.* Oxford University Press: Oxford. (Original work published in *Economica* 1937)

Cox, S., Alani, H., Glaser, H., & Harris, S. (2004). The Semantic Web as a semantic soup. In *Proceedings of the First Workshop on Friend of a Friend, Social Networking and the Semantic Web,* Galway, Ireland.

Cranor, L., Langheinrich, M., Marchiori, M., Presler-Marshall, M., & Reagle, J. (2002). The platform for privacy preferences 1.0 (P3P1.0) Specification. *World Wide Web Consortium.* Retrieved August 20, 2007, from http://www. w3.org/TR/P3P/

Downes, L., & Mui, C. (2000). *Unleashing the Killer App.* Harvard Business School Press.

Eischen, K. (2002). *The social impact of informational production: Software development as an informational practice* (CGIRS working paper 2002-1). Center for Global, International and Regional Studies, University of California, Santa Cruz.

Ellman, J. (2004, January/February). Corporate ontologies as information interfaces. *IEEE Intelligent Systems,* pp. 79-80.

Evans, P., & Wurster, T. (2000). *Blown to bits.* Harvard Business School Press.

Fensel, D., Bussler, C., Ding, Y., Kartseva, V., Klein, M., Korotkiy, M., Omelayenko, B., & Siebes, R. (2002). Semantic Web application areas. In *Proceedings of the 7th International Workshop on Applications of Natural Language to Information Systems (NLDB 2002),* Stockholm, Sweden.

Gilder, G. (1993, September 13). Metcalfe's Law and legacy. *Forbes ASAP.*

Glaser, H., Alani, H., Carr, L., Chapman, S., Ciravegna, F., Dingli, A., Gibbins, N., Harris, S., schraefel, M., & Shadbolt, N. (2004). CS AKTive space: Building a Semantic Web application. In *Proceedings of 1st European Semantic Web Symposium (ESWS 2004),* Crete, Greece (pp. 417-432).

Gruhl, D., Chavet, L., Gibson, D., Meyer, J., Pattanayak, P., Tomkins, A., & Zien, J. (2004). How to build a Webfountain: An architecture for very large-scale text analytics. *IBM Systems Journal, 43*(1), 64-76.

Harris, S., & Gibbins, N. (2003). 3store: Efficient Bulk RDF storage. In *Proceedings of the ISWC'03 Practical and Scalable Semantic Systems (PSSS-1),* Sanibel Island, FL.

Hendler, J. (2001, March/April). Agents and the Semantic Web. *IEEE Intelligent Systems,* pp. 30-37.

Kalfoglou, Y., Alani, H., Schorlemmer, M., & Walton, C. (2004). In On the emergent Semantic Web and overlooked issues: *Proceedings of the 3rd International Semantic Web Conference (ISWC 2004),* Hiroshima, Japan.

Kelly, K. (1998). *New rules for the new economy: 10 Radical strategies for a connected world.* Penguin Books.

Klein, M., & Fensel, D. (2001). Ontology versioning on the Semantic Web. In *Proceedings of the 1st International Semantic Web Working Symposium,* Stanford University, CA (pp. 75-91).

Markillie, P. (2005, March 31). Crowned at last. *The Economist.*

Marshall, C., & Shipman, F.M. (2003). Which Semantic Web? In *Proceedings of the 14th HyperText Conference (HT 2003),* Nottingham, UK (pp. 57-66).

McGuinness, D.L. (2002). Ontologies Come of Age. In D. Fensel, J. Hendler, H. Lieberman, & W. Wahlster (Eds.), *Spinning the Semantic Web: Bringing the World Wide Web to Its Full Potential.* MIT Press.

Mika, P. (2005). Flink: Semantic Web technology for the extraction and analysis of social networks. *Journal of Web Semantics, 3*(2-3), 211-223.

Nixon, L. (2004, December). *Prototypical business use cases* (FP6 IST NoE Deliverable D1.1.2, KnowledgeWeb EU NoE FP6 507482). Retrieved August 20, 2007, from http://www.starlab.vub.ac.be/research/projects/knowledgeweb/KWeb-Del112v1.pdf

Noy, N.F., Sintek, M., Decker, S., Crubezy, M., Fergerson, R.W., & Musen, M.A. (2001, March/April). Creating Semantic Web contents with Protégé-2000. *IEEE Intelligent Systems,* pp. 60-71.

O'Hara, K. & Shadbolt, N. (2008). *The spy in the coffee machine: The end of privacy as we know it.* Oneworld Publications.

schraefel, M.C., Preece, A., Gibbins, N., Harris, S., & Millard, I. (2004). Ghosts in the Semantic Web machine? In *Proceedings of the 1st Workshop on friend of a friend, social networking and the Semantic Web*, Galway, Ireland.

Schreiber, G., Akkermans, H., Anjewierden, A., de Hoog, R., Shadbolt, N, de Velde, W.V., & Wielinga, B. (1999). *Knowledge engineering and management: The CommonKADS approach.* MIT Press.

Shadbolt, N., O'Hara, K., & Crow, L. (1999). The experimental evaluation of knowledge acquisition techniques and methods: History, problems, and new directions. *International Journal of Human-Computer Studies (IJHCS), 51*, 729-755.

Shadbolt, N., schraefel, M., Gibbins, N., & Harris, S. (2003). CS AKTive Space: Or how we stopped worrying and learned to love the Semantic Web. In *Proceedings of the 2nd International Semantic Web Conference*, Florida.

Uschold, M. (2003). Where are the semantics in the Semantic Web? *AI Magazine, 24*(3), 25-36.

Uschold, M., & Gruninger, M. (2004). Ontologies and semantics for seamless connectivity. *SIGMOD Record, 33*(4).

Varian, H.R. (2004). Competition and Market Power. In H.R. Varian, J. Farrell, & C. Shapiro (Eds.), *The economics of information technology: An introduction* (pp. 1-47). Cambridge University Press: Cambridge.

Weitzner, D.J., Hendler, J., Berners-Lee, T., & Connolly, D. (2005). Creating a policy-aware Web: Discretionary, rule-based access for the World Wide Web. In E. Ferrari & B. Thuraisingham (Eds.), *Web and Information Security.* Idea Group Inc.

Wenger, E. (1998). *Communities of practice: learning, meaning and identity.* Cambridge: Cambridge University Press.

Wilks, Y. (2004). Are ontologies distinctive enough for computations over knowledge? *IEEE Intelligent Systems, 19*(1), 74-75.

Williamson, O.E. (1975). *Markets and hierarchies.* New York: Free Press.

Williamson, O.E. (1991). *The nature of the firm: origins, evolution and development.* Oxford: Oxford University Press.

ENDNOTES

[1] http://www.iabuk.net
[2] http://www.pwc.com/
[3] http://knowledgeweb.semanticweb.org/
[4] http://www.foaf-project.org/
[5] http://www.friendster.com/
[6] http://www.orkut.com/
[7] http://www.linkedin.com/
[8] http://www.thefacebook.com/
[9] http://www.songbuddy.com/
[10] http://www.autotrader.co.uk
[11] http://www.rightmove.co.uk
[12] http://challenge.semanticweb.org/
[13] A full list of submissions is available at http://www.cs.vu.nl/~pmika/swc/submissions.html
[14] Demonstration available online at http://triplestore.aktors.org/SemanticWebChallenge/
[15] Available online at http://www.aktors.org/ontology/
[16] Demonstration available online at http://prauw.cs.vu.nl:8080/flink/
[17] Demonstration available from http://www.confoto.org/home

Chapter XV
Semantic Location Modeling for Mobile Enterprises

Soe-Tsyr Yuan
National Chengchi University, Taiwan

Pei-Hung Hsieh
STPRIC, National Science Council, Taiwan

ABSTRACT

A location model represents the inclusive objects and their relationships in a space and helps engender the values of location based services (LBS). Nevertheless, LBS for enterprise decision support are rare due to the common use of static location models. This chapter presents for enterprises a framework of dynamic semantic location modeling that is novel in three ways: (1) It profoundly brings location models into enterprise business models; (2) with a novel method of dynamic semantic location modeling, enterprises effectively recognize the needs of the clients and the partners scattered in different locations, advancing existing business relationships by exerting appropriate service strategies through their mobile workforces; (3) through the location model platform of information sharing, enterprises are empowered to discover potential business partners and predict the values of their cooperation, gaining competitive advantages when appropriate partnership deals are made by enterprise mobile workforces. This proposed framework has been implemented with the J2EE technology and attained the positive evidences of its claimed values.

INTRODUCTION

With the advent of wireless communication technologies, the era of mobile enterprises unfolds. Many international enterprises like IBM, Sun, HP, and Microsoft are vying to develop mobile enterprise servers and solution architectures. According to a Cutter report, 57% of the employees in the enterprises worldwide were regarded as the "mobile workforce" in 2005 (Ericsson Enterprise,

2002). Accordingly, following the e-business trend, competitive advantages built on wireless technologies in dynamic mobile environments are now widely recognized by enterprises.

The conventional perception of mobile enterprises is that enterprise users are able to have personalized, seamless access to enterprise applications and services from anyplace and at anytime, regardless of the devices employed, in order to facilitate the tasks at hand (Bouwman et al., 2005; Ericsson Enterprise, 2002).

Subsequently, location is an inherent feature of many mobile services. Location-based services (LBS) are information services that exploit knowledge about where an information device user is located. According to Ovum, an analyst and consulting company, the market for LBS will grow to $12 billion by 2006. Existing LBS primarily rest on four categories of services (Varshney, 2000): (1) safety (e.g., emergency services, roadside assistance); (2) navigation and tracking (e.g., vehicle navigation, asset tracking, people tracking); (3) transactions (e.g., location-sensitive billing, zone-based traffic calming); and (4) information (e.g., yellow pages, location-based advertising). The main idea behind the former three categories is locating targeted objects for provision/consumption of certain external resources. The last category then focuses on targeted advertising, linking nearby consumers/buyers and providers/sellers to facilitate additional revenue generation (Polyzos, 2002; Ververidis & Yuan & Peng, 2004; Yuan & Tsao, 2003). LBS has been a hot area of research because mobility of information device users leads to the generation of user location information that subsequently drives a slew of new services.

Moreover, enterprise decision support (Bolloju, 2003) is often regarded as: (1) use of corporate data to derive and create higher level information and knowledge, (2) integration of organizational information to support all departments and end users, and (3) provision of tools to transform scattered data into meaningful business information. Enterprises utilizing geometrical data are often the likes of logistic companies of which LBS mainly rests on the provision of support on navigation and the tracking of their employees (shipping vehicles) or clients. For instance, logistic delivery planning locates shipping vehicles based on geometric models: static location models (MapInfo, http://www.mapinfo.com/products/Features.cfm; RITI Technology Inc., http://www.elocation.com.tw) to know all inventories in transit and enable efficient logistic deliveries (Varshney, 2000). Nevertheless, it is rare to perceive LBS as enterprise decision support in attaining higher level information and knowledge. *It naturally comes to a question of how to marry enterprise decision support with LBS so as to deeply utilize the business data together with the geometric data.* In searching for the answer to the question, there is a need to identify the reasons behind the limited extent of this marriage. (Afterwards, this sort of marriage is named *enterprise-based LBS.*)

In this research, we believe the possibilities behind this limitation are (1) the integration of enterprise business models and existing location models is difficult; and (2) the limitation of existing location models hinders additional development on enterprise-based LBS.

With the aforementioned suppositions, *this chapter aims to present a framework of dynamic semantic location modeling (DSLM) that shows certain integration of enterprise business models and the proposed location model (that surmounts the problems encountered in static location models), realizing enterprise-based LBS* (e.g., the location-sensitive decisions of potential strategic partners required in the expansion of enterprise alliance networks). The DSLM framework is believed to encourage the development of myriad research on enterprise-based LBS in the future. This chapter will first discuss the limitations of existing location models and then present the DSLM framework, followed by some evaluation results and conclusion.

LOCATION MODELING

Existing methods for location modeling are twofold (Domnitcheva, 2001): the first one is geometric modeling that is built upon the geometric coordinate system. The other is symbolic modeling that represents locations with symbols and symbol sets.

Each location modeling method has its pros and cons. Geometric modeling (static location models) has the advantages of high accuracy and easy communication between different kinds of platforms. However, geometric modeling requires reference points and mappings between information objects and geometric coordinate objects. On the other hand, symbolic modeling represents locations with location object names (e.g., 11th Park in Taipei), each of which unfolds as a set containing the objects residing in the designated location. Symbolic modeling accordingly is easy to comprehend, but requires effort in managing the naming of the location objects and the handling of the ranges and the overlaps of the location objects[1] (vLeonhardt, 1998).

While exerting geometric models (static location models) for enterprise-based LBS, there are two primary problems encountered:

- **Meaningless syntactic information:** A mobile enterprise application system can attain only *syntactic* information objects regarding a given location. For instance, when a salesperson queries the system for product sales information of a designated branch office, he may get numerous sales figures for the product at the designated location, but do not know whether these figures imply good sales or bad sales. For situations that salespeople are capable of judging the performance of these figures, the judgments cannot be wisely retained for facilitating subsequent relevant decision making (that however appreciates these *semantic* judgments).

- **No seamless information exchange/integration:** When the exchange or integration of location-sensitive information is intended by enterprises, this might give rise to the need of a middleware for the information translation when enterprises employ different static local models. The rationale is twofold: (1) the mapping between information objects and coordinate objects in a static location model is fixed (*static*), and thus it is hard to interoperate the information objects exchanged; (2) this fixed mapping also creates difficulties in the merging of the two static location models when tight enterprise relationships are attempted (i.e., the *dynamic* expansion of existing location models).

From the above discussion, there are two vital desired features for enterprise-based LBS: "semantic" and "dynamic." *"Semantic" indicates that an enterprise can define its own objects, object values, object relationships in a location model* (Pradhan, 2002). *"Dynamic" then denotes that a location model can grow and adapt with the enterprise interactions, building "dynamic links" between locations [6].* These two features drive the necessity of the development in a new method of location modeling[2] in order to shed light on advanced enterprise-based LBS.

This chapter presents DSLM that unfolds itself as a new location modeling method and is the first attempt integrating enterprise business models and the proposed location model so as to realize an advanced enterprise-based LBS. The contributions of DSLM are threefold (denoted by BOLM, PNLM, and LMP) and outlined in Table 1. BOLM, PNLM, and LMP differ with each other mainly in the scopes of their functions. BOLM and PNLM can endow enterprise mobile workforce with location-sensitive decision information about their clients or potential partner enterprises. On the other hand, LMP furnishes enterprises of an industry with a platform in which location-sensitive new potential partners can be identified. Their details will be described in Section 3.2, 3.3, and 3.4, respectively.

As for other existing enterprise location services, they mainly focus on the potentials of RFID by managing readers, filtering and aggregating raw RFID data, and facilitating data exchange among the supply chain partners (Bouwman, Haaker, & Faber, 2005). There is a need to search for advanced work for location intelligence (that combines spatial-data collection with advanced analysis and visualization methods to transform sitting to knowledge) (Grimes, 2005).

THE FRAMEWORK OF DSLM

DSLM aims to fulfill a certain integration of enterprise business models and location models in terms of the three DSLM solutions. These solutions involve enterprise clients, enterprise partners, and a platform enabling the search of new partners. Accordingly, interoperability and decision-support aid are the key characteristics of DSLM. This section starts with the description of the ontology employed in DSLM (that defines the semantics required to represent our location models and to enable location intelligence in enterprise interoperability) followed by the three DSLM solutions addressed in Table 1.

DSLM Ontology

DSLM fits in the category of symbolic modeling, but the relationships between symbols and symbol sets can be changed dynamically. The DSLM ontology is a shared ontology that is regarded as the interchange format, enabling common access to enterprise operational data (Jasper & Uschold, 1999). DSLM ontology defines objects, object relationships, and relationship measurements. The following subsections will detail these terms.

Table 1. The DSLM solutions in mobile enterprise decision support

Solution	Within Enterprise ***Business-Oriented Location Model (BOLM)***	Between Enterprises ***Partner-Network Location Model (PNLM)***	Within Industry ***Location Model Platform (LMP)***
Main Function	Assist an enterprise mobile workforce to understand the business relationships with clients in certain locations.	Endow an enterprise mobile manager with the knowledge of the benefits of the cooperation with potential partner enterprises in a certain location area so as to attain satisfactory cooperation contracts or deals.	Assist an enterprise to search for potential partner enterprises to cooperate in certain location areas (of different location regions).
Benefit	Employ proper location-sensitive strategies to better utilize the enterprise's resources.	1. Location-sensitively attain the cooperation relationships between enterprises. 2. Expand the service scope and range through cooperation between enterprises.	Realize an information-sharing platform between enterprises in different locations.

Objects

DSLM ontology defines four types of objects (original unit, business unit, client unit, and business-oriented location model) as shown in Figure 1 and defined in Definition 1:

- **Original unit (OU):** An entity in a map that is not at all referenced in the location model of an enterprise because of no business relationship between the enterprise and the entity. For instance, if there is no business relationship between a freight company A and a bookstore B, then B will be regarded

Figure 1. Objects in DSLM

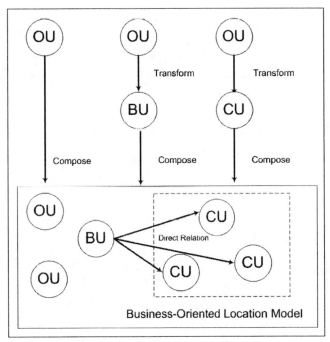

Definition 1.

OU(Y) : Y is the Original Unit in the DSLM

BU(C) : C is the Business Unit of the DSLM

CU(D,C) : D is the client of Business Unit C in the DSLM

$BOLM(C) = def \exists_{X_1, X_2, \ldots, X_n}$ is $CU(X_1, C)$, $CU(X_2, C)$, ...,$CU(X_n, C)$

$\exists_{Y_1, Y_2, \ldots, Y_m}$ is $OU(Y_1)$, $OU(Y_2)$,..$OU(Y_m) \in BOLM(C)$

$\bigwedge_{i=1}^{n} \bigwedge_{j=1}^{m} ((BU(C) \wedge CU(X_i, C) \wedge OU(Y_j)) \Big|_{DR(x_i,C)}$

as an OU in A's Business-Oriented Location Model.

- **Business unit (BU):** Upon the construction of a business-oriented location model for an enterprise, the OU representing the enterprise transforms into a BU.

- **Client unit (CU):** An OU representing a client of the enterprise (constructing its business-oriented location model) transforms into a CU.

- **Business-oriented location model (BOLM):** The BOLM of an enterprise is

Figure 2. Object relationships in DSLM

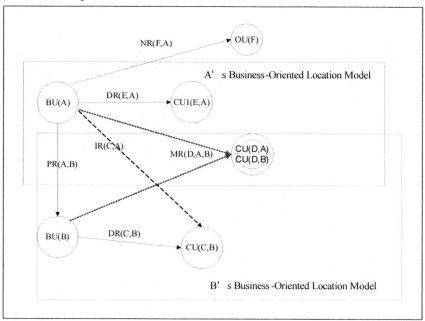

Definition 2.

```
A is a Business Unit in BOLM(A)
B is a Business Unit in BOLM(B)
F represents an Original Unit
E represents a Client Unit
```

NL(F, A) : F is not located in BOLM(A)

NR(F, A) : F has no relation with BU(A)

LI (E, A) : E is located in BOLM(A)

PR(A, B) : BU(A) and BU(B) has partner relation

$DR(x, A) = (LI(x, BOLM(A)) \wedge CU(x, A))$

$IR(y, A) = \exists_z (BU(z) \wedge (NL(y, BOLM(A)) \wedge PR(A, z) \wedge CU(y, z))$

$MR(z, A, B) = (LI(z, BOLM(A)) \wedge LI(z, BOLM(B) \wedge PR(A, B) \wedge CU(z, A) \wedge CU(z, B))$

comprised of the BU (representing the enterprise), the CUs (denoting all of the clients of the enterprise, and OUs (symbolizing the entities without business relationships with the enterprise). For instance, if a logistic company C has three clients (D, E, F), then C's BOLM is composed of 1 BU representing C and 3 CUs denoting D, E, F, and a couple of OUs. The relationship (DR relationship) between the enterprise and its clients will be detailed in the next subsection.

Object Relationship

Object relationships stand for the relationships between the BU, CUs, and the BOLM. There are a variety of relationships being modeled: direct relationship (DR), indirect relationship (IR), no relationship (NR), multiple relationship (MR), partner relationship (PR), located in (LI), and not-located in (NI) as shown in Figure 2 and defined in Definition 2:

- **Direct relationship (DR):** A relationship denoting the direct business relationship between a client and the enterprise such as DR(C, B) as shown in Figure 2 in which C is a client of the enterprise B.

- **Partner relationship (PR):** A business relationship between two enterprises such as PR(A, B) as shown in Figure 2.

- **Indirect relationship (IR):** A relationship between a client C (of the enterprise B) and the enterprise A that is formed because of a Partner Relationship between A and B such as IR(C, A) as shown in Figure 2.

- **Multiple relationship (MR):** A relationship between a client and multiple enterprises that have the partner relationship such as MR(D, A, B) as shown in Figure 2 in which the client D is a client of both A and B (that further have the partner relationship with each other).

- **No relationship (NR):** A relationship other than any of the aforementioned relationships.

- **Located in (LI):** An inclusive relationship between a BU (CU, or OU) and a BOLM.

- **Not-Located in (NI):** A non-inclusive relationship between a BU (CU, or OU) and a BOLM.

Relationship Measurement

In order to differentiate the relationships for the purpose of decision support, the relationship DR,

Algorithm 1. An exemplar of DR measurement

Function Direct_Relation_Measurement (BU, CU)

1. Select significant attributes A_i that characterizes the relationship between BU and CU, and transform their values to Semantic Levels SL_i according to BU's subjective judgment.

2. Assign weight W_i to all chosen attributes according to the levels of their significance to BU.

3. $DR = \sum_{i=1}^{n} SL_i * W_i$

Note : This algorithm only exemplifies a linear measurement. Non-linear measurements can be employed in Step 3 as well.

Algorithm 2. An exemplar of IR measurement

Function Indirect_Relation_Measurement (Source Enterprise BU, Target Enterprise BU, CU)

1. Select significant attributes A_i from the client CU's Source Enterprise BU, and transform their values to Semantic Levels SL_i according to Target Enterprise BU's subjective judgment.

2. Assign weight W_i to all chosen attributes according to the levels of their significance to Target Enterprise BU.

3. $IR = \sum_{i=1}^{n} SL_i * W_i$

Note : This algorithm only exemplifies a linear measurement. Non-linear measurements can be employed in Step 3 as well.

Algorithm 3. Example of MR measurement

Function Multiple_Relation_Measurement (Source Enterprise BU, Target Enterprise BU, CU)

1. Calculate DR(Source Enterprise BU, CU).

2. Source Enterprise BU calculates DR'(Target Enterprise BU, CU) by using the CU's attributes and data retained in Target Enterprise BU.

3. MR = DR' - DR

IR, and MR are associated with measurements. These measuring are based on the values of certain object attributes that an enterprise concerns such as distance[3] from the enterprise, average revenue and average order. Algorithm 1, Algorithm 2, and Algorithm 3 exemplify certain algorithms for calculating the relationship measurements:

- **DR measurement:** Between the direct clients (CU) of an enterprise (BU), DR measurements aim to differentiate the clients. Algorithm 1 exemplifies one possible way of such differentiation that is accomplished through the calculation of a weighted sum of the values of the client's attributes chosen by the enterprise.

- **IR measurement:** Between the indirect clients (CU) of an enterprise (Target Enterprise BU) because of its partnership with another enterprise (Source Enterprise BU),

IR measurements intend to distinguish the indirect clients by calculating a weighted sum of the CU's attribute values gathered from Source Enterprise BU.[4] However, the weights are assigned from the point view of Target Enterprise BU (instead of from Source Enterprise BU's as shown in Algorithm 2).

- **MR measurement:** Given a MR (in which a client CU is associated with Source Enterprise BU and Target Enterprise BU by the MR bindings), MR measurements aims to further discriminate these bindings in terms of different originating perspectives (i.e., from the perspective of Source Enterprise BU). Algorithm 3 shows the method for a MR measurement from the perspective of Source Enterprise BU. This MR measurement represents a strength difference between the DR measurement (of Source Enterprise

BU and CU) and the DR' measurement (of Target Enterprise BU and CU) for which the retrieval of CU's data retained in Target Enterprise BU is made). In other words, from the perspective of Source Enterprise BU, a MR measurement reveals an important message about the subjective relative strength (with respect to Target Enterprise BU) in regard to the relationship with the client CU. For instance, it manifests a stronger relationship that Source Enterprise BU has with CU than that of Target Enterprise BU when the MR measurement is less than zero.[5]

Mobile Enterprise Decision Support Using DSLM

This section describes the three DSLM solutions (BOLM, PNLM, LPM) mentioned in Table 1. Each of the solutions supplies relevant decision-support aids and leads to certain integration of enterprise business models and enterprise location models as described in Section 3.1.

BOLM

A business-oriented location model (BOLM) (as defined in Definition 1) represents a location model that is composed of the objects and the relationships that are embodied with semantics and are able to be dynamically expanded and updated as the myriad enterprise relationships develop with the clients. The construction of a BOLM for an enterprise involves the calculation of the DR measurements (i.e., the semantics perspective) with respect to the enterprise clients and evolves these DR measurements with continuous interactions between the enterprise and the current clients engaged (i.e., the dynamic perspective).

The application of a BOLM (e.g., a mobile workforce deciding the service priorities for clients in a certain location area) accordingly involves consulting these relationship measurements together with additional myriad considerations (attributes) of service requests (e.g., request distance, emergency, profit, etc.). For simplicity, a liner weighted scheme is exerted on these service attributes to attain the proportion of the significance share for a given service request (invoked by a given client) besides the other proportion of the significance share coming from relationship measurements, followed by another liner weighted scheme combining both significance shares.

The following exemplifies the BOLM application (that subsequently will be evaluated in Section 4.1):

- A logistic enterprise A has seven clients (spread over different regions of a given area): B, C, D, E, F, G, H (that simultaneously make requests to Enterprise A for its services by the temporal order of {C, F, D, E, H, B, G}. Assume their distances to a mobile workforce (shipping vehicle) of Enterprise A (arriving in this designated area) are increasingly ordered as follows: {C, E, F, H, D, B, G}.

- Based on the client attributes and their weights shown in Table 2, the DR measurements of BOLM for these clients are calculated as shown in Table 3 (with Algorithm 1).

- While the mobile workforce wirelessly accesses the enterprise's BOLM for the decision of an appropriate arrangement to serve the clients requests, additional myriad attributes of service requests (request distance, emergency, and profit) are taken into account as shown in Table 4. Subsequently, the DR measurements and the service request considerations are combined as shown in Table 5, and the decision of the priorities of the clients to serve is then determined.

- Table 6 contrasts the BOLM service-request arrangement with the others' (First-In-First-Out and Shortest-Distance-First),

Table 2. The attributes (and their corresponding weights) considered in DR measurements

DR Attributes	Average Shipment	Average Revenue	Average Positive Feedback
Weights	0.5	0.2	0.3

Note: The weights (0.5, 0.2, 0.3) represent the relative degrees of importance considered by Enterprise A when differentiating its clients in terms of the three chosen attributes.

Table 3. Results of DR measurements

	B	C	D	E	F	G	H
Average Shipment	7	5	4	7	2	1	3
Average Revenue	3	5	3	5	4	6	7
Average Positive Feedback	4	5	6	3	3	6	7
DR Measurement	5.3	5	4.4	5.4	2.7	3.5	5

Note: A SL value (e.g., a value ranging from 1 to 10) for a designated attribute denotes a subjective performances score with respect to the attribute from the viewpoints of Enterprise A.

Table 4. Service request considerations

		B	C	D	E	F	G	H
Request distance	0.5	7	1	5	1	2	7	2
Emergency	0.1	3	3	6	1	5	3	4
Profit	0.4	4	4	7	2	1	4	3
Weighted SUM		5.4	2.4	5.9	1.4	1.9	5.4	2.6

Table 5. The resulting service request arrangement by the BOLM method

		B	C	D	E	F	G	H
DR Measurement	0.8	5.3	5	4.4	5.4	2.7	3.5	5
Service Request Considerations	0.2	5.4	2.4	5.9	1.4	1.9	5.4	2.6
Weighted SUM		5.62	4.48	5.19	4.6	2.54	3.88	4.52
Order		1	5	2	3	7	6	4

Table 6. A contrast between the different service-request arrangements

	B	C	D	E	F	G	H
First-In-First-Out	6	1	3	4	2	7	5
Shortest-Distance-First	6	1	5	2	3	7	4
BOLM	1	5	2	3	7	6	4

manifesting that the BOLM method takes on a different service-request arrangement (that will be shown to outperform First-In-First-Out and Shortest-Distance-First in Section 4.1).

PNLM

Partner-network location model (PNLM) enables the realization of the benefits of cooperation between enterprises residing in different location regions, in terms of the expanded market share of clients or the increased relationships with clients. This realization is able to assist an enterprise mobile manager to negotiate with potential partner enterprises of a certain location area regarding their cooperation contracts or deals.

Suppose a PNLM is formed because of the cooperation between enterprises A and B. The PNLM from A's perspective is then defined as in Definition 3. A picturesque view of this PNLM is shown in Figure 3. (Figure 4 then shows that of

Definition 3.

$$PNLM(A,AB) = def \ \exists x_1, x_2, x_3, \ldots, x_m \ is \ CU(x_1,A), CU(x_2,A), \ldots, CU(x_m,A)$$

$$\exists y_1, y_2, y_3, \ldots, y_n \ is \ CU(y_1,B), CU(y_2,B), \ldots, CU(y_n,B)$$

$$\exists z_1, z_2, z_3, \ldots, z_p \ is \ CU(z_1,A), CU(z_2,A), \ldots, CU(z_p,A)$$

$$also \ is \ CU(z_1,B), CU(z_2,B), \ldots, CU(z_p,B)$$

A and B have PR(A, B)

$$\bigwedge_{i=1}^{m} \bigwedge_{j=1}^{n} \bigwedge_{k=1}^{p} (BU(A) \wedge BU(B) \wedge CU(x_i,A)\big|_{DR(x_i,A)} \wedge CU(y_j,B)\big|_{IR(y_j,A)} \wedge$$

$$CU(z_k,A)\big|_{MR(z_k,A,B)})$$

Figure 3. PNLM of enterprise A and B in A's point of view

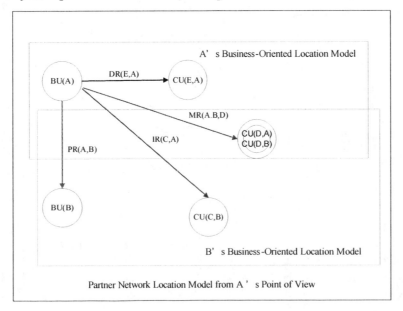

the PNLM from B's perspective.) In other words, PNLM is constructed out of a PR relationship between A and B, and subsequently IR and MR are generated. Since IR and MR are directional relationships, a PNLM accordingly is formulated as a directional model (i.e., from the perspective of A (or B)). The PNLMs from different perspectives differ with each other in terms of the different measurements calculated.

The benefits of exerting PNLM in an enterprise are exemplified by two scenarios as shown below:

- **Competitors cooperation scenario:** Two competitive enterprises (such as the former Compaq and HP) cooperate through PNLM that facilitates things such as expanding market share in myriad location regions, perceiving client relationships development, recognizing their services overlap, discerning the increase in their service scope, and so forth. This scenario emphasizes the importance of the number of clients increasing because of cooperation.

- **Vertical supply chain scenario:** A manufacture enterprise residing in southern Taiwan seeks a northern logistic enterprise to cooperate through PNLM for providing better services to the manufacture enterprise's clients in northern Taiwan. This scenario then emphasizes the increased relationship measurements because of the cooperation.

With the two aforementioned scenarios, a PNLM performance evaluation algorithm is provided (as shown in Algorithm 4) for evaluating the performance of the cooperation with a Target Enterprise BU from the perspective of a Source Enterprise BU. The performance evaluation is represented as a vector comprising a CAN value and a SIM value. *The CAN value denotes the increase in the number of the clients because of the cooperation between Source Enterprise BU and Target Enterprise BU, and the SIM value then stands for the increased relationship measurements because of the cooperation (i.e., the sum of the IR and MR measurements).*

The rationale behind this performance vector is twofold: (1) Different enterprises might have dif-

Figure 4. PNLM of enterprise A and B in B's point of view

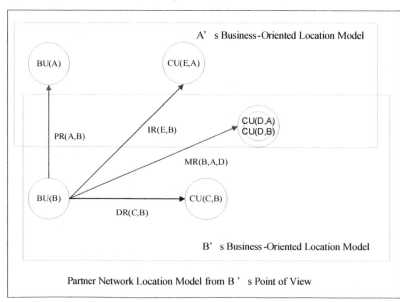

ferent objectives in the cooperation (as exemplified in the above scenarios) and thus the performance vector is unfolded as a vector of a CAN value and a SIM value (instead of a single scalar); (2) rendering different [CAN, SIM] vectors (corresponding to different Target Enterprises) on a 2-dimension space, it is easy to snatch the various strengths between different cases of enterprise cooperation (i.e., an exemplar of advanced location intelligence derived from dynamic semantic reasoning and computation).

Algorithm 4. PNLM performance evaluation

Function PNLM_Performance (PNLM, Source Enterprise BU, Target Enterprise BU)
1. From the Source Enterprise's perspective, calculate the increase in CU because of the given PNLM and give rise to a statistics named a CAN value.
2. From the Source Enterprise's perspective, calculate the increased amount of measurements in relationships because of IR and MR encountered. This amount is named a SIM value.
3. Set the PNLM performance vector with respect to the Target Enterprise BU as a vector of [CAN, SIM].

Note: If Source Enterprise BU cannot attain relevant client's attribute values (CU) from Target Enterprise BU during the calculation of the relationship measurements, then this CU would be considered as an OU. Source Enterprise BU subsequently calculates the relationship measurements in terms of the OU's attribute values.

Figure 5. (a) α's BOLM; (b) β's BOLM; (c) PNLM(α,β)

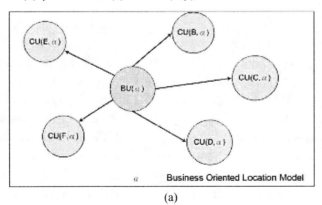

(a)

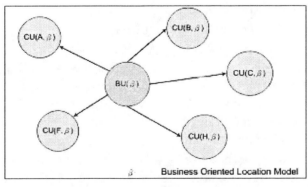

(b)

Figure 5. continued

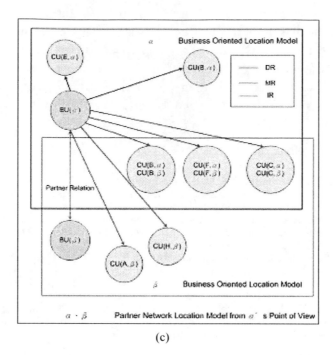

(c)

As follows is an exemplar regarding the computation of the CAN and SIM values:

- In Figure 5, (a) shows the picturesque view of Enterprise α of clients {B, C, D, E, F}; (b) shows that of Enterprise β of clients {A, B, C, F, H}; (c) then denotes the picturesque view of PNLM(α,β) that is a directional view from Enterprise α's perspective.
- Table 7 shows the DR measurements in Enterprise α's BOLM and in Enterprise β's BOLM; Table 8 then exhibits those IR measurement and MR measurement that are generated from the creation of PNLM(α,β).
- Since CAN denotes the increase in the number of the clients because of the cooperation between Source Enterprise α and Target Enterprise β, the CAN value of PNLM(α,β) accordingly equals to 2 (that arises because of the clients in {A, H} becoming the indirect

clients of Enterprise α due to the partnership with Enterprise β).

- Since SIM represents the increased relationship measurements (i.e., the sum of the IR measurements and the MR measurements) because of the cooperation, the SIM value of PNLM(α,β) accordingly equals to 5 (that is calculated by summing the values {4, 2, -4, 2, 1}[6] that denote those IR and MR measurements attained in enterprises {A, B, C, F, H} respectively).

LMP

Location model platform (LMP) is a platform for the exchange of BOLM abstracts. In other words, the shared BOLM abstracts empower the search of potential enterprises to cooperate without exposing enterprises' confidential and private information (i.e., another exemplar of advanced location intelligence derived from

Table 7. The DR measurements presumed for Enterprise αand Enterprise β

	A	B	C	D	F	H	I	J
α	N/A	2	7	4	3	N/A	N/A	N/A
β	6	4	3	N/A	5	1	N/A	N/A

Table 8. The IR and MR measurements presumed from the perspective of Enterprise α

	A		B		C		D		F		H		I		J	
	IR	MR	IR	MR	IR	MR	IR	MR	IR	MR	IR	MR	IR	MR	IR	MR
α	4	N/A	N/A	2	N/A	-4	N/A	N/A	N/A	2	1	N/A	N/A	N/A	N/A	N/A

Figure 6. Example of BOLM abstract

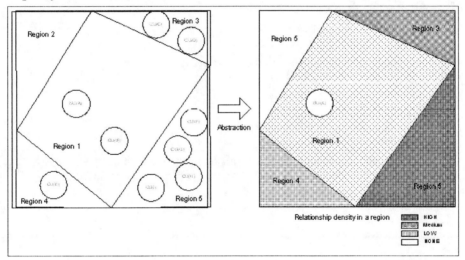

dynamic semantic reasoning and computation). Figure 6 shows an example of BOLM abstracts, and Algorithm 5 lists the algorithm for producing BOLM abstracts.

A BOLM abstract unfolds as a distribution of various business sizes on a designated geographical coverage that is composed of a certain number of geographical regions (as shown in Figure 6). The business size in a region is represented with a certain semantic label (High Density, Medium Density, or Low Density) that is allocated accord-ing to the relative strength of ongoing business unfolding in the designated region with respect to that developing in all of the regions. Ongoing business is measured by the size of the clients and the size of the relationship measurements.

EVALUATION

Our DSLM is implemented using the service-oriented architecture (Machiraju, 2001). J2EE

Algorithm 5. BOLM abstract construction algorithm

Function BOLM_Abstract_Construction (BOLM)

1. From BOLM, identify all CUs that have direct relationship (DR) to the business.

2. In each geographical region, sum up all DR measurements and multiply this sum with the number of CUs in the region, obtaining a scalar representing a Region Relationship (RR).

   $\forall$ Region $R_j \in$ BOLM(C), $j=1,...,n$; $CU(x_p,C) \subset R_j$, $p=1,...,m$;

   $$RR_j = (\sum_{p=1}^{m} DR(x_p,C)) * m$$

3. Total Region Relationship (TR) is the sum of all the RRs.

   $$TR = \sum_{j=1}^{n} RR_j$$

4. Region Relationship Percentage is defined as the percentage of a designated RR to TR.

   $$RRP_j = \frac{RR_j}{TR} \times 100\% , \ j = 1,.....,n;$$

5. Assign Semantic Levels (region density) to RRPs :

   - High Density : $100\% > RRP >= \dfrac{100\%}{total\ region}$

   - Medium Density : $\dfrac{100\%}{total\ region} > RRP >= \dfrac{100\%}{m \times total\ region}$

   - Low Density : $\dfrac{100\%}{m \times total\ region} > RRP > 0$

   - None   : RRP=0

   Note : m is a tuning parameter determined by the platform designer.

6. Label the region density (the semantic level of RRP) in every region.

7. Return the labeled abstract.

---

and Enterprise JavaBeans technology are used to develop the DSLM system (as shown in Figure 7). The Model View Controller (MVC) pattern is exerted to hinge on a clean separation of objects into one of three categories: **models** for maintaining data (BOLM EJB Entity Bean and PNLM EJB Entity Bean), **views** for displaying all or a portion of the data (JSP Javabeans), and **controllers (EJB Session Beans)** for handling events that affect the model or view(s). Because of this separation, multiple views and controllers can interface with the same model. Even new types of views and controllers that never existed

before can interface with a model without forcing a change in the model design.

In this section, different sets of experiments are employed to realize the claimed contributions of the three solutions (BOLM, PNLM and LMP) in Section 4.1, 4.2 respectively. Section 4.3 then provides a short discussion of the evaluation results. Although the full-scope justifications won't be available until fielded experiments (i.e., attaining long-term observation of the DSLM performance in relevant enterprises/industries) are constructed, these results anew shed light on future enterprise LBS.

BOLM Evaluation

This evaluation unfolds itself by exerting a logistic enterprise BOLM example on the task of service request arrangement (as shown in Figure 8) in order to show the increased values brought by the decision support of BOLM.

In this logistic enterprise BOLM example, six types of clients (that are commonly perceived as shown in Table 9) dynamically generate requests to the enterprise (of which its mobile workforce with their shipping vehicles are responsible for fulfilling the requests of the clients). Each request is composed of a variety of attribute values (such as request distance, request unit price, and request quantity) that are also dynamically generated (given the assumption that there are limits set for request distance and request quantity in this example).

This example compares three different methods for the task of service request arrangement in terms of the average resulting value to the enterprise:

- **First-In-First-Out:** Serving requests by the order of the request sequence.
- **Far-Distance-Based:** Serving requests by the decreasing order of the request distances.[7]
- **BOLM:** Serving request by the order of client relationship measurements.

In Figure 8, Broker Agent pools 10 clients[9] requests (forming a request set) and sends them to the enterprise service arrangement method periodically. The request-sending magnitude of a client controls how often this client will post requests to the enterprise. Clients will tune their request-sending magnitude in the following ways: *if the enterprise rejects a client's request, the client will tune down the request-sending magnitude, but will raise this magnitude vise versa.*[10] The

Figure 7. DSLM J2EE implementation architecture

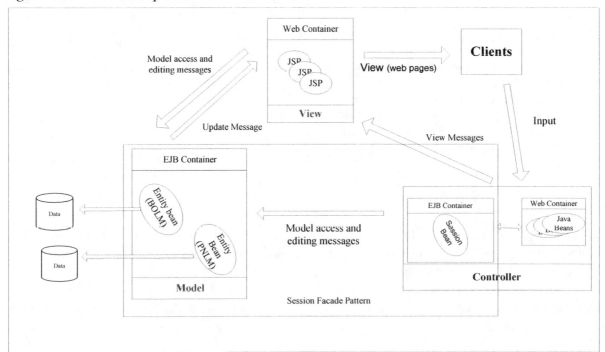

Figure 8. BOLM experiment system architecture

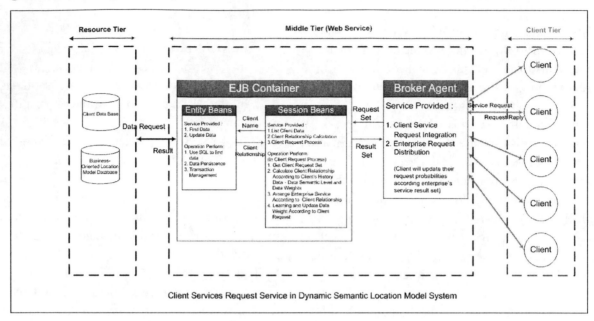

Table 9. Client request type in experiment environment

Attribute Client Type	Request Distance	Request Quantity	Request Unit Price
1	Long	High	High
2	Short	High	High
3	Long	Low	High
4	Short	Low	High
5	Long	High	Low
6	Short	High	Low

*(Note: Request Distance: "Long" represents the range from 500 to 1300 and "short" from 100 to 499;
Request Quantity: "High" represents the range from 50 to 130 and "Low" from 10 to 49;
Request Revenue = Request Quantity * 1.5 (with High Request Unit Price);
Request Revenue = Request Quantity (with Low Request Unit Price);
Request Revenue is also multiplied by 3.5[8] while Request Distance is Long)*

value of a request set (i.e., the 10 pooled client request per period) will be calculated with equation (1) (in which the value of a request set for the logistic enterprise is proportional to the revenue received but reverse proportional to the distance transported and the quantity carried).

$$Request\ Set\ Value = Total\ Request\ Revenue / (Total\ Request\ Distance * Total\ Request\ Quantity)$$

$$(1)$$

A reply set is the arrangement results (with respect to a given request set) returned by the enterprise service arrangement method.[11] Equation (2) computes the value of the reply set.

Reply Set Value = Total Reply Revenue / (Total Reply Distance * Total Reply Quantity)

(2)

Request Set Value *and* Reply Set Value *are two metrics employed to evaluate the performance of the service arrangement methods.* High Request Set Value *indicates the continuity of intensive business opportunities, and high* Reply Set Value *then denotes quality arrangement between service requests.*

Distinguished from First-In-First-Out *and* Far-Distance-Based, *the BOLM method employs LMS weight update rule [4,17] to* evolve the weights of the service-request attributes (as shown in Table 4) for the purpose of adaptively serving clients in light of the dynamic magnitudes of their service requests. *This adaptation aims at adjusting the weights toward the direction of high* Request Set Value *and high* Reply Set Value. *The weight learning equation is shown in equation (3).*

Weight = Weight + learning rate * (Request set Value / Reply set Value)*Xi

(3)

*Note: learning rate = 0.1; If the weight of the distance attribute is under tuning, then Xi represents the sum of distance in reply set.

With the evaluation experiment setup addressed, evaluation results show that the choice of the request arrangement method will affect the magnitudes of client requests, request set values, and reply set values. On the other hand, *a good request arrangement method should be able to stably generate competitively high client request set values and enterprise reply set values by continuously arranging the requests of the clients to serve prosperously.*

We gradually experiment up to 1,000 client request sets (an exemplar of a dynamically generated request set is shown in Table 10). That is, there are 10,000 client requests in total sent to each request arrangement method. An investigation of which method can stably generate higher client request set values and enterprise reply values is explored. *Figure 9 then shows the evaluation results in which the BOLM method outperforms the other two methods (FIFO and Far-Distance-Based) throughout the whole experiment process (i.e., from a small number of request sets to a great number of request sets). Furthermore, the two values stay quite stable throughout the experiment process with the BOLM method.*

Figure 9. The evaluation results for the three request arrangement methods

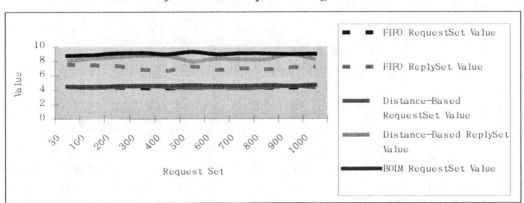

Table 10. Exemplars of a request set and the reply set obtained by the BOLM method

```
Client Request Set are :
The No.0 request set is :client = 2; distance = 404; revenue = 189; quantity = 126
The No.1 request set is :client = 3; distance = 1156; revenue = 182; quantity = 52
The No.2 request set is :client = 1; distance = 608; revenue = 309; quantity = 59
The No.3 request set is :client = 6; distance = 143; revenue = 61; quantity = 41
The No.4 request set is :client = 5; distance = 817; revenue = 73; quantity = 21
The No.5 request set is :client = 1; distance = 1069; revenue = 346; quantity = 66
The No.6 request set is :client = 6; distance = 139; revenue = 58; quantity = 39
The No.7 request set is :client = 4; distance = 181; revenue = 66; quantity = 66
The No.8 request set is :client = 2; distance = 319; revenue = 169; quantity = 113
The No.9 request set is :client = 6; distance = 142; revenue = 42; quantity = 28
request distance = 4978.0
request revenue = 1495.0

request quantity = 611.0
request set value = 4.915244095295898
++++++++++++++++++++++++++++++++++++++++++++++++++++++++++++++
The Reply Set after Enterprise process are :
The request set Enterprise accepted : client = 5; distance = 817; revenue = 73; quantity = 21
The request set Enterprise accepted : client = 1; distance = 608; revenue = 309; quantity = 59
The request set Enterprise accepted : client = 1; distance = 1069; revenue = 346; quantity = 66
The request set Enterprise accepted : client = 2; distance = 404; revenue = 189; quantity = 126
The request set Enterprise accepted : client = 2; distance = 319; revenue = 169; quantity = 113
The request set Enterprise rejected : client = 3; distance = 1156; revenue = 182; quantity = 52
The request set Enterprise accepted : client = 6; distance = 143; revenue = 61; quantity = 41
The request set Enterprise accepted : client = 6; distance = 139; revenue = 58; quantity = 39
The request set Enterprise accepted : client = 6; distance = 142; revenue = 42; quantity = 28
The request set Enterprise rejected : client = 4; distance = 181; revenue = 66; quantity = 66
reply distance = 3641.0
reply revenue = 1247.0
reply quantity = 493.0
reply value = 6.947024896198523
```

Figure 10. Client request magnitude changes in the FIFO method

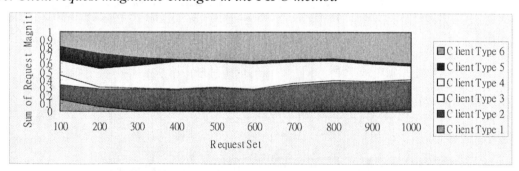

Figures 10, 11, and 12 show the client request magnitude changes (i.e., the dynamics of the distribution of the requests) for the three request arrangement methods. In the FIFO method, the client types 2, 4, and 6 that make Low distance requests are the higher request magnitude ones (i.e., with wider ranges along the dimension of Sum of Request Magnitude). In contrast, the client types 1, 3, and 5 are the higher magnitude ones in the Far-Distance-Based method. However, in

Figure 11. Client request magnitude changes in the far-distance-based method

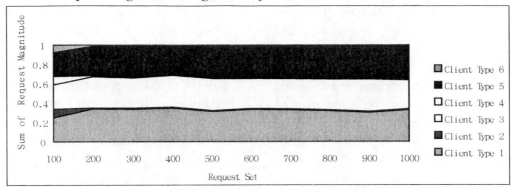

Figure 12. Client request magnitude changes in the BOLM method

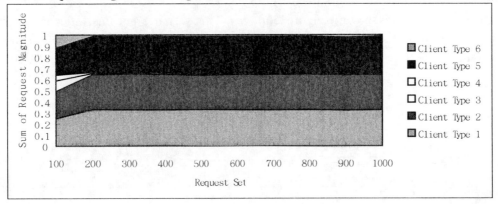

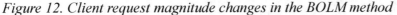

the BOLM method, the client types 1, 2, and 5 are the higher magnitude ones. This is due to request revenue being computed by request distance and request quantity. The client types 1 and 2 have High request quantity and the client type 5 has Higher distance compensation than the client type 6. In other words, the BOLM method is able to come up with valuable clients to serve.

Evaluation of PNLM and LMP

The application of PNLM and LMP involves factors considered in the decisions of seeking and evaluating enterprises for intercooperation or interoperability (that usually induces monumental

policies, practices, contracts on enterprise alliance). However, this section aims at delivering certain prospects realized by the existence of PNLM and LMP in terms of two scenarios and their positive evaluation results (that will be detailed in Section 4.2.1 and 4.2.2, respectively).

These scenarios unfold themselves around a common geographical setting and circumstance, detailed as follows:

* As shown in Figure 13, in the geographical setting of Taipei city there are five geographical regions (identified by Region1 to Region5) covering the 36 smaller geographical areas (those in a 6*6 coordinated plane).

- Four enterprises (Enterprise1 to Enterprise4) are exerted for manifesting the two scenarios. These enterprises are presumed to be cooperative when conditions are met. Figure 14 exemplifies a fragmented portion of the client data of an enterprise. In this figure, each client record is composed of its location values (Location, LX, LY)[12] and its attribute values (PROP_A, PROP_B, PROP_C, PROP_D). Without loss of generality, these attributes are merely represented by symbols (A, B, C, D) (that can be tailored to a customized set of attributes in an industry and an enterprise of this industry can exert only a subset of the attributes for measuring its client relationship as shown in Figure 15(a) to (b), and the attribute values are randomly generated.

- A set of general principles underlying the selection of enterprises to cooperate [1,7,10] is employed to evaluate the performance of PNLM and LMP. The following are the principles: compatible (conforming to the needs), complementing (complementing with each other either in functions or in operational locations), analytic (measuring the merits of the cooperation), and feasible (examining the feasibility of the cooperation), and homogeneous (sharing the same competitive goal).

Scenario 1

This scenario goes as follows: Enterprise1 is a logistic delivery company not residing in Taipei city and is searching for good regional operation representatives in Taipei.

Figure 13. An exemplar of the geographical setting

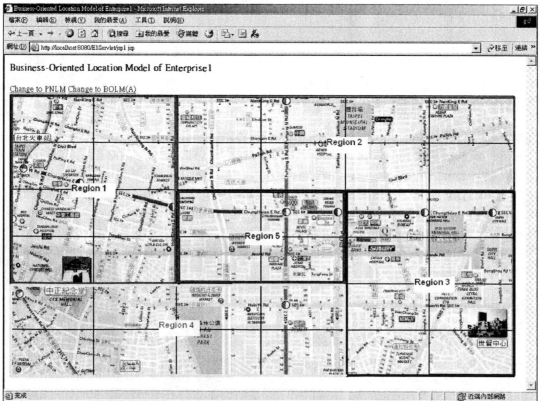

Figure 14. A fragmented portion of the client data for an enterprise

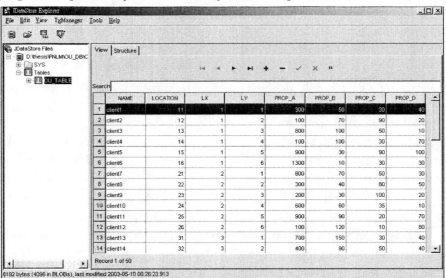

Figure 15. (a) shows the selection of the attributes (together their weights) in measuring the client relationship for Enterprise3; (b) then shows the mappings to semantic labels from the attribute values (e.g., the semantic label of 2 is assigned when the values of the attribute A are less than 600 hut greater than 300).

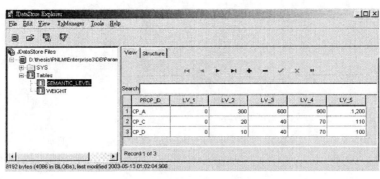

(a)

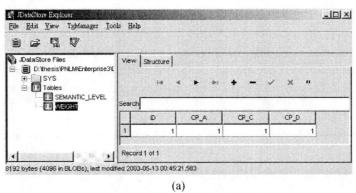

(b)

This search and evaluation can be facilitated with LMP and PNLM and unfold as follows:

- Attain a set of candidates of regional representatives from LMP based on the needs and the conditions. For instance, Enterprise1 requires a representative that fairly engages in commercial activities (exhibiting its adequate business connection) in Region3 and Region5. This need can be fulfilled by looking up LMP (as shown in Figure 16) for the enterprises that are of mediocre densities of clients (e.g., densities roughly larger than 0.1 but less than 0.2) in Region3 and Region5, and the set of candidates accordingly comprises Enterprise3 and Enterprise4.

- Analyze the candidates with the measurements of CAN and SIM attained from PNLM. In other words, assess the values (the quantity of the indirect client size and the quality of their clients in terms of the relationship sum) brought to Enterprise1 through the partnership with Enterprise3 (Enterprise4). Figure 17 shows PNLMs associated with Enterprise1&3 and Enterprise1&4, and exhibits the

(CAN, SIM) value vector of (20, 188[13]) for the partnership with Enterprise3 and (20, 182) for Enterprise4. Accordingly, Enterprise3 outperforms Enterprise4 from the perspective SIM. This information can furnish Enterprise1's mobile managers with a valuable starting point to continue alliance contract negotiation with Enterprise3 or Enterprise4 together with the consideration of additional alliance criteria.

- In this scenario, PNLM and LMP fully (partly) satisfy the following cooperative principles: compatible (LMP assists in generating the cooperation candidates in designated regions), analytic (PNLM measures the values CAN and SIM), and feasible (PNLM's ClickPoint function equips the evaluating task a close look of the to-be-clients).

Scenario 2

This scenario proceeds as follows: Enterprise1 is a logistic delivery company and has clients spread over the Taipei geographical regions. Enterprise2, Enterprise3, and Enterprise4 simultaneously are soliciting the partnership with Enterprise1.

Figure 16. An exemplar of LMP

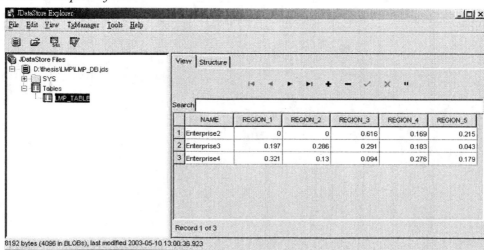

Figure 17. (a) Exemplifies the PNLM data associated with Enterprise1&3 and (b) then exemplifies that of Enterprise1&4

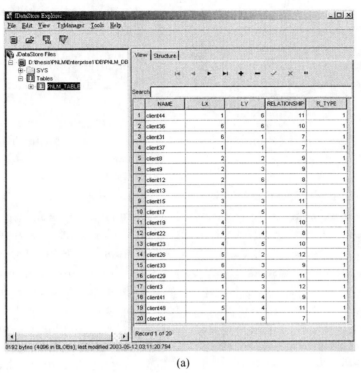

(a)

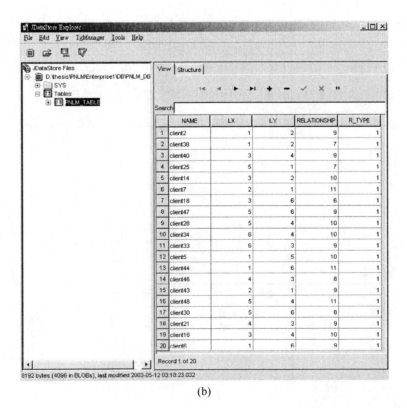

(b)

Due to the limited resource, Enterprise1 has to select the one partner among them. Furthermore, Enterprise1 prefers a complement partner in Region3. This scenario differs from the previous one in Enterprise2-4 initiating the requests and thus PNLM sufficing to assist the decision making. Moreover, this scenario investigates the application of the cooperation criteria of the compliment principle.

This investigation with PNLM unfolds as follows:

- In each BOLM of Figure 18, yellow-colored lines indicate the existence of clients in designated geographical areas colored nonblack/white (in contrast to the black/white geographical areas representing none-client areas).

- Figure 18(a) shows the clients of Enterprise1 spread over Region1&2, while Figure 18(b) exhibits those of Enterprise2 unfolding in Region3&5. The PNLM of Enterprise1&2 manifests clients spreading over Region1&2&3&5 (Region3&5 circled to indicate the addition of regions imposed by the partnership with Enterprise2 as shown in Figure 18(c), and enables the Click Point enquiries rendered on the colored areas for the details of selected clients (as shown in Figure 19).

- Figure 20(a) shows the fragmented data associated with the PNLM of Enterprise1&2 that manifests the (CAN, SIM) value vector of (18, 163) based on the BOLMs of Enterprise1 and Enterprise2 as shown in Figure 20(b) and Figure 20(c), respectively.

- The aforementioned steps are repeated for attaining the PNLMs of Enterprise1&3 and Enterprise1&4 (as shown in Figure 21) and the resulting values of CAN and SIM are presumed in Table 11.

- Table 11 summarizes the (CAN, SIM) value vectors for Enterprise2-4, enlightening certain clues to the decision making. That

Figure 18. The picturesque views of (a) Enterprise1's BOLM; (b) Enterprise2's BOLM; (c) PNLM of Enterprise1&2

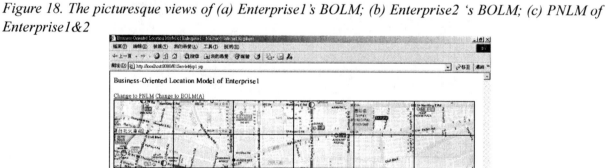

(a)

continued on following page

Figure 18. continued

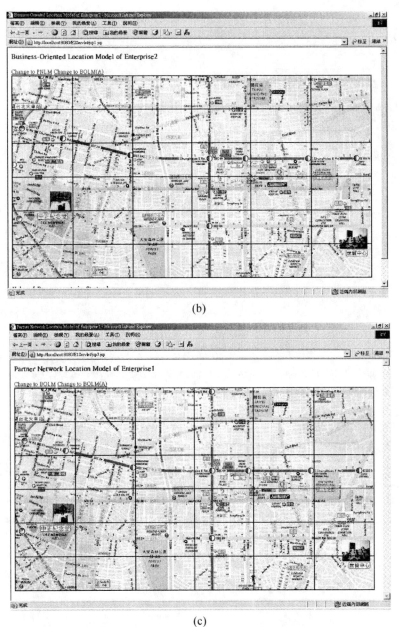

(b)

(c)

is, Enterprise2 prosperously serves the needs better than Enterprise3-4 because of its higher SIM in Region3 (for complementing Enterprise1) besides its overall high values of CAN and SIM. This analysis endows Enterprise1's mobile managers with valuable decision knowledge while negotiating contract deals with Enterprise2 (or Enterprise3/Enterprise4).

- In this scenario, PNLM fully (partly) satisfy the following cooperative principles: compatible (PNLM assists in the selection of

Figure 19. Click point enquiry (of the details of a client) and its reply window

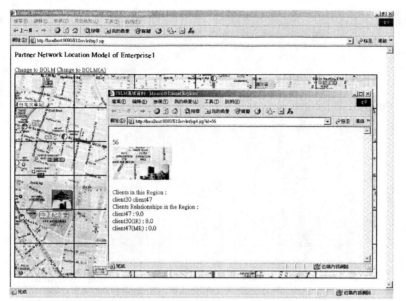

Figure 20. (a) shows the fragmented data associated with the PNLM of Enterprise1&2; (b) and (c) then show the data associated with the BOLMs of Enterprise1 and Enterprise2 respectively.

	NAME	LX	LY	RELATIONSHIP	R_TYPE
1	client15	3	3	11	1
2	client16	3	4	10	1
3	client40	3	4	9	1
4	client21	4	3	9	1
5	client46	4	3	8	1
6	client22	4	4	-2	2
7	client17	3	5	5	1
8	client42	3	5	7	1
9	client18	3	6	6	1
10	client23	4	5	10	1
11	client24	4	6	7	1
12	client29	5	5	11	1
13	client30	5	6	8	1
14	client47	5	6	0	2
15	client35	6	5	11	1
16	client36	6	6	10	1
17	client28	5	4	10	1
18	client48	5	4	11	1
19	client34	6	4	10	1
20	client33	6	3	3	2
21	client27	5	3	9	1

Record 1 of 21

8192 bytes (4096 in BLOBs), last modified 2003-05-14 04:27:36.681

(a)

continued on following page

Figure 20. continued

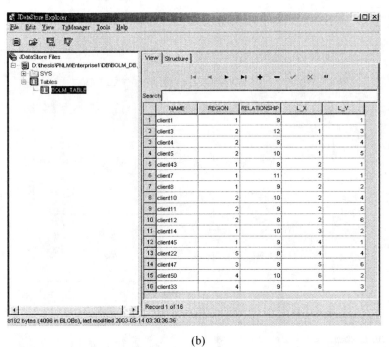

(b)

(c)

Figure 21. The picturesque views of (a) PNLM of Enterprise1&3; (b) PNLM of Enterprise1&4

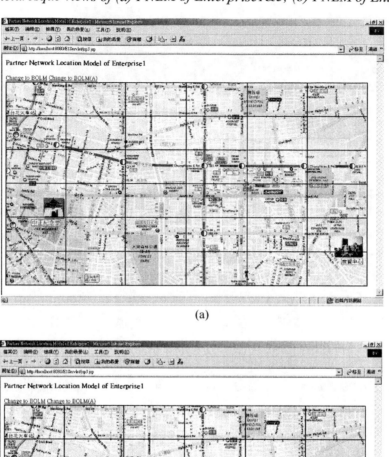

(a)

(b)

a partner), complementing (PNLM locates a complement partner in Region3), analytic (PNLM measures the values of CAN and SIM), and feasible (PNLM's ClickPoint function equips the evaluating task a close look of the to-be-clients).

Discussion

The aforementioned evaluation results aim to justify the contributions of the DSLM framework by exemplifying certain integration of enterprise business models and the proposed location model,

Table 11. The (CAN, SIM) value vectors between Enterprise2-4

	(CAN)	(SIM)
Enterprise2	18	163 (75)*
Enterprise3	15	132 (43)*
Enterprise4	14	125 (13)*

*Note: * indicates the SIM value attained only from the clients in Region3*

driving the enterprise LBS research a step further. We conclude our evaluation with a short discussion as follows:

- Among the two aspired features for enterprise-based LBS ("*semantic*" and "*dynamic*"), BOLM enables to semantically and dynamically define for an enterprise its clients, client values, and client relationships in a location model, while PNLM empowers an enterprise to *merge* its location model with the other enterprise's due to their inter-enterprise cooperation. LMP, on the other hand, facilitates this inter-enterprise cooperation in a given industry. That is, DSLM bestows enterprise mobile workforces high-level location-sensitive decision information so as to properly serve the clients or justifiably negotiate contracts with other enterprises.

- The results of Section 4.1 show that *semantic* feature of the location model (i.e., the linking between the enterprise business models and the enterprise location model in BOLM) endows the enterprise higher values than those of *static* location models (e.g., First-In-First-Out and Far-Distance-Based), and is capable of adapting to the varying service needs of the clients.

- Section 4.2 evinces that the merits of the semantic feature in BOLM can be extended to PNLM. That is, PNLM fully (or partly) enables the valuation of inter-enterprise cooperation/inter-operability in terms of the

myriad principles of enterprise cooperation (compatible, complementing, analytic, and feasible). However, the scenarios have not yet addressed the homogeneous principle (sharing the same competitive goal) that awaits further investigation.

- Scenario 1 of Section 4.2 demonstrates that LMP moderates the creation of PNLMs between enterprises, precipitating the diffusion of the merits of the extended semantic values.

- DSLM can be applied to myriad kinds of enterprises with appropriate ontology modeling: enterprises of operations sensible to locations (e.g., transportation, logistics, touring, etc.), enterprises of majority mobile workforces (e.g., insurance, estate agencies, etc.), and enterprises of clients spreading over various regions (e.g., newspaper, online merchants, etc.).

- DSLM primarily intends to bring about cooperation between enterprises. However, this framework can reverse its functions by serving as a tool for competitive analysis (that is, substituting clients data with competitors data in order to perform competitive analysis).

- The practical implication of DSLM is that BOLM, PNLM, and LMP case be used in location-based enterprise decision support for maximizing enterprise profits or engaging enterprise geographical expansion (or cooperation).

CONCLUSION

The contribution of dynamic semantic location modeling devised in this chapter is the first attempt in integrating enterprise business models with location models (that have been playing a very important role in mobile commerce). DSLM advances the former location modeling methods by embodying the dynamic and semantic features. The DSLM is a kind of symbolic model that includes business oriented location models (BOLM) as objects. BOLM describes different business units and their relationships. For location-based enterprise decision support, our chapter presents a framework of three different deployment of the DSLM (business-oriented location model, location model platform, and partner network location model).

Enterprises can build up their business-oriented location model first and then search for their potential partners in location model platform. Finally, if the advanced cooperation between two enterprises is possible, the partner network location model can be constructed with their business-oriented location models. In short, the dynamic semantic location model provides certain solutions to enterprise-based LBS that take into account enterprise business models, bestowing enterprise mobile workforce location-sensitive decision information so as to properly serve the clients or justifiably negotiate contracts with other enterprises.

We have evaluated the partner network location model and location model platform and created a few innovative scenarios about the integration of enterprise business models and the proposed location model. These scenarios imply some suggestions about how the proposed framework could be utilized along with a variety of situations of enterprises having different perspectives and strategies in assessing their clients, partners, and business contexts. We hope our work can shed light on further integration of enterprise business models and location models for advanced mobile enterprise applications.

REFERENCES

Bolloju, N. (2003). Extended role of knowledge discovery techniques in enterprise decision support environments. In *Proceedings of the 34th Annual Hawaii International Conference on System Sciences (HICSS-34)*, Maui, HI.

Bouwman, H., Haaker, T., & Faber, E. (2005). Developing mobile services: Balancing customer and network value. In *Proceedings of the 2nd IEEE International Workshop on Mobile Commerce and Services (WMCS'05)*, Munich, Germany.

Bruce, B., Cross, M., Duncan, T., Hoey, C., & Wills, M. (2000). *e.Volution* (pp. 81-96). Prestoungrange University Press.

Domnitcheva, S. (2001). *Location modeling: State of the art and challenges.* Paper presented at the UbiComp Workshop on Location Modeling for Ubiquitous Computing, Atlanta, GA.

Doz, L. Y. (2002). *Managing partnerships and strategic alliances.* Insead. Retrieved August 22, 2007, from http://www.insead.edu/executives

Ericsson Enterprise. (2002). *The Path to the mobile enterprise.* Retrieved August 22, 2007, from http://www.ericsson.com/products/whitepapers_pdf/whitepaper_mobile_enterprise_rc.pdf

Example for Reinforcement Learning: Playing Checkers. (2001). Retrieved August 22, 2007, from http://www.cs.wustl.edu/~sg/CS527_SP02/lecture2.html.

Fetnet. http://enterprise.fetnet.net/event /Special _02.htm.

Grimes, S. (2005). Location, location, location. *intelligent enterprise.* Retrieved August 22, 2007, from http://www.intelligententerprise.com/toc/?day=01&month=09&year=2005

Hiramatsu, H. (2001). A spatial hypermedia framework for position-aware information delivery systems. *Lecture Notes in Computer Science, 2113*, 754-763.

Hynes, N., & Mollenkopf, D. (1998). *Strategic alliance formation: Developing a framework for research.* Paper presented at the Australia New Zealand Academy of Marketing Conference, Otago, New Zealand.

Jasper, R., & Uschold, M. (1999). *A framework for understanding and classifying ontology applications.* Paper presented at the 12th Workshop on Knowledge Acquisition, Modeling and Management, Banff, Canada.

Machiraju, V. (2001). *Service-oriented research opportunities in the in the world of appliances (Tech. Rep.).* HP Software Technology Lab.

Mitchell, T. M. (1997). *Machine learning* (pp. 10-11). McGraw-Hill.

Pradhan, S. (2002). *Semantic location.* Retrieved August 22, 2007, from http://cooltown.hp.com/dev/wpapers /semantic/semantic.asp

Rohs, M., & Roduner, C. (2006). Towards an enterprise location service. In *Proceedings of the International Symposium on Applications and the Internet Workshops*, Phoenix, AZ.

Varshney, U. (2000). Recent advances in wireless networking. *IEEE Computer, 33*(6), 100-103.

Ververidis, C., & Polyzos, G. (2002). Mobile marketing using location based services. In *Proceedings of the 1st International Conference on Mobile Business*, Athens, Greece.

vLeonhardt, U. (1998). *Supporting location: awareness in open distributed systems.* Unpublished doctoral thesis, Imperial College, Department of Computing, London.

Yuan, S. T., & Peng, K. H. (2004). Location based and customized voice information service for mobile community. *Information Systems Frontiers, 6*(4), 297-311.

Yuan, S. T., & Tsao, E. (2003). A recommendation mechanism for contextualized mobile advertising. *Expert Systems with Applications, 24*(4), 399-414.

ENDNOTES

[1]   The efforts require the capability of interfacing with geometric models when handling the ranges and the overlaps of the location objects.

[2]   The new modeling method is grounded on symbolic modeling (in which a location is considered as a set containing the objects residing in the designated location) so as to be appropriately extended as shown in the later sections.

[3]   In a BOLM, the information of the distance between a CU and a BU is captured as an attribute of the CU (whenever required) that subsequently enables location-based commerce as addressed in Section 1.

[4]   Averaged attribute values are used if there are multiple Source Enterprises.

[5]   Please note that for the same pair of BUs the MR measurement is different if the source BU exchanges with the target BU.

[6]   For simplicity, in Table 8 the IR values are presumed and the MR values are attained by respectively calculating the difference of the correspondent DR values. (The complete process of calculating these DR, IR, MR values involves the awareness of the attribute weights and the attribute values before the application of Algorithm 2 and Algorithm 3.)

[7]   Farther-distance service requests are presumed to be of higher values (than those of shorter-distance service requests) for a

logistic delivery company (if its charges take into account the distances of the service requests).

8   The purpose of additionally tuning Request Revenue by 1.5 (3.5) is for making the comparison of the resulting values (for the six types of clients) more perceivable.

9   Without loss of generality, a request set of size 10 is used in our experiment settings (i.e., the size could be equal to any other number).

10   This tuning originates from an intuition that acceptance of requests implicitly encourages the occurrence of subsequent requests (in contrast to the situation that rejection of requests usually dismays the succeeding).

11   Due to the limited resources of the enterprise, there might be client requests that cannot be served and hence are not taken into account in the client reply set.

12   Location represents the region number, and LX and LY indicate the coordinates of the 6*6 coordinated plane.

13   In Figure 17, clients of the IR are indicated with the R_Type of value 1 (while clients of MR are indicated with R_Type of value 2) and thus the SIM value is the sum of these relationship measurements.

Chapter XVI
Service Composition Approaches for Ubiquitous and Pervasive Computing:
A Survey

Mohamed Bakhouya
The George Washington University, USA

Jaafar Gaber
Universite de Technologie de Belfort-Montbeliard, France

ABSTRACT

This chapter describes and classifies service composition approaches according to ubiquitous and pervasive computing requirements. More precisely, because of the tremendous amount of research in this area, we present the state of the art in service composition and identify key issues related to the efficient implementation of service composition platforms in ubiquitous and pervasive computing environments.

INTRODUCTION

Ubiquitous and pervasive computing (UPC) are new paradigms with a goal to provide computing and communication services all the time and everywhere. In **ubiquitous computing** (UC), the objective is to provide users the ability to access services and resources all the time irrespective of their location (Gaber, 2000; Weiser, 1996).

Pervasive computing (PC), often considered the same as **ubiquitous computing** in the literature, is a related concept that can be distinguished from **ubiquitous computing** in terms of environment conditions (Gaber, 2006). We can consider that the aim in UC is to provide any mobile device an access to available services in an existing network all the time and everywhere while the main objective in PC is to provide spontaneous services created on the fly by mobiles that interact by ad hoc connections (Gaber, 2000, 2006)

In **ubiquitous and pervasive computing**, **service composition** plays a fundamental role, and automation is essential to improve speed and efficiency of users' responses and benefits. **Service composition** is the act of taking several component products or services, and handling them together to meet the needs of a given user (Chakraborty, 2001). For example, in an online business process of reservation of air tickets, a reservation service carries out three distinct services: ticket availability check, credit card check, and updating the required database to reserve a ticket for the user. Therefore, these three services must be integrated together to serve numerous user requests (Oprescu, 2004).

In UPC, automatic **service composition** requires dealing with four major research issues: service matching and selection, scalability, fault tolerance, and adaptiveness. The service matching and selection is the first step in creating any composite service and requires a **service discovery** system. The role of **service discovery** system is to locate the service components that provide the functionality to be placed in the new service. It allows the selection of services providing functionalities (i.e., capabilities) that match the requested functionalities. More precisely, **service discovery** systems should also be able to find out all services conforming to a particular functionality, irrespective of the way of invocation. To achieve this requirement, semantic level reasoning of describing the functionality of service is required. This is why the **Web service** community has developed a number of languages to formally describe services in order to facilitate their discovery.

Recently, **Web services** are becoming the most predominant paradigm for distributed computing and electronic business. In other words, the Web has become the platform through which many companies communicate with their partners, interact with their back-end systems, and perform electronic commerce transactions. Examples of **Web services** include bill payment, customized online newspapers, or stock trading services. As pointed out in Hu (2003), a **Web service** is a software system designed to support interoperable machine-to-machine interaction over a network. More precisely, **Web services** are self-contained, modular units of application logic that provide business functionality to other applications via an Internet connection (Bucchiarone & Gnesi, 2006). Several XML-based standards are proposed to formalize the specification of **Web services** to allow their discovery, composition, and execution (Baget, Canaud, Euzenat, & Saïd-Hacid, 2003; Zeng, Benatallah, Ngu, Dumas, Kalagnanam, & Chang, 2004). These standards are primarily syntactical; **Web service** interfaces are like remote procedure call and the interaction protocols are manually written. On the other side, the **Semantic Web** community focuses on reasoning about Web services by explicitly declaring their preconditions and effects with terms precisely defined in ontologies.

The **service discovery** system should also be scalable across large-scale networks and adaptable to dynamic changes especially when services dynamically join and leave the network. To implement this process, most **service discovery** systems like SLP (Guttman, 1999), Jini (Jini, 2000; Robert, 2000), and SSDS (Xu, Nahrstedt, & Wichadakul, 2001) require that service components must be stored in component directories that can be accessed at runtime. These centralized systems cannot meet the requirements of both scalability and adaptability simultaneously. Several decentralized systems are proposed to address these issues. A survey and classification of **service discovery** systems proposed in the literature are presented in Bakhouya and Gaber (2006b).

Service coordination and management is the second issue to be addressed in automatic **service composition**. More precisely, composition platforms must have one or some brokers that coordinate and manage the different services involved in the composition. The problem of coordination and management becomes difficult when the entities are distributed across the network and poses a scalability problem, especially when numerous users are concurrently making composite service requests. For example, if elementary services are distributed across the network, the use of a centralized broker requires a lot of central processing unit (CPU) cycles to process users' composition requests.

Since a composite service is dependent on many distributed elementary services, fault tolerance is another important issue to be included in **service composition** platforms in order to ensure their proper functioning. For example, if an elementary service shuts down, the request processing is hindered (Chakraborty & Joshi, 2001). In this case, the platform should be able to detect and restore it. It should be noted also that in dynamic networks, where services are coming up and going down frequently, the **service composition** platform should be able to adapt its composition by taking maximum advantage of the currently available services. This increases the composite service availability in a dynamically changing environment.

These important issues can be classified into two major directions in dynamic **service composition** research. The first direction addresses languages to specify and describe services including complex planning mechanisms that utilize these descriptions to generate composite services. In other words, this research direction is trying to define languages to specify services, invocation mechanisms, and composite services. The second direction aims to develop architectures that enable scalability, fault tolerance, and adaptive **service composition**.

In this chapter, approaches proposed in the literature for **service composition** are presented. The aim is to identify key issues related to the efficient implementation of a **service composition** platform for UPC. The first section will present the service-oriented architecture and **Web service**

generalities. In the second section, we will start from the current interest in **Web services** and compare the two major approaches: the **syntactical Web approach** and the **Semantic Web approach**. In the third section, we will then explore the Web **service composition** methodologies and present the most **service composition** platforms known so far in UPC. It should be noted that automatic or dynamic **service composition** must involves the automatic selection, composition, and cooperation of appropriate services to perform and satisfy user task, given a high-level description of the task's objective. While in manual **service composition**, the user must select the required services, manually specify the composition, and ensure their interoperation in order to satisfy its request.

Service-Oriented Architecture (SOA) and Web Services

The current technological architecture for **Web services**, as depicted in Figure 1, is structured around three entities: the requesters, providers, and registries (Akram, Medjahed, & Bouguettaya, 2003; Dustdar & Schreiner, 2005; Wu & Chang, 2005). This architecture involves three major standards: web service definition language (WSDL), universal description, discovery, and integration (UDDI), and simple object access protocol (SOAP) (Maamar, Mostefaoui, & Yahyaoui, 2005). The aim is to define mechanisms that describe, advertise, bind, and trigger Web services. The service provider first publishes its WSDL services description in a UDDI registry. The service requester then searches in the UDDI for a **Web service** that matches the given criteria. If the required service is found, the requester invokes the service provider using SOAP messaging protocol.

UDDI is a repository, similar to a Jini repository (Robert, 2000), that provides mechanisms for service providers to publish their services and for clients to search required Web services. More

Figure 1. Web service entities with their roles and operations

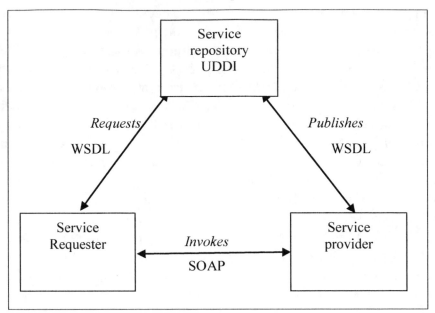

precisely, using a UDDI interface, clients can dynamically look up as well as discover services provided by business partners. SOAP is a message layout specification that defines a uniform way of passing XML-encoded data. WSDL defines services as collections of network endpoints.

This **Web service** architecture is sufficient for some simple interaction needs; it is not sufficient for integration of business processes that involve multiple services (Padhye, 2004). More precisely, they do not deal with the dynamic composition of existing services (Serin, Hendler, & Parsia, 2003). For example, WSDL specifies only the syntax of messages that enter or leave a computer program (i.e., service), but not the order of messages (i.e., composition flow) that have been exchanged between services. A more challenging problem is to compose services dynamically in order to resolve unexpected user requests. This raises the need for Web **services composition** that provides the mechanism to fulfill the complexity of business processes execution. In other words, composition of **Web service** is needed to support

business-to-business or enterprise application integration (Srivastava & Koehler, 2003). This is why the business world has developed a number of **syntactical XML-based standards** to formalize the specification of composition and execution of composite services. However, these **Web service** flow specification languages, such as BPEL4WS (Curbera, Goland, Klein, Leymann, Roller, Thatte, & Weerawarana, 2002), are syntactical and do not deal with the semantic behavior of services. Today, Web services need to be described with additional semantic annotation to create a **Web semantic service-oriented architecture,** as depicted in Figure 2.

The **Web semantic** (Bucchiarone & Gnesi, 2006) is an extension of the current Web by marking up Web content, properties, and relations, in a reasonably expressive markup language with well-defined behaviors, for example, by declaring their preconditions and effects with terms defined in ontologies. In general, **ontology** is a shared conceptualization based on the semantic proximity of terms in a specific domain of interest (Majithia, Walker, & Gray, 2004; Medjahed, 2004). In other words, the aim of ontologies is to extend the existing Web by adding a semantic level to its content. **Semantic ontology's languages** developed for this purpose are Darpa agent markup language for service description (DAML) and **ontology** Web language (OWL). Using these **semantic languages**, the information necessary to select, compose, and respond to services is encoded at the service Web site. The main objective is to develop **semantic languages** that will perform automatic composition by describing and exploiting services' capabilities and user constraints and preferences.

In the rest of this chapter, Web services description languages together with **service composition** models are presented. We start with the definitions of syntactic Web services and **semantic Web** services and then we present the **service composition** approaches.

SERVICES DESCRIPTION AND MODELING

Syntactic Description Languages

The **Web service** model evolves from three main XML-based technologies: UDDI, SOAP, and WSDL, as described above. WSDL is an **XML-based language** for describing network services as a set of endpoints operating on messages containing either document-oriented or procedure-oriented information (Milanovic & Malek, 2004). More precisely, it specifies the location of the service and communications that the service will perform. Four types of communication, called endpoints, are defined involving a service's operations (Bucchiarone & Gnesi, 2006): the endpoint receives a message (one-way) and sends a message (notification), the endpoint receives a message and sends a correlated message

Figure 2. Web service protocols and languages

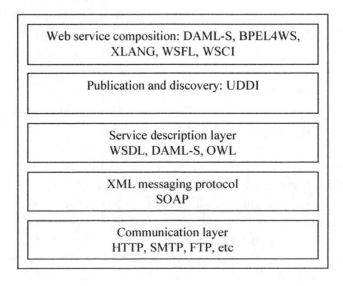

(request-response), and it sends a message and receives a correlated message (solicit-response). Operations are grouped into port types, which describe abstract endpoints of a **Web service** such as a logical address under which an operation can be invoked (WSDL).

WSDL does not support semantic description of services like the definition of logical constraints between its input and output parameters. Generally, the major limitation of **XML-based languages**, such as WSDL, is their lack of explicit semantics that limit the capability of matching Web services. A semantic knowledge, using **ontology**, would help in the identification of the most suitable services for a particular request (Abela & Slanki, 2003). Ontologies play a key role in the semantic description extending syntactic description by providing precisely defined terms of services (Bucchiarone & Genesi, 2006). In other words, semantic description using ontologies will facilitate the automation of **Web service** tasks, including the automated Web **service discovery**, the execution, the composition, and the interoperation. The next section presents the most known semantic service description languages, DAML-S and OWL.

Semantic Description Languages

DAML-OIL, proposed by the Darpa agent markup language (DAML) committee, was one of the first languages designed for expressing ontologies. The service coalition of DAML has also developed a DAML-based **Web service ontology** to semantically describe **Web service**, namely DAML-S (Ankolekar et al., 2002). More precisely, DAML-S is a DAML-based **Web service ontology**, which supplies **Web service** providers with a core set of markup language constructs for describing the properties and capabilities of their services in unambiguous, computer-interpretable form. DAML-S is proposed primarily with the intention to provide the automatic Web **service discovery**, the automatic **Web service** invocation, the auto-

matic Web **service composition** and interoperation, and the automatic **Web service** execution monitoring. Automatic Web **service discovery** involves the automatic location of Web services that provide a particular service and that adhere to requested constraints. To realize this task, DAML-S provides declarative advertisements of service properties and capabilities that can be used for automatic **service discovery** (Narayanan & McIlraith, 2002). Automatic **Web service** invocation involves the automatic execution of an identified **Web service** by a computer program or a software agent. DAML-S provides declarative APIs for Web services that are necessary for automated **Web service** invocation. Automatic Web **service composition** and interoperation processes involve the automatic selection, composition, and interoperation of Web services to perform some tasks, given a high-level description of an expected objective. This task is realized by providing declarative specifications of the prerequisites and consequences of individual service use that are necessary for automatic **service composition** and interoperation. To realize the automatic **Web service** execution monitoring, DAML-S provides descriptors for the execution of services.

More precisely, to facilitate automatic Web **service discovery**, invocation, and composition execution monitoring tasks, DAML-S provides one possible representation through the class profile. This profile describes a service as a function of three basic types of information: the organization providing the service, the function that the service computes, and the service's characteristics. The provider information refers to the entity that provides the service (e.g., service name, contact information). The functional description of the service specifies the inputs (e.g., credit card number and expiration date) required by the service and the outputs (e.g., receipt) generated. It describes also the precondition required (e.g., valid credit card) by the service and the expected effects (e.g., card is charged) that result from the execution of the service. Finally, characteristics of the service allow the description of the host properties that are used to describe features of the service (e.g., price).

In DAML-S, a service's process model, that describes the flow and data-flow involved in using a service, is also proposed. In this model, a service can be viewed as a process. A process has any number of inputs representing the information that is required for its execution. Under some condition, the process is executed to provide some number of outputs. In order for the process to be invoked, a number of preconditions must hold to generate a number of effects. In this way, a particular subclass of service model called process model is defined. Two leader components of a process model are also defined (DAML). The first, called process **ontology**, describes a service in terms of its inputs, outputs, preconditions, and effects. The second, called process control **ontology**, enables planning, composition, and service interpretation.

When defining processes using DAML-S, there are three tasks that need to be started for the process model to be successful. The first task relates process inputs to process' conditions, outputs or effects, including its precondition, the conditions governing conditional effects and conditional outputs, and the effects and outputs themselves. The second task relates inputs and outputs of a composite process to the inputs and outputs of its various subprocesses. The third task relates the inputs and the outputs of elements of a composite process definition to the parameters of other process components.

Ontology web language (OWL) has been developed by the W3C (World Wide Web Consortium). OWL has been inspired by DAML-OIL and provides three increasingly expressive sublanguages: OWL Lite, OWL DL (description logic), and OWL Full. OWL Full can be viewed as an extension of resource description framework (RDF), while OWL Lite and OWL DL can be viewed as extensions of the restricted view of RDF. Recall that RDF is one of the first languages for representing information about resources in the World Wide Web developed by W3C. The main difference between OWL Full and OWL DL lies in restrictions on the use of some of those features and on the use of RDF features. OWL Full allows free mixing of OWL with RDF Schema, like RDF Schema, and does not enforce a strict separation of classes, properties, individuals, and data values. OWL DL puts constraints on the mixing with RDF and requires classes, properties, individuals, and data values (OWL). To the contrary, OWL Lite is a minimal sublanguage of OWL DL, more practical, and easier to use for tools developers since it supports only a subset of the language constructs.

SERVICE COMPOSITION MODELING

As stated in the second section, the current **Web service** model, using WSDL, UDDI, and SOAP, enables the **service discovery**; however, it does not consider the automatic and dynamic integration and composition of services. More precisely, this model specifies only services and operations that perform, but not the order of, a flow specification of exchanged messages between services (Abela & Solanki, 2003). Several **Web service** flow specification languages like BPEL4WS (Business processes execution languages for web services), XLANG, WSFL (Web Service Flow Language), and DAML-S have proposed to describe the order of messages to be exchanged between services (Akram et al., 2003; Curbera et al., 2002; Leymann, 2001; Van der Aalst, Dumas, & ter Hofstede, 2003).

In this section, **service composition** approaches are presented. As depicted in Figure 3, we can classify them into three categories: workflow-based approaches, **artificial intelligence (AI) planning-based approaches**, and **learning-based approaches**. More precisely, the purpose of this section is to provide a review of these approaches, to help better understand how a Web services composition can be accomplished so far.

Workflow-Based Approaches

Recall that a **service composition** mechanism describes how different services can be composed into a coherent global service to satisfy the user request. As depicted in Figure 3, workflow-based approaches are classified into two categories: **orchestration methods** and **choreography methods** (Beek Bucchiarone, & Gnesi, 2006; Hu, 2003; Peltz, 2003a). **Orchestration methods** use a central coordinator (i.e., orchestrator) to combine and invoke Web services. More precisely, orchestration represents control from one party's perspective and interactions occuring at the message level. The **choreography methods** do not assume the exploitation of a central coordinator. These methods define the conversation that should be undertaken by each participant. In other words, the composition is achieved via

peer-to-peer interactions among the collaborative services. Hence, **choreography methods** are more collaborative than **orchestration methods** and allow each involved party to describe its part in the interaction (Hu, 2003).

Orchestration Methods

Several research projects regarding **service composition**, based on the orchestration principle, have been proposed. For example, the dynamic **service composition** called software hot-swapping (Mennie & Pagurek, 2000) at Carleton University, eFlow (Casati, Ilnicki, Jin, Krishnamoorthy, & Shan, 2000) from HP Laboratories, and Ninja (Feng, 1999) **service composition** at the University of California Berkeley. These approaches typically address the fault-tolerance, scalability, and management of **service discovery** and composition requirements. They do not explicitly address languages issues to describe services.

EFlow, for example, is an e-commerce service platform that provides techniques for integrating

Figure 3. Classification of services composition approaches

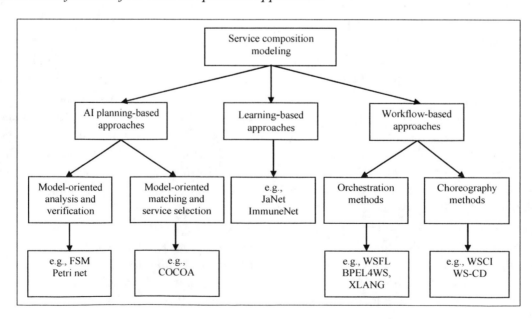

various e-services to compose complex e-commerce transactions. The eFlow architecture consists of three entities (Chakraborty & Joshi, 2001): the eFlow composition, the **service discovery** system, and the elementary services. The role of eFlow engine is to schedule and maintain the state of every running composite service and coordinate the different events between e-services. The discovery system is responsible for finding required e-services. In eFlow, a composite service is modeled as a graph, which defines the execution order among the services. The graph is created manually and consists of service nodes that represent simple or composite services and event nodes or decision nodes that represent rules which control the execution flow between service nodes (Su & Rao, 2004). Each service is specified using XML language providing some details such as URL to contact and input/output information.

More precisely, the eFlow is based a centralized broker which manages the **service composition** process, like path creation, **service discovery**, appropriate combination of different services, and management of the information flow between composed services. The drawback of a controlized approach is that if a huge number of users attempts to access variety and increasing number of services distributed over the network, the broker could quickly becomes a bottleneck.

Another interesting platform developed at Carleton University focuses on a dynamic **service composition** called hot-swapping (Feng, 1999; Mennie & Pagurek, 2000). Their work on **service composition** follows their research in the field of dynamic component up-gradation at runtime. More precisely, software hot-swapping is defined in Mennie and Pagurek (2000) as a process of upgrading software components at runtime in systems which cannot be brought down easily, cannot be switched off-line for long periods of time, or cannot wait for software to be recompiled once changes are made. The hot-swapping differs from other forms of software composition

since it deals exclusively with network services. Network services are individual software components, which can be distributed within a network environment, that provide a specific set of well-defined operations. This platform uses Jini (Jini, 2000) discovery architecture to locate services to be composed. **XML-based language** is used to describe capabilities, constraints, inputs, outputs, and dependencies of each service.

To carry out heterogeneous **service composition**, two techniques are proposed in Feng (1999): the interface fusion technique and the stand-alone technique. The first involves the formation of composite services that can be accessed by a common interface. The advantage of this technique is the speed at which a composite can be created and exported since it does not need to be constructed dynamically. With the second technique, a service is created by dynamically assembling the available services, for example, in the form of pipes and filters. In this configuration, the input to the composite service is sent to the first service component, which in turn sends its output to the input of the next service component and so on, along a chain. As stated by Feng (1999), the main advantage of this technique is that all service components do not need to share state information and they are not aware or dependent on other service components.

The Ninja platform has been developed to enable and to address dynamic **service composition** (Mao, Brewer, & Katz, 2001). It addresses the composition of arbitrarily complex services from simpler ones over a wide area network. The Ninja platform architecture is composed of the following five entities (Chakraborty & Joshi, 2001): Internet service provider (ISP), network service providers (NSP), automatic path creation (APC), service discovery service provider (SDSP), and end clients (EC). ISPs are providers of resources that can be requested by clients. NSPs provide connection to the various devices through varying types of networks. APC is the central entity that coordinates the **service composition** process. It enables service access, dynamic **service composition**, data flow optimization, and adaptability. SDSP is a discovery system similar to Jini or SLP directory systems responsible for locating required services available in the network using XML description language. ECs are machines, either fixed or mobiles, connected to the Internet. The **service composition** mechanism used in the Ninja platform is as follows. When an APC entity gets a request for a particular composite service, it figures the set of operators and connectors that help in transporting the data from one network operator to another. The APC creates a logical path by shortest path techniques (Mao et al., 2001). Then it creates, instantiates, and executes the corresponding physical path.

The Ronin agent framework (RAF) is a composition agent- and service-oriented architecture for deploying dynamic distributed systems (Chen, 1999, 2004). The main objective of this work regarding **service composition** is to enable mobile users to access all sorts of information from their mobile devices. The information is assumed to be available from static information providers residing in the fixed network infrastructure.

The main entity in the RAF is the notion of the Ronin agent and its corresponding deputy. The agent deputy acts as a front-end interface for the other agents in the system. To discover the different information about agents available in the system, an enhanced Jini Discovery framework, called Xreggie (Chakraborty, Perich, Avancha, & Joshi, 2001), is used. The Ronin agent framework includes an agent description facility based on the **syntactical language** XML to describe a service by defining its capability, and system feature on which it is running. Based on this description, the enhanced Jini lookup finds out a service that matches a request by comparing the XML attributes of each service and selecting one to the request.

Recently, the Xreggie system was enhanced using DAML language to describe the services and their capabilities. Consequently, this en-hanced **service discovery** allows more flexibility in finding out different types of services and reasoning about their capabilities. Recall that DAML is a **semantic language** that is more powerfully expressive than XML and is seen as a very important language to enable the service selection and matching. More precisely, the main advantage is that it allows the discovery system to locate services by semantically reasoning about the request and service descriptions. However, the system suffers from the scalability imposed by the centralized Jini architecture. The **service composition** also depends on the centralized broker that represents the central point of failure. Also, the state information of each agent has to be maintained and the use of broadcast technique to disseminate the state information poses a scalability problem (Chakraborty, 2001).

Chackraborty and Joshi (2001) present a survey related to **service composition** platforms and evaluate criteria for judging protocols used to enable such composition. Their work claims that many of the current technologies still do not cover all these aspects in their implementation and deal with a centralized broker that manages the **service composition** process. The drawback is that if a huge number of users attempts to access a variety and increasing number of services distributed over the network, the broker quickly becomes a bottleneck, and in addition, it represents a central point of failure. Therefore, achieving coordination in collaborative applications that consist of composed Web services is difficult. It should be noted that the platforms presented above focus primarily on local collaborative applications without the emphasis on service description languages issues explicitly.

In parallel to these works which focus mainly on developing architectures that enable scalability, fault tolerance, and adaptive **service composition**, several service description languages that allow developers to create, execute, and combine Web services into complex services using **Web service** model and language standards are proposed, for example, a Web service flow language (WSFL) proposed by IBM to specify how to implement a business process model using the **Web service** architecture. In other words, WSFL is integrated with UDDI and WSDL for dynamic selection of Web services. Service providers must properly describe their services in order to be classified to handle a specific activity in the business process. More precisely, a business process is created using WSDL and is represented by a XML flow model. The flow model defines activities that are implemented in the form of Web services. It defines also the flow data sequences from an activity to another (Abela & Solanki, 2003).

XLANG is another flow composition language developed initially by Microsoft for the creation of business processes and the interaction between **Web service** providers (Abela, 2003; Satish, 2001). The specification provides support for sequential, parallel, and conditional process control flow. It also included a robust exception handling facility, with support for long-running transactions through compensation. XLANG uses WSDL as a means to describe the service interface of a process. More precisely, it extends WSDL language by adding some behavior capabilities. A behavior defines the list of actions that belong to the service and the order in which these actions must be performed. The execution order of XLANG actions is defined through control processes (e.g., sequence, while loop, etc.).

Recently, the Web services workflow specifications outlined by XLANG and WSFL have been superseded by a new specification from IBM, Microsoft, and BEA called BPEL4WS (Peltz, 2003a, b). BPEL4WS is a flow language that allows the specification of multiple Web services coordination. More precisely, BPEL4WS provides a notation for describing interactions of Web services as business process (Charif & Sabouret, 2005; Curbera et al., 2002). The definition of such business protocols involves the precise specification of the mutually visible message exchange behavior of each of the parties involved in the protocol,

without revealing their internal implementation. In this way, the BPEL4WS process defines how multiple service interactions are coordinated to achieve a business goal, as well as the state and the logic necessary for this coordination.

BPEL4WS assumes that services are described using XML-based specifications like WSDL (Peltz, 2003a, b). The interaction between services is described in a business protocol specification language. For this purpose, BPEL4WS provides a programming language-like constructs (e.g., sequence, switch, while) as well as graph-based links that represent additional ordering constraints on the constructs (Srivastava & Koehler, 2003). It also provides a set of primitive activities like Invoke, Receive, and Wait. BPEL4WS is extended with two significant and complementary specifications: WS-Coordination and WS-Transaction for providing protocols that coordinate the actions of distributed applications (Abela & Solanki, 2003). More precisely, the WS-Transaction specification allows a composite service to monitor the success or failure of each individual, and the coordinated activity. The WS-Coordination defines a framework through which the composite service can work from a shared coordination context. BPEL4WS use of WSDL port information for service description is quite expressive with respect to process modeling constructs.

It is worth noting that these service flow languages focus on representing compositions where services, flow of the process, and the bindings between services are static and known *a priori* (Su et al., 2004; Sirin et al., 2003; Beek & Rao, 2006).

Choreography Methods

Choreography is defined in Kavantaz, Budett, Ritzinger, Fletcher, Lafon, and Bareto (2005) as a mechanism that allows constructing compositions of **Web service** participants by explicitly asserting their common observable behaviors. More precisely, choreography defines the rules and interactions of collaboration between two or more services. Choreography is described from the perspectives of all parties and defines the complementary observable behavior between participants of collaboration.

WS-CDL (Kavantaz et al., 2005), for example, is a **choreography language** for specifying peer-to-peer protocols where each party is autonomous without hierarchy among them, and there is no centralization point like in **orchestration approaches**. More precisely, a WS-CDL choreography description is a collection of activities that may be performed by one or more participants. There are three types of activity in WS-CDL, namely control-flow activities (e.g., Sequence, Parallel, and Choice), work unit activities, and basic activities. Control-flow activities are similar to the basic control-flow constructs found in typical imperative programming languages. It can also be seen that these activities correspond to the Sequence, Flow, While, Switch, and Pick activities in BPEL4WS. A work unit describes the conditional and, possibly, repeated execution of an activity. The third type of WS-CDL activities, basic activities, describes points in a choreography where one role performs no action or performs an action behind the scenes that does not affect the rest of the choreography (Barros, Dumas, & Oaks, 2005).

To enable static validation and verification of choreographies to ensure that the runtime behavior of participants conforms to the choreography plan, WS-CDL must be based on a formal language that provides these validation capabilities (Barros et al., 2005). In addition, descriptions that abstractly specify behavior at a higher level, in terms of capability, would allow runtime selection of participants able to fulfill that capability, rather than restricting participation in the choreography to participants based on their implementation of a specific WSDL interface or WSDL operations.

Web service choreography interface (WSCI) is another XML-based description language proposed by BEA Systems, Intalio, SAP AG, and Sun Microsystems for choreographing message flow between Web services (Abela & Solanki, 2003; Arkin, Askary, Fordin, Jekeli, Kawaguchi, Orchard, et al., 2002). WSCI describes the flow of messages exchanged by a **Web service** in a particular process, and also describes the collective message exchange among interacting Web services, providing a global view of a complex process involving multiple Web services (Arkin et al., 2002). WSCI is a construct-based language that deals with the external observable rather than the internal definition of service behavior. This behavior is expressed in terms of temporal and logical dependencies among the exchanged messages, featuring sequencing rules, correlation, exception handling, and transactions. The advantage of this description is to enable developers, architects, and tools to describe and compose a global view of the dynamic of the message exchanged by understanding the interactions with Web services. However, it does not expose any form of semantics and therefore does not facilitate the process of automated composition.

Self-Serv framework (Sheng, Benatallah, Dumas, & Mak, 2002) can compose Web services and the resulting composite service can be executed in a decentralized dynamic environment (Benatallah, Dumas, Fauvet, & Paik, 2001). The execution of a composite service is not dependent on a central scheduler like Eflow (Abela, 2003), but rather on software components hosted by each of the providers participating in a **service composition**. More precisely, service providers participating in a composition process collaborate in a peer-to-peer manner to ensure that the schema and the control flow of the composition are respected.

Recall that to describe the control-flow of composite services, there are several existing process-modeling languages as described above. In Self-Serv, a subset of statecharts has been adopted to express the control-flow of the composite service. States can be simple or composite. A simple state corresponds to the execution of a service, whether elementary or composite. Compound states on the other hand, contain one or several entire statecharts within them, thereby providing a decomposition mechanism. When a composite state is entered, its initial state(s) become(s) active. The execution of a composite state is considered complete when it reaches (all) its final state(s).

This approach clearly provides greater scalability and availability than a centralized one where the execution of a service depends on a central scheduler. Specifically, the responsibility of coordinating providers participating in a composite service execution is distributed across several lightweight software components hosted by the providers themselves.

SpiderNet is a platform that differs from the above works by providing fully decentralized efficient **service composition** solution (Xiaohui, Nahrstedt, & Yu, 2004). It focuses on addressing the challenge of scalable QoS and resource management issues, which is important for composing QoS sensitive distributed applications. More precisely, SpiderNet is a service oriented P2P system called P2P service overlay where peers can provide not only media files but also a number of application service components such as media transcoding and data filtering as well as application-level data routing.

In SpiderNet, new services can be flexibly composed from available service components based on the user's function and quality-of-service requirements. A service is defined as a self-contained application unit providing certain functionality. Service components are represented by a service graph, which collectively deliver advanced composite services to the end user. The link in the service graph is called service link that can be mapped on an overlay network path.

SpiderNet platform implements the decentralized **service discovery** based on the Pastry distributed hash table (DHT) to efficiently locate services. The composition protocol is described as flows. Given the user's QoS/resource require-ments, the source first generates a composition probing message, called probe. Peers process a probe independently and in parallel until the probe arrives at the destination in order to locate multiple candidate service graphs. When the destination collects the probes for a request, it then selects the best qualified service graph based on the resource and QoS and sends an acknowledge message along the reversed selected service graph to confirm resource allocations and initialize service components at each intermediate peer. Finally, the application sender starts to stream application data units along the selected service graph.

Another framework (Basu, Ke, & Little, 2002) addresses a distributed platform based on a hierarchical task-graph approach to enable **service composition** in mobile ad hoc networks. In this work, a composite service is represented as a task-graph with leaf nodes representing logical services and edges between nodes represent required data flows between corresponding services. A service is a functionality provided by a single device or by a federation of cooperating devices. The main purpose of this framework is to achieve the following goals. The first allows the construction of complex distributed services from simpler services, while the second provides a runtime discovery of devices and instances of services that are most suitable for executing the larger distributed application. The third goal provides the rapid adaptation to node and link failures due to the dynamic changes of the environment (e.g., the users' mobility).

For instantiating hierarchical task graphs and for handling disruptions in services due to mobility of devices, distributed algorithms have been proposed (Basu et al., 2002). **Service composition** is coordinated and managed by the source of the request. More precisely, this approach selects the source of the request as the composition manager for itself (Chakraborty Perich, Joshi, Finin, & Yesha, 2002; Chakraborty, Yesha, Finin, & Joshi, 2005). After a service is composed on demand and used, its components retain their associations for a certain interval of time. If another user requests the service, he does not have to compose it on demand. However, this approach employs global broadcasting techniques that can increase load traffic in the network.

In Abela and Solanki (2003), authors have identified several requirements for Web **service composition** languages. The most important requires that a composition language has to be adequately expressive, and it has to have a well-defined and robust formal model in order to facilitate the automated composition of services. In other words, in order to dynamically compose Web services, languages modeling flow between services needs to have well-defined semantics. Flow composition languages like XLANG, WSFL, BPEL4WS, WS-CDL, and WSCI may be syntactically sound using WSDL, however they lack semantics and expressiveness. More precisely, they do not expose any form of semantics and therefore do not facilitate the process of automated and dynamic composition, which is suitable for **ubiquitous and pervasive environments**.

A more challenging problem, when a functionality that cannot be realized by the existing services is required, the existing services can combined together on demand to fulfill the request (Fujii & Suda, 2004; Maamar et al., 2005; Sirin et al., 2003). More precisely, problems related to **Web service** are how to specify them in an expressive manner, how to dynamically discover and compose them, and how to ensure their correctness (Beek et al., 2006).

Recently, artificial intelligence based approaches have been proposed in the literature. They take two major research directions to address these issues. The first concerns composite services behavior correctness. Indeed, interactions between service components come up with several problems like messages that are never received and the behavior incompatibility of interacted services. Therefore, interoperation

of the independent communicating component services should be granted; we will denote this issue in what follows by the first requirement, that is, services behavior correctness. The second research area addresses approaches that enable dynamic, runtime semantic **service discovery**, interaction and composition across the Web; we will denote this issue by the second requirement. These approaches, grouped into the AI planning approaches category depicted in Figure 3, are the subject of the next section.

AI Planning-Based Approaches

Recently, several techniques with unambiguous and formal semantics have been proposed in order to verify that a **service composition** process works properly. These techniques, known as **formal methods**, are generally used for the specification and the verification of complex systems. Among them, a variety of approaches based on state-action models (e.g., labeled transition system, timed automata, and Petri nets) and process models (e.g., p-calculus) are used to formally describe and reason about Web services conversation and composition (Beek et al., 2006). However, **formal methods** allow simulating and verifying the behavior of Web **service composition** at design time. Thus, the verification enables the detection and the correction of errors as early as possible, but they do not address the second requirement related to dynamic composition because they do not deal with the location of services or the recognition or the selection of those that match to the target service requested, thus creating the suitable composite service (Charif & Sabouret, 2005). Approaches using ontologies and augmented with formal techniques have been developed specifically to describe flow of **service composition**. These approaches allow the description of Web services' conversation, which is an important requirement to achieve dynamic **service composition**.

Model-Oriented Analysis and Verification

There are several methods to translate **service composition** descriptions (e.g., OWL-S or DAML-S, BPEL4WS) to formal description using **formal methods** like Petri net and labeled state transition that have been proposed. For example in Hamadi and Benatallah (2003) and Narayanan and McIlraith (2002), a Web **service composition** using Petri net algebra is proposed. A service behavior is considered as an ordered set of operations and has an associated Petri net that describes it. Operations are modeled by transitions, and the states of the service are modeled by places. The arrows between places and transitions are used to specify causal relations. After specifying composition with a Petri net, some algebraic properties, such as absence of deadlocks, could be proved.

Colored Petri Net (CP-net) is also used to tackle the reliability of **service composition** as proposed in Yang, Tan, and Xiao (2005). Generally, formal approaches aim to introduce much simpler descriptions and to model services in order to ensure verification of properties such as safety and liveness. In other words, describing services in such an abstract way lets us reason about the composition's correctness.

An approach proposed in Ankolekar et al. (2002) uses the situation calculus to model the composition process with their inputs, outputs, preconditions, and effects. Axioms in the situation calculus are mapped onto Petri net representation, which are then used to describe the semantics of the DAML-S control and constructs. In Foster, Uchitel, Magee, and Kramer (2003) modeling and verifying the composition of Web services workflows using the finite state processes (FSP) notation is proposed. To illustrate how these compositions are verified, workflow scenarios described by BPEL4WS using message sequence charts, together with a model-checking tool to interactively verify the workflow properties are

constructed. Recall that BPEL4WS is a language used to specify interactions between Web services.

An approach proposed in Koshikina (2003) describes the mapping rules for translating a BPEL4WS process model to a finite state automaton. The aim of this approach is to analyze composition processes in order to detect possible deadlocks. To achieve this, a process algebra called the BPE-calculus is introduced. It is a small language which captures all the BPEL features relevant to the analysis. This process algebra is modeled using a labeled transition system with a verification tool called the Concurrency Workbench. This tool allows to verify many properties of BPE-calculus processes specified in a logic called μ-calculus (Koshikina, 2003).

Other approaches to verify business process are based on model checking techniques. For example, in Beek et al. (2005) authors describe how to use the SPIN model checker to verify **Web service orchestration** using Web Services Flow Language. In order to do the verification using SPIN, business processes are first translated into Promela, the specification language provided by SPIN.

It is worth noting that these approaches simulate and verify the behavior of Web **service composition** at design time. However, a few of these methods address automated composition where the end user or application developer specifies a goal and an "intelligent" composition engine selects adequate services and offers the composition transparently to the user (Mokhtar, Georgantas, & Issarny, 2005). More precisely, unlike the specification and the correctness verification of the composition process, a main issue is how to identify candidate services, compose them, and verify how closely they match a request. In other words, the issue involved in the composition is how to select the most suitable services among a lot of services in a distributed manner. Recently, interesting formal based approach classified as model-based matching and service selection strategy is proposed in Mokhtar, Georgantas, and Issarny (2006). The following subsection focuses on the description of this strategy.

Model-Oriented Matching and Service Selection

Recall that dynamic composition allows the integration on the fly of a set of required services to perform a user request or a given target task. More precisely, the aim of dynamic composition is how to identify and select candidate services, compose them, and verify how closely they match a user request.

It should be noted that matching services is an important function of dynamic **service composition**. It allows the selection of services providing capabilities that are semantically equivalent to the capabilities requested by the target user task (Mokhtar et al., 2005). Several matching algorithms have been proposed in the literature (Ankolekar et al., 2002; Sycara, Paolucci, Ankolekar, & Srinivasan, 2003). These algorithms are classified in Mokhtar et al. (2005) into two categories: interface-level matching algorithms and process-level matching algorithms. In the first category, services and requests are described as a set of provided outputs and required inputs. Matching between a service and a request consists in matching all outputs and inputs of the request against all outputs and inputs of the service, respectively. This is the most used algorithm for matching **Semantic Web** services at the interface level in the literature (Kavantzas et al., 2005; Paolucci, Kawmura, Payne, & Sycara, 2002; Sycara et al., 2003). In contrary, the second category of matching algorithms, based on conversation description at the process level, provides a more precise matching, since the conversation description is richer than the interface description. More precisely, service conversation description provides more information about the service's behavior.

Recently, a COnversation-based **service composition** middleware (COCOA) using OWL/

DAML-S language that supports dynamic **service composition** by using both interface-level matching (i.e., signatures matching) and process-level approaches (i.e., behaviors matching) to construct semantic composite services, is proposed in Mokhtar et al. (2006).

It should be noted, as described above, that DAML-S profile, process and grounding provide only the specification of **Semantic Web** services. More precisely, DAML-S, which defines a clear semantics for services description, is not provided with a tool or a means for composing dynamically the specific functionalities or actions desired. Indeed, the composite process description must be given *a priori* and cannot be built at runtime (Charif & Sabouret, 2005). In COCOA, this specification is complemented on one hand by an execution model that preserves the DAML-S semantics and, on the other hand, by an implemented computational architecture that enables and ensures dynamic, runtime semantic **service discovery**, interaction, interoperation, and composition across the Web.

COCOA is composed of two main algorithms, a **service discovery** algorithm (COCOA-SD) and a conversation integration algorithm (COCOA-CI). COCOA-SD allows the discovery and the selection of services as candidates to the composition. COCOA-CI performs the dynamic composition of the selected services. Using these algorithms, COCOA allows a user with a task description to execute it on the fly without any previous knowledge about the networked services (Mokhtar et al., 2006).

The first step in COCOA is the selection of the most suitable services to respond to the user requirements using COCOA-SD algorithm. More precisely, this algorithm that allows the construction of a task's behavior by using fragments of the service behaviors is performed in two phases as follows. The first phase concerns the semantic operation matching that allows the selection of services that may be integrated together to compose the user task. To perform this phase, both the requested capabilities and those provided by services, are described using OWL-S as a set of IOPEs (inputs, outputs, preconditions, effects). More precisely, the requested and advertised capabilities are represented by a set of provided inputs and preconditions and a set of outputs and effects. COCOA-SD is used to match a requested service capability with a set of advertised ones. Two levels of matching have been defined, exact matching and weak matching, as proposed in Mokhtar et al. (2005, 2006). There is an exact or weak matching between a requested and a provided capability if all the inputs of the provided capability are matched against inputs of the requested one, and all the outputs of the requested capability are matched against outputs of the provided one. The difference between the two matching mechanisms is that in weak matching mechanism subsumption is supported between matched inputs and outputs. In addition, the weak matching is recognized between the two capabilities, if preconditions of the provided capability can be inferred from preconditions of the requested one, and effect of the requested capability can be inferred from effects of the provided one. In the exact matching mechanism, preconditions and effects of both capabilities should be equivalent.

The objective of the semantic matching phase is to compare semantically described operations involved in the task's conversation with those involved in the services' conversations. After matching the capabilities of the target task with those of the networked services, the **service discovery** algorithm selects those ones that will be useful for the composition.

The second phase concerns the conversation matching that filters and selects the most suitable services by comparing the structure of the task's conversation with those of selected services in the first phase. To do this, the mapping rules for translating an OWL-S process model to a finite state machine are defined in this method and formalized as follows. An automaton is represented by the 5-tuple $\langle Q, S, \delta, S_0, F \rangle$, where Q is a finite set of states, S is a finite set of symbols that define the alphabet of the language the automaton accepts, δ is the transition function; that is $\delta: Q \times S \to Q$, S_0 is the start state, the state in which the automaton is when no input has been processed yet, and F a set of final states, a subset of Q (i.e., $F \subset Q$).

In COCOA, the symbols correspond to the atomic processes involved in the conversation. The initial state corresponds to the root composite process, and a transition between two states is performed when an atomic process is executed. Each process, either atomic or composite, that is involved in the OWL-S conversation, is mapped to an automaton and linked together with the other ones in order to build the conversation automaton. This is achieved following the OWL-S process description and mapping rules as described in Mokhtar et al. (2006).

COCOA-SD uses conversation descriptions to select service capabilities that are semantically equivalent to required capabilities of the user task. To perform this process, regular expressions that represent languages generated by the user task automaton and automatons of selected services are used. More precisely, consider L the language generated by the extracted regular expression and by L_1, L_2, ..., L_n the languages generated by the automata of the selected services S_1, S_2, ..., S_n respectively. COCOA-SD selects all service S_i such that $L \cap L_i \neq \varphi$.

Once the filtration phase is achieved by CO-COA-SD, COCOA-CI integrates all the automata of selected services in one global automaton and selects one sub-automaton that correspond to task's conversation automata. More precisely, the COCOA-CI algorithm is involved into three steps. The first step is to connect the selected services' automata to form a global automaton by adding a new initial state. ε-transitions (ε is the empty symbol) are also added on one hand to link this state to each of the initial states of the selected services and, on the other hand, to link each final state of the selected services with the new initial state.

The next step of the conversation matching phase is to parse each state of the task's automaton by starting with the initial state and by following the automaton transitions. The objective of this step is to find at each step of the parsing process an equivalent state to the current one in the task. More precisely, equivalence is detected between a task's automaton state and a global automaton state, when for each input symbol of the former, there is at least a semantically equivalent input symbol of the latter. The result of this step gives a list of sub-automata of the global automaton that behave like the task automaton. The third step consists of finding a sub-automaton, among this list, that behaves like the task's automaton. More precisely, this step verifies that for each transition set that corresponds to an atomic conversation, there is no ε-transition going to the initial state before this conversation is finished. ε-transitions that connect final states to the initial state of the global automaton mark the end of a service conversation. This step allows eliminating a list of sub-automata that do not verify the atomic conversation constraints and select arbitrarily one of those that behaves as the user task. An executable description of the user task that includes references to required services is generated and sent to an execution engine that executes this description by invoking the appropriate service operations.

This approach permits to the arbitrary selection of the resulting **service composition** as they all conform to the target user task (Mokhtar et al., 2005, 2006). However, it does not allow the selection of the most effective composition among the eligible ones. More precisely, it does not allow the selection of the optimal or the best one. In general, most research to date in **service composition** are based on a broker that chooses the composite service after calculating all the candidates. The client/server strategy based on a broker is not scalable to large networks due to the large amount of available services. To set up **self-adaptive service composition systems, reinforcement learning** mechanism could be used (Gaber, 2000, 2006). The next section presents **reinforcement learning**-based approaches that permit the selection and the emergence of the most suitable composite service in a distributed manner without any central controller.

Learning-based Approaches for Service Composition and Emergence

In 2000, Gaber (2000) has pointed out that most research to date in **service discovery** and **composition** is based on the **traditional client-server paradigm** (CSP). He states that this paradigm is impractical in **ubiquitous and pervasive environments** and does not meet their related needs and requirements of adaptability to a dynamic environment, self-organization, and emergence. Therefore, Gaber (2000, 2006) has proposed two alternative paradigms to the traditional **client to server interaction paradigm** to design and implement **ubiquitous and pervasive applications**: the **adaptive services-to-client paradigm** (SCP) and the **spontaneous service emergence paradigm** (SEP). The **adaptive services/client paradigm** can be considered the opposite of CSP. Indeed, with the traditional CSP paradigm, the user initiates a request, should know *a priori* that the required service exists, and should be able to provide the location of a server holding that service. With the alternative SCP paradigm, it is the service that comes to the user. In other words, in this paradigm, a decentralized and **self-organizing middleware** should be able to provide services to users according to their availability and the network status. As pointed out in Gaber (2000), such a middleware can be inspired, for example, from a biological system like the natural immune system. More precisely, unlike the classical client/server approach, each user request is considered as an attack launched against the global network. An **immune networking middleware** reacts like the natural immune system against pathogens that have entered the body. It detects the infection (i.e., user request) and delivers a response to eliminate it (i.e., satisfy the user request).

The second alternative SEP paradigm to the client/server one is more adequate for **pervasive applications**. It involves the concept of **spontaneous emergence service composition** that suits **pervasive environments**. More precisely, spontaneous services can be created on the fly and be provided by mobiles that interact by ad hoc connections (Gaber 2000, 2006). The **spontaneous service emergence paradigm** (SEP) could also be implemented by a natural system that involves self-organizing and emergence behaviors (Gaber, 2000, 2006). Natural and biological systems like the human immune system (Jerne, 1994; Watanabe, Ishiguro, & Uchkawa 1999) has a set of organizing principles such as scalability, adaptability, and availability that are useful for developing a networking model in a highly dynamic and instable setting (Gaber, 2000).

Recently, **agent-based approaches**, with self-adapting and self-organizing capabilities, have been proposed to implement SCP and SEP, respectively (Bakhouya, 2005; Bakhouya & Gaber, 2006a, b; Gaber, 2006). More precisely, these approaches, inspired by biological and natural systems, provide scalable and adaptive **service discovery** and **composition** systems for **ubiquitous and pervasive environments**. Approaches to implement SCP and SEP are presented in the next section.

JaNet

Itao, Nakamura, Matsuo, Suda, and Aoyama (2002a) have proposed a platform called Ja-Net for service **emergence** in large-scale networks. Ja-Net Architecture is motivated by the observation that the above desirable properties, such as scalability and adaptability, have already been

realized in various large-scale biological systems like the bee colony. More precisely, Ja-Net defines a framework for developing large-scale, distributed, heterogeneous, and dynamic network applications. In this framework, a service is implemented by a collection of **distributed agents**, called cyber-entities. Ja-Net achieves built-in capabilities to create/emerge services adaptively according to user preferences (Itao, Nakamura, Matsuo, Suda, & Aoyama, 2002b). This is analogous to a bee colony (a network application) consisting of multiple bees (cyber-entities). Each cyber-entity implements a functional component related to its service or application. In addition, cyber-entity has simple behaviors such as migration, replication, reproduction, relationship establishment and death, and implements a set of actions related to a service that the cyber-entity provides.

Applications are provided through interactions of its cyber-entities. To provide an application, cyber-entities first establish relationships with each other and then choose cyber-entities to interact with based on their relationships. Strength of a relationship indicates the usefulness of the partner and dynamically adjusted based on the level of satisfaction indicated by a user who received the application.

In Ja-Net, a cyber-entity consists of three main parts: attributes, body, and behaviors. Attributes carry information regarding the cyber-entity (e.g., cyber-entity ID, service type, keywords, age, etc.). The cyber-entity body implements the service provided. Cyber-entity behaviors implement related actions such as migration, replication, relationship establishment, and death. Attributes, body, and behaviors of cyber-entities are described by XML and communicate using Speech Act-based Ja-Net ACL (Agent Communication Language). However, Ja-Net is not fault-tolerant to node failures. This is because, if a node fails or leaves the network, cyber-entities residing in it are destroyed and their state information is also removed. Consequently, cyber-entities could be misinformed about the presence of some services at particular nodes. Thus, since cyber-entities collaboration requires global information state maintenance, this platform does not address dynamic environments such as mobile ad hoc networks where nodes are mobiles and could be partially connected.

Recall that **pervasive computing**, often considered the same as **ubiquitous computing** in the literature, is a related concept that can be distinguished from **ubiquitous computing** in terms of environment conditions. We can consider that the aim in UC is to provide any mobile device access to available services in an existing network all the time and everywhere while the main objective in PC is to provide spontaneous services created on the fly by mobiles that interact by ad hoc connections (Bakhouya & Gaber, 2007; Gaber, 2000, 2006).

ImmuneNet

Recently, **agent-based approach**, with self-adapting and self-organizing capabilities, has been proposed in Bakhouya (2005); Bakhouya and Gaber (2006b); and Gaber (2006) to implement the SCP paradigm for UC applications. This approach, inspired by the human **immune system**, provides scalable and adaptive **service discovery** and **composition systems** for **ubiquitous environments**. In this approach, servers are organized into decentralized multi-agent communities. More precisely, a community represents a composite service. A component service is composed of a set of hardware or software resources that users need to discover and select. Semantic languages, such as DAML-S, that enable their dynamic discovery and their composition, could describe these services. For example, for a punctual need, a user who would like to open video files or create video CD might need the following services: a video player, format transcoding software, the MPEG-4 codec (for his wireless laptop), video effect or edge detection algorithms, and so forth. Since he does not know from which hosts he should get these resources throughout his current session, the user submits his multimedia composed service as a request, eventually with QoS requirements, to the resource discovery system that will process it according to the network status at that instant.

In this approach, the communities' organization based on the federation of **autonomous agents** is required to be self-organizing with inherent support for scalability and adaptability to user requirements and network changes. To address scalability, service agents (Sagent) that represent the resources of their corresponding servers in the network are organized into communities. More precisely, service agents establish relationships between them based on their affinity. Affinity corresponds to the adequacy with which two service agents could bind to create a composed service or to point out a similar service. To address the adaptability issue, affinity relationships between service agents are dynamic; the affinity values can be adjusted at runtime to cope with changes in the network.

The resource discovery and composition system proceeds as follows. It consists of two processes: service dissemination process and user request resolution process. The resource dissemination is a mechanism for server agents to learn about the existence of each other in order to create affinity relationships. In this process, when a server joins the network, it creates a **mobile agent**, to publish its service. A **mobile agent** initiates a random walk and moves randomly in the network. **Agents** can create replications (i.e., clones) of themselves, and consequently, they will learn about the existence of the all other Sagents in the network. In particular, when two Sagents detect an interest in common, a relationship is established and the two Sagents will group themselves into the same community. More generally, two Sagents have interest in common if their respective resources allow the creation of a composed service or if the Sagent holds a similar resource or service. It should be noted that, regarding the agent cloning operation, the distributed algorithm proposed in Bakhouya and Gaber (2006b) is used to regulate and control dynamically the number of clones spawned in the network.

The request resolution process is the process by which the user or an application will be provided by the required service information. In this approach, to locate a single service or a composed service (i.e., a community), the user creates a **mobile agent**, called request agent. This agent initiates a random walk in the network until it meets an appropriate community that can resolve the request (i.e., an initial entry point of a community). In this community, the request agent uses and **adjusts affinity relationships** in three stages: request forwarding stage, result backtracking stage, and service execution stage. During the forwarding stage, the request agent guided by the affinity relationships seeks server agents from inside the community that can provide the required resources. By the **affinity adjustments** during this stage, a selected path from within the community graph will emerge as a response to the request. When all the required resources are discovered, the request agent starts the backtracking phase. During this phase, the path computed between an endpoint in the community and the initial entry point will be reinforced globally by secondary **affinity adjustments**.

After the result backtracking phase, the requester node has all the discovered service paths and can choose one of them based on its affinity values and initiates the request agent that starts the service execution phase. In this phase, the request agent moves from the initial point of the service path, via the intermediate nodes, to the final point of the community and links elementary services together in order to form a composite service. During this phase, request agent triggers the service methods provided by each node in the service path and transports the result to other nodes until it meets the last node. If one elementary service fails during the service session, the Sagent that represents it in the preceding node will detect it and find a substitute and recompose the

service path. Since the Sagent has several affinity relationships to other Sagents that are similar or complementary, it selects a required one and sends a request agent to link it up.

It is worth noting also that this multi-organization based on dynamic affinities supported by relationships provides a highly decentralized system while remaining adaptive in a dynamic environment. More precisely, this decentralized architecture offers a high degree of resilience against servers leaving the network. For example, when a server leaves the network, all peer relationships with other servers are removed without additional overhead since it does not rely on any hard overlay control structure. Also, a leaving server or a communication failure does not have impact on the resolution of user requests.

The second alternative paradigm to the client/server one for **service composition** that suits **pervasive environments** involves the concept of spontaneous emergence. This paradigm can be carried out also by an inspired natural immune middleware that allows the emergence of ad hoc services on the fly according to dynamically changing context environments such as computing context and user context (Gaber, 2000). In this model, organizations or groups of autonomous agents represent ad hoc or composed services. More precisely, **agents** correspond to the **immune system** B-cells. **Agents** establish relationships based on affinities to form groups or communities to provide composite services. A community of **agents** corresponds to the idiotypic network in the human immune system (Gaber, 2006).

The service emergence model is the following (Gaber, 2000). A service is represented by an agent. The service is characterized by the set of functionality that is provided, while the agent represents the coherent behavior and activities of the service and its capability of performing tasks as a component of an ad hoc network (i.e., an ad hoc coalition). More precisely, a service agent is described by a set of states such as C = Initialization or context updating, M = Mature composition, I = Immature composition, A = Annihilated composition, E = Execution, T = Termination. The emergence and the behavior evolution of composed services depend on the agents' internal states, the agent-to-agent interactions, and the context sensed by the agents during their executions (Gaber, 2000). More precisely, agents interactions, or B-cells, act like the idiotypic network of the immune system by stimulation/suppression chain to provoke the emergence of an immune response against antigens, that is, the ad hoc context (e.g., Bakhouya & Gaber, 2006c).

More generally, **agents** together with their affinity relationships as a whole form a **propitient multi-agent system** (PMAS) (Bakhouya & Gaber, 2006a). A **propitient system** is a system with the ability to self-organize in order to adapt towards the most appropriate agent organization structures according to unpredictable changes in the environment. This emergent behavior is delivered as a result of agents-to-agents and agent-to-environment interactions that adapt until the system hits a most suitable affinity network (Gaber & Bakhouya, 2006). In other words, a **propitient multi-agent system** implements the **service emergence paradigm** (SEP) for **pervasive applications**.

CONCLUSION

The design and development of **ubiquitous and pervasive applications** require alternative operational models to the traditional **client-to-server interaction paradigm**. **Adaptive services-to-client paradigm** and **spontaneous service emergence paradigm** are more adequate to **ubiquitous and pervasive computing environments**. **Service composition** systems based on these three paradigms and proposed in the literature are presented with emphasis on automatic and dynamic **service composition** approaches and on self-organization and self-adaptation approaches to implement SCP and SEP. Self-adaptation, self-

organization, and emergence are crucial issues in systems that operate in an open and dynamic environment.

FUTURE WORK

Service composition is an important and active area of research and has been addressed widely in the fields of **Web service** and **service-oriented architecture**. As stated above, research in **service composition** can be divided into two directions: the service model description languages and the **service discovery** and **composition platforms**.

Several languages such as DAML-S have been developed to address the first requirement. Indeed, the **Web service** community concentrates its efforts mainly on developing Web services with rich semantic annotation to create a **Web semantic service-oriented architecture**. Hence, several service description languages that enable developers to create, execute, and combine Web services into complex ones using the standard **Web service** model have been proposed. It should be noted that the standard Web-based services model has a centralized architecture based on the traditional client-server paradigm (CSP). However, to meet UPC requirements, new design approaches for Web-**service discovery** and composition are required and should be scalable and adaptable to continuously changing UPC environments. Therefore, languages with a higher semantic level based on ontologies to facilitate services discovery and composition should be addressed.

In parallel to this issue, the service-oriented architecture community has focused mainly on developing architectures that enable scalable, fault tolerance, and adaptive **service discovery** and composition to fulfill the second requirement, but the development of semantic languages is not addressed. Moreover, most of the proposed systems are based on the client to server paradigm (CSP) based on the systematic usage of repositories and registries. However, the ability to maintain, allocate, and access a variety of continuously increasing heterogeneous resources and services distributed over a network is difficult to achieve with the traditional client-server approaches (Gaber, 2000, 2006). More precisely, these architectures cannot meet simultaneously the requirements of scalability and adaptability suitable for **ubiquitous and pervasive environments**.

As mentioned in this chapter, an appropriate model, the **adaptive services-to-client paradigm** for **service discovery** and composition in UC has been presented. In this paradigm, decentralized and **self-organizing middlewares** should be developed with the ability to provide services to users according to their availability and according to the network status at the user request moment. This kind of middleware should be intelligent, scalable, adaptable, and available in highly dynamic and instable environments.

The **spontaneous service emergence paradigm** is the alternative paradigm to the traditional client-server one that suits **pervasive environments**. This paradigm involves the concept of **spontaneous emergence of services** without *a priori* planning. This paradigm can be carried out by middleware that allows the emergence of ad hoc services on the fly according to dynamically changing context environments such as computing context and users' context.

In UPC environments, approaches dedicated to identify and select candidate services, to compose them, and verify how closely they match a user request, should emphasize self-organizing and self-adapting principles that are crucial issues for systems that operate in an open and dynamic environment. To achieve these requirements, approaches and languages proposed by Web services and **service-oriented architecture** communities should be combined together with the main objective to develop highly flexible and scalable approaches and platforms with self-organizing capabilities in order to cope with dynamically changing contexts and networks environments.

REFERENCES

Abela, C. (2003). *Semantic Web services composition.* Retrieved August 23, 2007, from http://www.citeseer.ist.psu.edu/735292.html

Abela, C., & Solanki, M. (2003). *A landscape of markup languages for Web services composition NetObject.* Retrieved August 23, 2007, from http://citeseer.ist.psu.edu/solanki03landscape.html

Akram, M. S., Medjahed, B., & Bouguettaya, A. (2003). Supporting Dynamic Changes in Web Service Environments. In *Proceedings of the International Conference on Service Oriented Computing* (LNCS 2910, pp. 319-334, ISBN: 978-3-540-20681-1).

Ankolekar, A., Burstein, M., Hobbs, J. R., Lassila, O., Martin, D., McDermott, D., McIlraith, S. A., Narayanan, S., Paolucci, M., Pyne, T., & Sycara, K. (2002). DAML-S: Web service description for the Semantic Web. In *Proceedings of the First International Semantic Web Conference (ISWC).* Retrieved August 23, 2007, from http://www.cs.cmu.edu/~softagents/daml.html

Ankolekar, A., Huch, F., & Sycara, K. (2002). Concurrent execution semantics for DAML-S with subtypes. In *Proceedings of the 1st International Semantic Web Conference on The Semantic Web* (LNCS 2342, pp. 318-332, ISBN: 3-540-43760-6).

Arkin, A., Askary, S., Fordin, S., Jekeli, W., Kawaguchi, K., Orchard, D., Pogliani, S., Riemer, K., Struble, S., Takacsi-Nagy, P., Trickovic, I., & Zimek, S. (2002). *Web service choreography interface (WSCI).* Retrieved August 23, 2007, from http://www.w3.org/TR/wsci/

Baget, J. F., Canaud, E., Euzenat, J., & Saïd-Hacid, M. (2003). Les Langages du Web Sémantique. *INRIA Rhône-Alpes.* Retrieved August 23, 2007, from http://www.inrialpes.fr/exmo/cooperation/asws/ASWS-Langages.pdf

Bakhouya, M. (2005). *Self-adaptive approach based on mobile agent and inspired by human immune system for service discovery in large scale networks.* Unpublished doctoral thesis, Universite de Technologies de Belfort-Montbeliard (UTBM).

Bakhouya, M., & Gaber, J. (2006a). Self-organizing approach for emergent multi-agent structures. In *Proceedings of the Workshop on Complexity through Development and Self-Organizing Representations (CODESOAR'06) at GECCO'06.* Seattle: ACM Press.

Bakhouya, M., & Gaber, J. (2006b). Adaptive Approaches for Ubiquitous Computing. In (Eds.), *Mobile Networks and Wireless Sensor Networks* (pp. 129-163). Hermes Science.

Bakhouya, M., & Gaber, J. (2006c). Adaptive Approach for the Regulation of a Mobile Agent Population in a Distributed Network. In *Proceedings of the 5th International Symposium on Parallel and Distributed Computing (ISPDC'06),* Timisoara, Romania. IEEE Press.

Bakhouya, M., & Gaber, J. (2007). Ubiquitous and Pervasive Applications Design. In D. Taniar (Ed.), *Encyclopedia of Mobile Computing & Commerce.* Hershey, PA: Idea Group Publishing.

Barros, A., Dumas, M., & Oaks, P. (2005). A critical overview of the Web services choreography description language (WS-CDL). *Business Process Trends.* Retrieved August 23, 2007, from http://www.bptrends.com

Basu, P., Ke, W., & Little, T.D.C. (2002). Scalable service composition in mobile ad hoc networks using hierarchical task graphs. In *Proceedings of the 1st Annual Mediterranean Ad Hoc Networking Workshop (Med-Hoc-Net 2002),* Sardegna, Italy. Retrieved August 23, 2007, from http://hulk.bu.edu/pubs/publications.html

Beek, M., Bucchiarone, A., & Gnesi, S. (2006). *A survey on service composition approaches: From industrial standards to formal methods* (Tech. Rep. Bib. Code 2006-TR-15). Retrieved August 23, 2007, from http://dienst.isti.cnr.it/

Benatallah, B., Dumas, M., Fauvet, M. C., & Paik, H. Y. (2001). *Self-coordinate, self-traced composite services with dynamic provider selection* (Tech. Rep. No. UNSW-CSE-TR-0108). The University of New South Wales Department of Computer Science and Engineering. Retrieved August 23, 2007, from http://sky.fit.qut.edu.au/~dumas/

Bucchiarone, A., & Gnesi, S. (2006). *A survey on services composition.* Retrieved August 23, 2007, from http://www.selab.isti.cnr.it/ws-mate/program.html

Casati, F., Ilnicki, S., Jin, L., Krishnamoorthy, V., & Shan, M.-C. (2000). Adaptive and dynamic service composition in eFlow. In *Proceedings of the Conference on Advanced Information Systems Engineering* (pp. 13-31). Retrieved August 23, 2007, from http://citeseer.ist.psu.edu/casati-00adaptive.html

Chakraborty, D. (2001). *Service composition in ad-hoc environments* (Tech. Rep. No. TR-CS-01-20). University of Maryland, Department of Computer Science and Electrical Engineering, Baltimore County, MD. Retrieved August 23, 2007, from http://citeseer.ist.psu.edu/529431.html

Chakraborty, D., & Joshi, A. (2001). *Dynamic service composition: State-of-the-art and research directions* (Tech. Rep. No. TR-CS-01-19). University of Maryland, Department of Computer Science and Electrical Engineering, Baltimore County, MD. Retrieved August 23, 2007, from http://citeseer.ist.psu.edu/chakraborty01dynamic.html

Chakraborty, D., Perich, F., Avancha, S., & Joshi, A. (2001). Dreggie: Semantic service discovery for m-commerce applications. In *Proceedings of the Workshop on Reliable and Secure Applica-*

tions in Mobile Environment, 20th Symposium on Reliable Distributed Systems (pp. 28-31). Retrieved August 23, 2007, from http://citeseer. ist.psu.edu/chakraborty01dreggie.html

Chakraborty, D., Perich, F., Joshi, A., Finin, T., & Yesha, Y. (2002). A reactive service composition architecture for pervasive computing. In *Proceedings of the 7th Personal Wireless Communications Conference (PWC 2002)* (pp. 53-62, ISBN: 1-4020-7250-3).

Chakraborty, D., Yesha, Y., Finin, T., & Joshi, A. (2005). Service composition for mobile environments [Special issue: Mobile Services]. *Journal on Mobile Networking and Applications, 10*(4), 435-451.

Charif, Y., & Sabouret, N. (2005). An overview of Semantic Web services composition approaches. In *Proceedings of the International Workshop on Context for Web Services (CWS-05), ENTCS, 146*(1), 33-41.

Chen, H. (1999). *Developing a dynamic distributed intelligent agent framework based on the jini architecture.* Unpublished master's thesis, University of Maryland, Baltimore County. Retrieved August 23, 2007, from http://gentoo. cs.umbc.edu/ronin/doc/

Chen, H. (2004). *Ronin agent framework.* Retrieved August 23, 2007, from http://gentoo. cs.umbc.edu/ronin/

Curbera, F., Goland, Y., Klein, J., Leymann, F., Roller, D., Thatte, S., & Weerawarana, S. (2002). *Business process execution language for Web services* (Version 1.0). Retrieved August 23, 2007, from http://www-106.ibm.com/developerworks/library/ws-bpel/.

DAML-S and Related Technologies. Retrieved August 23, 2007, from http://www.daml.org/services/daml-s/0.9/survey.pdf

Dustdar, S., & Schreiner, W. (2005). A survey on Web services composition. *International Journal on Web and Grid Services, 1*(1), 1-30. Inderscience Enterprises Ltd.

Feng, N. (1999). *S-module design for software hot swapping technology.* Unpublished master's thesis, Carleton University, System and Computer Engineering Department. Retrieved August 23, 2007, from http://citeseer.ist.psu. edu/feng99smodule.html

Foster, H., Uchitel, S., Magee, J., & Kramer, J. (2003). Model-based verification of web service compositions. In *Proceedings of the IEEE International Conference on Automated, Software Engineering* (ISBN: 0-7695-2035-9).

Fujii, K., & Suda, T. (2004). Dynamic service composition using semantic information. In *Proceedings of the 2nd ACM International Conference on Service Oriented Computing (ICSOC '04)* (pp. 39-48, ISBN: 1-58113-871-7).

Gaber, J. (2000). *New paradigms for ubiquitous and pervasive computing* (Research Rep. No. RR-09-00). Université de Technologies de Belfort-Montbéliard (UTBM), France.

Gaber, J. (2006). New paradigms for ubiquitous and pervasive applications. In *Proceedings of the 1st Workshop on Software Engineering Challenges for Ubiquitous Computing*, Lancaster, UK.

Gaber, J., & Bakhouya, M. (2006). An affinity-driven clustering approach for service discovery and composition for pervasive computing. In *Proceedings of the ACS/IEEE International Conference on Pervasive Services ICPS'06* (pp. 277-280, ISBN: 1-4244-0237-9).

Guttman, E. (1999). Service location protocol: Automatic discovery of IP network services. *IEEE Internet Computing, 3*(4), 71-80.

Hamadi, R., & Benatallah, B. (2003). A Petri-net-based model for Web service composition. In *Proceedings of the 14th Australasian Database Conference on Database Technologies, 17*, 191-200. ISSN: 1445-1336.

Hu, M. (2003). Web services composition, partition, and quality of service in distributed system integration and re-engineering. In *Proceedings of the XML Conference*, Philadelphia, PA. IDEAlliance. Retrieved August 23, 2007, from http://www.idealliance.org/papers/dx_xml03/papers/05-05-04/05-05-04.pdf

Itao, T., Nakamura, T., Matsuo, M., Suda, T., & Aoyama, T. (2002a). Adaptive creation of network applications in the jack-in-the-net architecture. *IFIP Networking*, pp. 129-140.

Itao, T., Nakamura, T., Matsuo, M., Suda, T., & Aoyama, T. (2002b). Service emergence based on cooperative interaction of self-organizing entities. In *Proceedings of the IEEE Symposium on Applications and the Internet* (pp. 194-203), Nara, Japan.

Jerne, N. (1974). Towards a network theory of the immune system. *Ann. Immunol, 125*, 125-373.

Jini. (2000). *Jini network technology: Specifications* (Version 1.1 Beta). Sun Microsystems Inc. Retrieved August 23, 2007, from http://www.sun.com/software/jini/specs/.

Kavantzas, N., Budett, B., Ritzinger, G., Fletcher, T., Lafon, Y., & Bareto, C. (2005). Web services choreography description language Version 1.0. *World Wide Web Consortium*. Retrieved August 23, 2007, from http://www.w3.org/TR/2004/WD-ws-cdl-10-20041217/

Koshkina, M. (2003). *Verification of business processes for web services.* Unpublished master's thesis, York University, Departement of Computer Science, Toronto. Retrieved August 23, 2007, from http://www.cse.yorku.ca/~franck/research/students/maria.pdf

Leymann, F. (2001, May). *Web services flow language* (Rep. No. WSFL10). IBM Software Group.

Maamar, Z., Mostefaoui, S. K., & Yahyaoui, H. (2005). Toward an agent-based and context-oriented approach for web services composition. *IEEE Transactions on Knowledge and Data Engineering, 17*(5), 686-697.

Majithia, S., Walker, D. W., & Gray, W. A. (2004). A framework for automated service composition in service-oriented architecture. *LNCS 3053*, 269-283.

Mao, M. Z., Brewer, E. A., & Katz, R. H. (2001). *Fault-tolerant, scalable, wide-area Internet service composition* (Tech. Rep. No. UCB//CSD01-1129). University of California-Berkeley. Retrieved August 23, 2007, from http://www.cs.berkeley.edu/zmao/Papers/techreport.ps.gz

Medjahed, B. (2004). *Semantic Web enabled composition of Web services.* Unpublished doctoral dissertation. Virginia Tech, Department of Computer Science. Retrieved August 23, 2007, from http://www-personal.engin.umd.umich.edu/~brahim/

Mennie, D., & Pagurek, B. (2000). An architecture to support dynamic composition of service components, systems and computer engineering, Carleton University. In *Proceedings of the 5th International Workshop on Component-Oriented Programming (WCOP 2000)*. Retrieved August 23, 2007, from http://citeseer.ist.psu.edu/mennie00architecture.html

Milanovic, N., & Malek, M. (2004). Current solutions for Web service composition. *IEEE Internet Computing, 8*(6), 51-59.

Mokhtar, S. B., Georgantas, N., & Issarny, V. (2005). Ad hoc composition of user tasks in pervasive computing environments. *LNCS, 3628*, 31-46.

Mokhtar, S. B., Georgantas, N., & Issarny, V. (2006). COCOA: Conversation-based service composition in pervasive computing environments. In *Proceedings of the IEEE International Conference on Pervasive Services (ICPS'06)* (pp. 29-38).

Narayanan, S., & McIlraith, S. A. (2002). Simulation, verification and automated composition of Web services. In *Proceedings of the 11th International World Wide Web Conference (WWW-11)*, Honolulu, HI (pp. 77-88).

Oprescu, J. (2004). *Découverte et Composition de Services dans des Réseaux Ambiants.* Unpublished doctoral thesis, Ecole Doctorale Mathematiques, Sciences et Technologie de l'Information, laboratoire LSR–IMAG. Retrieved August 23, 2007, from http://drakkar.imag.fr/article.php3?id_article=175

OWL. Ontology Web Language. Retrieved August 23, 2007, from http://www.w3.org/TR/owl-ref/

Padhye, M. (2004). *Coordinating heterogeneous Web services through handhelds using SyD's wrapper framework.* Unpublished doctoral thesis, Georgia State University, College of Arts and Sciences. Retrieved August 23, 2007, from http://etd.gsu.edu/theses/

Paolucci, M., Kawmura, T., Payne, T., & Sycara, K. (2002). Semantic matching of Web services capabilities. In *Proceedings of the 1st International Semantic Web Conference on The Semantic Web* (LNCS 2342, pp. 333-347, ISBN:3-540-43760-6).

Peltz, C. (2003a). Web services orchestration and choreography. *Computer, 36*(10), 46-52.

Peltz, C. (2003b). Web services orchestration: A review of emerging technologies, tools, and standards. *Hewlett Packard Co.* Retrieved August 23, 2007, from devresource.hp.com/drc/technical_white_papers/WSOrch/WSOrchestration.pdf

Robert, E. M. (2000). *Discovery and its discontents: Discovery protocols for ubiquitous computing* (Research Rep. No. UIUCDCS-R-2000-2154). Department of Computer Science, University of Illinois Urbana-Champaign, Urbana. Retrieved August 23, 2007, from http://citeseer.ist.psu.edu/mcgrath00discovery.html

Satish, T. (2001). *XLANG Web services for business process design.* Retrieved August 23, 2007, from http://www.gotdotnet.com/team/xml_wsspecs/xlang-c/default.htm

Sheng, Q. Z., Benatallah, B., Dumas, M., & Mak, E. (2002). SELFSERV: A platform for rapid composition of web services in a peer-to-peer environment. In *Proceedings of the 28th Very Large DataBase Conference (VLDB'2002)*, Hong Kong, China. Retrieved August 23, 2007, from http://sky.fit.qut.edu.au/~dumas/

Sirin, E., Hendler, J., & Parsia, B. (2003). Semi-automatic composition of web services using semantic descriptions. In *Proceedings of the Web Services, Modeling, Architecture and Infrastructure Workshop in conjunction with ICEIS 2003.* Retrieved August 23, 2007, from http://citeseer.ist.psu.edu/sirin02semiautomatic.html

SOAP. Simple Object Access Protocol. Retrieved August 23, 2007, from http://www.w3.org/TR/soap12-part0/

Srivastava, B., & Koehler, J. (2003). Web service composition: Current solutions and open problems. In *Proceedings of the ICAPS 2003 Workshop on Planning for Web Services* (pp. 28-35). Retrieved August 23, 2007, from http://citeseer.ist.psu.edu/srivastava03Web.html

Su, X., & Rao, J. (2004). A survey of automated Web service composition methods. In *Proceedings of the 1st International Workshop on Semantic Web services and Web process composition.* Retrieved August 23, 2007, from http://www.cs.cmu.edu/~jinghai/

Sycara, K., Paolucci, M., Ankolekar, A., & Srinivasan, A. (2003). Automated discovery, interaction and composition of Semantic Web services. *Journal of Web Semantics, 1*(1), 27-46.

UDDI. Universal Discovery Description and Integration Protocol. Retrieved August 23, 2007, from http://www.uddi.org/

Van der Aalst, W. M. P., Dumas, M., & ter Hofstede, A. H. M. (2003). Web service composition languages: Old wine in new bottles? In *Proceedings of the 29th EUROMICRO Conference* (pp. 298-305).

Watanabe, Y., Ishiguro, A., & Uchkawa, Y. (1999). Decentralized behavior arbitration mechanism for autonomous mobile robot using immune system. In *Artificial Immune Systems and Their Applications*. Springer-Verlag.

Weiser, M. (1996). *Hot topics: Ubiquitous computing.* IEEE Computer.

WSDL. Web Services Description Language. Retrieved August 23, 2007, from http://www.w3.org/TR/wsdl

WSCI. *Web Service Choreography Interface 1.0 Specification.* Intalio, Sun Microsystems, BEA Systems, SAP.

Wu, C., & Chang, E. (2005). State-of-the-art Web services architectural styles. In *Proceedings of the 3rd IEEE European Conference on Web Services (ECOWS)*, Vaxjo, Sweden. Retrieved August 23, 2007, from http://wscc.info/p51561/files/paper54.pdf

Xiaohui, G., Nahrstedt, K., & Yu, B. (2004). SpiderNet: An Integrated peer-to-peer service composition framework. In *Proceedings of the 13th IEEE International Symposium on High Performance Distributed Computing (HPDC)* (pp. 110-119).

Xu, D., Nahrstedt, K., & Wichadakul, D. (2001). Qos-aware discovery of wide-area distributed services. In *Proceedings of the 1st IEEE/ACM International Symposium on Cluster Computing and the Grid (CCGrid)*, Brisbane, Australia (pp. 92-99).

Yang, Y., Tan, Q., & Xiao, Y. (2005). Verifying Web services composition based on hierarchical colored petri nets. In *Proceedings of the 1st International Workshop on Interoperability of Heterogeneous Information Systems* (pp. 47-54, ISBN: 1-59593-184-5).

Zeng, L., Benatallah, B., Ngu, A., Dumas, M., Kalagnanam, J., & Chang, H. (2004). QoS-aware middleware for Web services composition. *IEEE Transactions on Software Engineering, 30*(5), 311-327.

ADDITIONAL READING

Abela, C., & Solanki, M. (2003). A landscape of markup languages for Web services composition. *NetObject*. Retrieved August 23, 2007, from http://citeseer.ist.psu.edu/solanki03landscape.html

Bakhouya, M., & Gaber, J. (2006). Self-organizing approach for emergent multi-agent structures. In *Workshop on Complexity through Development and Self-Organizing Representations (CODE-SOAR'06) at GECCO'06.* Seattle: ACM Press.

Bakhouya, M., & Gaber, J. (2007, February). Ubiquitous and pervasive applications design. In D. Taniar (Eds.), *Encyclopedia of Mobile Computing & Commerce*. Idea Group Publishing.

Peltz, C. (2003). Web services orchestration and choreography. *Computer, 36*(10), 46-52.

Weiser, M. (1996). *Hot topics: Ubiquitous computing.* IEEE Computer.

Compilation of References

Abbink, K., Irlenbusch, B., & Renner, E. (1999). An experimental bribery game. *The Journal of Law, Economics and Organization, 18*(2), 428-454.

Abela, C. (2003). *Semantic Web services composition.* Retrieved August 23, 2007, from http://www.citeseer.ist.psu.edu/735292.html

Abela, C., & Solanki, M. (2003). *A landscape of markup languages for Web services composition NetObject.* Retrieved August 23, 2007, from http://citeseer.ist.psu.edu/solanki03landscape.html

Admati, A., & Perry, M. (1987). Strategic delay in bargaining. *Review of Economical Studies, 54,* 345364.

Akram, M. S., Medjahed, B., & Bouguettaya, A. (2003). Supporting Dynamic Changes in Web Service Environments. In *Proceedings of the International Conference on Service Oriented Computing* (LNCS 2910, pp. 319-334, ISBN: 978-3-540-20681-1).

Alani, H., Dasmahapatra, S., O'Hara, K., & Shadbolt, N. (2003). Identifying communities of practice through ontology network analysis. *IEEE Intelligent Systems, 18*(2), 18-25.

Alderson, A., & Liu, K. (2000). Reverse requirements engineering: The AMBOLS approach. In P. Henderson (Ed.), *Systems Engineering for Business Process Change* (pp. 196-208). London: Springer.

Alexander, C. (1979). *The timeless way of building.* Oxford University Press.

Amin, M., & Ballard, D. (2000, July-August). Defining new markets for intelligent agents. *IT Pro,* pp. 29-35.

Andreasen, A. (1965). Attitudes and customer behavior: A decision model. In L. Preston (Ed.), *New research in marketing.* California Institute of Business and Economics Research.

Ankolekar, A., Burstein, M., Hobbs, J. R., Lassila, O., Martin, D., McDermott, D., McIlraith, S. A., Narayanan, S., Paolucci, M., Pyne, T., & Sycara, K. (2002). DAML-S: Web service description for the Semantic Web. In *Proceedings of the First International Semantic Web Conference (ISWC).* Retrieved August 23, 2007, from http://www.cs.cmu.edu/~softagents/daml.html

Ankolekar, A., Huch, F., & Sycara, K. (2002). Concurrent execution semantics for DAML-S with subtypes. In *Proceedings of the 1st International Semantic Web Conference on The Semantic Web* (LNCS 2342, pp. 318-332, ISBN: 3-540-43760-6).

Applegate, L. M., Holsapple, C. W., Kalakota, R., Radermacher, F. J., & Whinston, A. B. (1996). Electronic commerce: Building blocks of new business opportunity. *Journal of Organizational Computing & Electronic Commerce, 6*(1), 1-10.

Aranguren, M. (2005). *Ontology design patterns for the formalisation of biological ontologies.* Master's Thesis, Manchester University.

Ardissono, L., Barbero, C., et al. (1999). An agent architecture for personalized Web stores. In *Proceedings of the Conference on Autonomous Agents* (pp. 182-189). ACM.

Aridor, Y., & Lange, D. (1998). Agent design patterns: Elements of agent application design. In *Proceedings of the Conference on Autonomous Agents* (pp. 108-115). ACM.

Arkin, A., Askary, S., Fordin, S., Jekeli, W., Kawaguchi, K., Orchard, D., Pogliani, S., Riemer, K., Struble, S., Takacsi-Nagy, P., Trickovic, I., & Zimek, S. (2002). *Web service choreography interface (WSCI)*. Retrieved August 23, 2007, from http://www.w3.org/TR/wsci/

Aron, R., Sundararajan, A., & Viswanathan, S. (2006). Intelligent agents in electronic markets for information goods: Customization, preference revelation and pricing. *Decision Support Systems, 41*, 764-786.

Ashish, N., & Knoblock, C. A. (1997). *Semi-automatic wrapper generation for Internet information sources.* Paper presented at the Second IFCIS Conference on Cooperative Information Systems, Kiawah Island, SC.

AuctionWatcher (2002). Retrieved August 14, 2007, from http://www.auctionswatch.info

Avgeriou, P., & Zdun, U. (2005). Architectural patterns revisited: A pattern language. In *Proceedings of the European Conference on Pattern Languages of Programs.*

Axelrod, R. (1987). The evolution of strategies in the iterated prisoner's dilemma. In L. Davis (Ed.), *Genetic algorithms and simulated annealing.* London: Pittman.

Baget, J. F., Canaud, E., Euzenat, J., & Saïd-Hacid, M. (2003). Les Langages du Web Sémantique. *INRIA Rhône-Alpes.* Retrieved August 23, 2007, from http://www.inri-alpes.fr/exmo/cooperation/asws/ASWS-Langages.pdf

Bailey, J., & Bakos, Y. (1997). An exploratory study of the emerging role of electronic intermediaries. *International Journal of Electronic Commerce, 1*(3), 7-20.

Bailey, M. N., & Lawrence, R. L. (2001). Do we have a new economy? *American Economic Review, 91*(2), 308-312.

Baker, M.J. (1999). *The marketing book.* Woburm, MA: Butterworth-Heinemann.

Bakhouya, M. (2005). *Self-adaptive approach based on mobile agent and inspired by human immune system for service discovery in large scale networks.* Unpublished doctoral thesis, Universite de Technologies de Belfort-Montbeliard (UTBM).

Bakhouya, M., & Gaber, J. (2006a). Self-organizing approach for emergent multi-agent structures. In *Proceedings of the Workshop on Complexity through Development and Self-Organizing Representations (CODESOAR'06) at GECCO'06.* Seattle: ACM Press.

Bakhouya, M., & Gaber, J. (2006b). Adaptive Approaches for Ubiquitous Computing. In (Eds.), *Mobile Networks and Wireless Sensor Networks* (pp. 129-163). Hermes Science.

Bakhouya, M., & Gaber, J. (2006c). Adaptive Approach for the Regulation of a Mobile Agent Population in a Distributed Network. In *Proceedings of the 5th International Symposium on Parallel and Distributed Computing (ISPDC'06)*, Timisoara, Romania. IEEE Press.

Bakhouya, M., & Gaber, J. (2007). Ubiquitous and Pervasive Applications Design. In D. Taniar (Ed.), *Encyclopedia of Mobile Computing & Commerce.* Hershey, PA: Idea Group Publishing.

Bakos, Y. (1998). The emerging role of electronic marketplaces on the internet. *Communications of the ACM.*

Barbash, A. (2001). Mobile computing for ambulatory health care: Points of convergence. *Journal of Ambulatory Care Management, 24*(4), 54-60.

Barbuceanu, M., & Lo, W. (2000). A multiattribute utility theoretic negotiation architecture for eletronic commerce. In *Proceedings of the 4th International Conference on Autonomous Agents,* Barcelona, Spain (pp. 239-247).

Barjis, J., Chong, S., Dietz, J.L.G, & Liu, K. (2002, September). Development of agent-based e-commerce systems using semiotic approach and DEMO transaction concept. *International Journal of Information Technology and Decision Making, 1*(3), 491-510.

Barnes, S. J. (2002). The mobile commerce value chain: Analysis and future developments. *International Journal of Information Management, 22*(2), 91-108.

Barros, A., Dumas, M., & Oaks, P. (2005). A critical overview of the Web services choreography description language (WS-CDL). *Business Process Trends*. Retrieved August 23, 2007, from http://www.bptrends.com

Basu, P., Ke, W., & Little, T.D.C. (2002). Scalable service composition in mobile ad hoc networks using hierarchical task graphs. In *Proceedings of the 1st Annual Mediterranean Ad Hoc Networking Workshop (Med-Hoc-Net 2002)*, Sardegna, Italy. Retrieved August 23, 2007, from http://hulk.bu.edu/pubs/publications.html

Baumohl, B. (2000, December 11). Can you really trust those bots? *TIME Magazine, 156*.

Bazerman, M., & Carroll, J. (1987). Negotiator cognition. In B. Staw & L.C. (Eds.), *Research in organizational behavior* (9). Greenwich, CT: JAI Press.

Beck, J. C., & Fox, M. S. (1994, May 15). *Supply chain coordination via mediated constraint relaxation*. Paper presented at the First Canadian Workshop on Distributed Artificial Intelligence, Banff.

Becker, G. S. (1968). Crime and punishment: An economic approach. *Journal of Political Economy, 76*, 169-217.

Becker, R., Holland, O. E., & Deneubourg, J. L. (1994). From local actions to global tasks: Stigmergy in collective robotics. In R. Brooks & P. Maes (Eds.), *Artificial Life IV*. Cambridge, MA: MIT Press.

Beek, M., Bucchiarone, A., & Gnesi, S. (2006). *A survey on service composition approaches: From industrial standards to formal methods* (Tech. Rep. Bib. Code 2006-TR-15). Retrieved August 23, 2007, from http://dienst.isti.cnr.it/

Benatallah, B., Dumas, M., Fauvet, M. C., & Paik, H. Y. (2001). *Self-coordinate, self-traced composite services with dynamic provider selection* (Tech. Rep. No. UNSW-CSE-TR-0108). The University of New South Wales Department of Computer Science and Engineering. Retrieved August 23, 2007, from http://sky.fit.qut.edu.au/~dumas/

Berners-Lee, T. (1998). *The Semantic Web road map*. Retrieved August 20, 2007, from http://www.w3.org/DesignIssues/Semantic.html

Berners-Lee, T. (1998-2005). *The fractal nature of the Web*. Retrieved August 20, 2007, from http://www.w3.org/DesignIssues/Fractal.html

Berners-Lee, T. (2003). Semantic Web: Where to direct our energy? Keynote speech at the *2nd International Semantic Web Conference (ISWC2003)*, Florida. Retrieved August 20, 2007, from http://www.w3.org/2003/Talks/1023-iswc-tbl/

Berners-Lee, T. (2004). WWW 2004 keynote. In *Proceedings of the 13th International World Wide Web Conference*, New York. Retrieved August 20, 2007, from http://www.w3.org/2004/Talks/0519-tbl-keynote/

Berners-Lee, T., Hall, W., Hendler, J.A., O'Hara, K., Shadbolt, N., & Weitzner, D.J. (2006). A framework for Web science. *Foundations and Trends in Web Science, 1*(1), 1-129.

Berners-Lee, T., Hendler, J., & Lassila, O. (2001, May). The Semantic Web. *Scientific American*.

Bernhardt, D., & Miao, J. (2004). Informed trading when information becomes stale. *The Journal of Finance, LIX*(1).

Bertsekas, D. P. (1995). *Dynamic programming and optimal control*. Belmont, MA: Athena Scientific.

Bettman, J. (1979). *An information processing theory to consumer choice*. AddisonWesley.

Bilenko, M., Mooney, R., Cohen, W., Ravikumar, P., & Fienberg, S. (2003, September/October). Adaptive name matching in information integration. *IEEE Intelligent Systems*, pp. 2-9.

Binmore, K., & Vulkan, N. (1999). Applying game theory to automated negotiation. *Netnomics, 1*(1).

Bird, S. D. (1993). Towards a taxonomy of multi-agent systems. *International Journal of Man-Machine Studies, 39*, 689-704.

Bird, S. D., & Kasper, G. M. (1995). Problem formalization techniques for collaborative systems. *IEEE Transactions on Systems, Man, and Cybernetics, 25*(2), 231-242.

Blum, C., & Roli, A. (2003). Metaheuristics in combinatorial optimization: Overview and conceptual comparison. *ACM Computing Surveys, 35*(3), 268-308.

Bohoris, C., Pavlou, G., & Cruickshank, H. (2000). Using mobile agents for network performance management. In *Proceedings of Network Operations and Management Symposium "The Networked Planet: Management Beyond 2000"* (pp. 637-652).

Bolloju, N. (2003). Extended role of knowledge discovery techniques in enterprise decision support environments. In *Proceedings of the 34th Annual Hawaii International Conference on System Sciences (HICSS-34)*, Maui, HI.

Bolton, F. (2002). Pure CORBA: A code intensive premium reference. Indianapolis: Sams Publishing.

Bonarini, A., & Trianni, V. (2001). Learning fuzzy classifier systems for multi-agent coordination. *Information Sciences, 136*(1-4), 215-239.

Bonatti, P., Kraus, S., & Subrahmanian, V. S. (2003). Secure agents. *Annals of Mathematics and Artificial Intelligence, 37*(1-2), 169-235.

Bouwman, H., Haaker, T., & Faber, E. (2005). Developing mobile services: Balancing customer and network value. In *Proceedings of the 2nd IEEE International Workshop on Mobile Commerce and Services (WMCS'05)*, Munich, Germany.

Brenner, W., Zarnekow, R., & Wittig, H. (1998). *Intelligent software agents: Foundations and applications.* Springer.

Bresnahan, J. (1998). Supply chain anatomy: The incredible journey. *CIO Enterprise Magazine*, August 15. Retrieved on March 12, 2006 from http://www.cio.com site

Brewer, P. J., Huang, M., Nelson, B., & Plott, C. R. (1999). *On the behavioral foundations of the law of supply and demand: Human convergence and robot randomness* (Social Science Working Paper 1079). California Institute of Technology.

Bruce, B., Cross, M., Duncan, T., Hoey, C., & Wills, M. (2000). *e.Volution* (pp. 81-96). Prestoungrange University Press.

Brynjolfsson, E., & Smith, M. (2000). Frictionless commerce? A comparison of Internet and conventional retailers. *Management Science, 46*(4).

Brynjolfsson, E., & Smith, M. (2000, July). *The great equalizer? Consumer choice behavior at Internet shopbots* (MIT Working Paper). Retrieved August 16, 2007, from http://ebusiness.mit.edu/erik/TGE%202000_08_12.html

Bucchiarone, A., & Gnesi, S. (2006). *A survey on services composition.* Retrieved August 23, 2007, from http://www.selab.isti.cnr.it/ws-mate/program.html

Buckingham Shum, S. (2004, January/February). Contentious, dynamic, information-sparse domains ... and ontologies? *IEEE Intelligent Systems,* pp. 80-81.

Bugliesi, M., & Castagna, G. (2001). Secure safe ambients. In *Proceedings of the 28th ACM Symposium on Principles of Programming Languages*, London (pp. 222-235).

Burke, G., & Vakharia, A. (2002). Supply chain management. In H. Bidgoli (Ed.), *Internet Encyclopedia*, New York: John Wiley.

Cabri, G., Leonardi, L., & Zambonelli, F. (2001). Mobile agent coordination for distributed network management. *Journal of Network & Systems Management, 9*(4), 435-456.

Cabri, G., Leonardi, L., & Zambonelli, F. (2002). Engineering mobile agent applications via context-dependent coordination. *IEEE Transactions on Software Engineering, 28*(11), 1039-1055.

Cai, D., Luo, Z., Qian, K., & Gao, Y. (2005, November 14-16). Towards efficient selection of Web services with reinforcement learning process. In *Proceedings of the 17th IEEE International Conference on Tools with Artificial Intelligence* (pp. 372-376). Los Alamitos: IEEE Computer Society Press.

Carlson, D. (2001). *Modeling XML applications with UML: Practical e-business applications*. Addison-Wesley.

Carmel, D., & Markovitch, S. (1996). Learning models of intelligent agents. In *Proceedings of the 13th National Conference on Artificial Intelligence and the 8th Innovative Applications of Artificial Intelligence Conference.*

Carroll, J. S. (1978). A Psychological approach to deterrence: The evaluation of crime opportunities. *Journal of Personality and Social Psychology, 36*(12), 1512-1520.

Carroll, J. S., & Weaver, F. (1986). Shoplifters' perceptions of crime opportunities: A process-tracing study. In D. B. Cornish & R. V. Clarke (Eds.), *The Reasoning Criminal: Rational Choice Perspectives on Offending* (pp. 19-38). Springer.

Carroll, J.J., Dickinson, I., Dollin, C., Reynolds, D., Seaborne, A., & Wilkinson, K. (2003). *Jena: Implementing the Semantic Web recommendations* (Tech. Rep. No. HPL-2003-146). HP Laboratories Bristol.

Casati, F., Ilnicki, S., Jin, L., Krishnamoorthy, V., & Shan, M.-C. (2000). Adaptive and dynamic service composition in eFlow. In *Proceedings of the Conference on Advanced Information Systems Engineering* (pp. 13-31). Retrieved August 23, 2007, from http://citeseer.ist.psu.edu/casati00adaptive.html

Castelfranchi, C. (1995). Guarantees for autonomy in cognitive agent architecture. In Wooldrige, M. and Jennings, N. R. (Eds.), *Intelligent Agents: Theories, Architectures, and Languages*, 890, pp. 56-70. Heidelberg, Germany: Springer-Verlag.

Castro-Schezl, J. J., Jennings, N. R., Luo, X., & Shadbolt, N. R. (2004). Acquiring domain knowledge for negotiating agents: a case of study. *International Journal of Human-Computer Studies, 61*, 3-31.

Census.gov. (2004). Retail e-commerce sales in second quarter 2004 were $15.7 billion. Retrieved August 15, 2007, from http://www.census.gov/mrts/www/current.html

Chaib-Draa, B. (1995). Industrial applications of distributed artificial intelligence. *Communications of the ACM, 38*(11), 49-53.

Chaib-Draa, B., & Mandiau, R. (1992). Distributed artificial intelligence: An annotated bibliography. *SIGART Bulletin, 3*(3), 20-37.

Chakraborty, D. (2001). *Service composition in ad-hoc environments* (Tech. Rep. No. TR-CS-01-20). University of Maryland, Department of Computer Science and Electrical Engineering, Baltimore County, MD. Retrieved August 23, 2007, from http://citeseer.ist.psu.edu/529431.html

Chakraborty, D., & Joshi, A. (2001). *Dynamic service composition: State-of-the-art and research directions* (Tech. Rep. No. TR-CS-01-19). University of Maryland, Department of Computer Science and Electrical Engineering, Baltimore County, MD. Retrieved August 23, 2007, from http://citeseer.ist.psu.edu/chakraborty01dynamic.html

Chakraborty, D., Perich, F., Avancha, S., & Joshi, A. (2001). Dreggie: Semantic service discovery for m-commerce applications. In *Proceedings of the Workshop on Reliable and Secure Applications in Mobile Environment, 20th Symposium on Reliable Distributed Systems* (pp. 28-31). Retrieved August 23, 2007, from http://citeseer.ist.psu.edu/chakraborty01dreggie.html

Chakraborty, D., Perich, F., Joshi, A., Finin, T., & Yesha, Y. (2002). A reactive service composition architecture for pervasive computing. In *Proceedings of the 7th Personal Wireless Communications Conference (PWC 2002)* (pp. 53-62, ISBN: 1-4020-7250-3).

Chakraborty, D., Yesha, Y., Finin, T., & Joshi, A. (2005). Service composition for mobile environments [Special issue: Mobile Services]. *Journal on Mobile Networking and Applications, 10*(4), 435-451.

Chamberlin, E. H. (1948). An experimental imperfect market. *Journal of Political Economy, 56*(2), 95-108.

Chan, H., Lee, R., Dillon, T., & Chang, E. (2001). *E-commerce fundamentals and applications*. West Sussex, England: John Wiley & Sons.

Chang, M.K., Cheung, W., & Lai, V. (2005). Literature derived reference models for the adoption of online shopping. *Information & Management, 42*, 543-559.

Chappell, D.A. (2004). *Enterprise service bus.* O'Reilly.

Charif, Y., & Sabouret, N. (2005). An overview of Semantic Web services composition approaches. In *Proceedings of the International Workshop on Context for Web Services (CWS-05), ENTCS, 146*(1), 33-41.

Chavez, A., & Maes, P. (1996). *Kasbah: An agent marketplace for buying and selling goods.* Paper presented at the Conference on Practical Applications of Intelligence Agents and Multi-Agent Technology.

Chen, H. (1999). *Developing a dynamic distributed intelligent agent framework based on the jini architecture.* Unpublished master's thesis, University of Maryland, Baltimore County. Retrieved August 23, 2007, from http://gentoo.cs.umbc.edu/ronin/doc/

Chen, H. (2004). *Ronin agent framework.* Retrieved August 23, 2007, from http://gentoo.cs.umbc.edu/ronin/

Chen, J.H., Chao, K.M., Godwin, N., Reeves, C., & Smith, P. (2002). An automated negotiation mechanism based on co-evolution and game theory. In *Proceedings of the ACM Symposium on Applied Computing*, Madrid, Spain (pp. 63-67).

Chen, L. (2005). *Optimal information acquisition, inventory control, and forecast sharing in operations management.* Dissertation thesis. Stanford, CA: Stanford University.

Cheng, F., Ryan, J.K., & Simchi-Levy, D. (2000). Quantifying the 'bullwhip effect' in a supply chain: The impact of forecasting, lead times, and information. *Management Science, 46*(3), 436-444.

Cheung, C., Chan, G., & Limayem, M. (2005). A critical review of online consumer behavior: Empirical research. *Journal of Electronic Commerce in Organizations, 3*(4), 1-19.

Cheung, W.K., Liu, J., Tsang, K.H., & Wong, R.K. (2004, September 20-24) Dynamic resource selection for service composition in the Grid. In *Proceedings of the 2004 IEEE/WIC International Conference on Web Intelligence* (pp. 412-418). Los Alamitos: IEEE Computer Society Press.

Choi, J. (2001). A customized comparison-shopping agent. *IEICE TRANS. COMMUN., E84-B*(6), 1694-1696.

Chong, S. (2001). *DEON: A semiotic approach to the design of agent-mediated e-commerce systems.* Unpublished doctoral thesis, Staffordshire University.

Chong, S., & Liu, K. (2000a). A semiotic approach to the design of agent-mediated e-commerce. In E.D. Falkenberg, K. Lyytinen, & A.A. Verrijn-Stuart (Eds.), *Information Systems Concepts: An Integrated Discipline Emerging* (pp. 95-114). Boston: Kluwer Academic Publishers.

Chong, S., & Liu, K. (2000b). A semiotic approach for distinguishing responsibilities in agent-based systems. In K. Liu, R. Stamper, R. Clarke, & P. Andersen (Eds.), *Organization Semiotics* (pp. 173-186). Kluwer Academic Press.

Chopra, S., & Meindl, P. (2003). *Supply chain management: Strategy, planning, and operation* (2nd ed.).

Christensen, C.M. (1997). *The innovator's dilemma.* Harvard Business School Press.

Ciravegna, F., Dingli, A., Guthrie, D., & Wilks, Y. (2003). Integrating information to bootstrap information extraction from Web sites. In *Proceedings of Workshop on Information Integration, IJCAI'03*, Acapulco, Mexico.

Cliff, D. (1997). *Minimal-intelligence agents for bargaining behaviors in market-based environments* (Tech. Rep. No. HPL-97-91). Hewlett-Packard Laboratories.

Cliff, D. (2001). *Evolution of market mechanism through a continuous space of aution-types* (Tech. Rep. No. HPL-2001-326). Hewlett-Packard Laboratories.

Cliff, D. (2003). Explorations in evolutionary design of online auction market mechanism. *Electronic Commerce Research and Applications, 2*, 162-175.

Cliff, D., Walia, V., & Byde, A. (2003). Evolved hybrid auction mechanisms in non-ZIP trader marketplaces. In *Proceedings of the IEEE International Conference on Computational Intelligence for Financial Engineering* (CIFEr03), Hong Kong, China.

Coase, R. (1991). The Nature of the Firm. In E.O. Williamson & S.G. Winter (Eds.), *The nature of the firm: Origins, evolution and development.* Oxford University Press: Oxford. (Original work published in *Economica* 1937)

Collins, J., Arunachalam, R., Sadeh, N., Eriksson, J., Finne, N., & Janson, S. (2005). *The supply chain game for the 2006 trading agent competition, competitive benchmarking for the trading agent community.* Retrieved August 18, 2007, from http://www.sics.se/tac/tac06scmspec_v16.pdf

Collis, J., & Lee, L. (1999). Building electronic marketplaces with the ZEUS agent tool-kit. *Lecture Notes in Artificial Intelligence, 1571,* 1-24.

Cooper, S., & Taleb-Bendiab, A. (1998). CONCENSUS: Multi-party negotiation support for conflict resolution in concurrent engineering design. *Journal of Intelligent Manufacturing, 9*(2), 155-159.

Corby, O., Dieng-Kuntz, R., Faron-Zucker, C., & Gandon, F. (2006). Searching the Semantic Web: Approximate query processing based on ontologies. *IEEE Intelligent Systems, 21*(1), 20-27.

Cournot, A. (1838). *Researches into the mathematical principles of the theory of wealth.* New York: Macmillan.

Coursaris, C., & Hassanein, K. (2002). Understanding m-commerce. *Quarterly Journal of Electronic Commerce, 3*(3), 247-271.

Cox, S., Alani, H., Glaser, H., & Harris, S. (2004). The Semantic Web as a semantic soup. In *Proceedings of the First Workshop on Friend of a Friend, Social Networking and the Semantic Web*, Galway, Ireland.

Cranor, L., Langheinrich, M., Marchiori, M., Presler-Marshall, M., & Reagle, J. (2002). The platform for privacy preferences 1.0 (P3P1.0) Specification. *World Wide Web Consortium.* Retrieved August 20, 2007, from http://www.w3.org/TR/P3P/

Cross, Gary J. (2000). How e-business is transforming supply chain management. *Journal of Business Strategy, 21*(2), 36-39.

Crowley, J. L., Coutaz, J., & Bérard, F. (2000). Perceptual user interfaces: Things that see. *Communications of the ACM, 43*(3), 54-64.

Crowston, K. (1997). *Price behavior in a market with Internet buyer's agents.* Paper presented at the International Conference on Information Systems, Atlanta, GA.

Curbera, F., Goland, Y., Klein, J., Leymann, F., Roller, D., Thatte, S., & Weerawarana, S. (2002). *Business process execution language for Web services* (Version 1.0). Retrieved August 23, 2007, from http://www-106.ibm.com/developerworks/library/ws-bpel/.

Cyert, R. M., & DeGroot, M. H. (1987). *Bayesian analysis and uncertainty in economic theory.* New York: Rowman & Littlefield.

DAML-S and Related Technologies. Retrieved August 23, 2007, from http://www.daml.org/services/daml-s/0.9/survey.pdf

Das, R., Hanson, J. E., Kephart, J. O., & Tesauro, G. (2001). *Agent-human interactions in the continuous double auction.* Proceedings of the IJCAI-2001, Seattle, USA, 2001.

Dastani, M., Jacobs, N., Jonker, C.M., & Treur, J. (2005). Modelling user preferences and mediating agents in electronic commerce. *Knowledge-Based Systems, 18,* 335-352.

Davenport, T.H., & Harris, J.G., (2005). Automated decision making comes of age. *MIT Sloan Management Review, 46*(4), 83-89.

Davis, M. (1988). Time and punishment: An intertemporal model of crime. *Journal of Political Economy, 96,* 383-390.

Davulcu, H., Kifer, M., Pokorny, L. R., Ramakrishnan, C. R., Ramakrishnan, I. V., & Dawson, S. (1999). Modeling and analysis of interactions in virtual enterprises. In *Proceedings of the 9th International Workshop on Research Issues on Data Engineering: Information Technology for Virtual Enterprises (RIDE 1999)* (pp. 12-18). IEEE Computer Society.

Debenham, J. (2000). Supporting strategic process. *Proceedings of the 5th International Conference on the Practical Application of Intelligent Agents and Multi-Agents* (pp. 237-256).

Debenham, J. (2003). An eNegotiation framework. *Proceedings of the 23rd International Conference on Innovative Techniques and Applications of Artificial Intelligence, AI'2003* (pp. 79-92).

Debenham, J. (2004). Bargaining with information. *Proceedings of the 3rd International Conference on Autonomous Agents and Multi Agent Systems AAMAS-2004* (pp. 664-671).

DeLoach, S.A. (2001). Analysis and design using MaSE and agentTool. In *Proceedings of the 12th Midwest Artificial Intelligence and Cognitive Science Conference (MAICS 2001).* Dietz, J.L.G. (1994). Business modeling for business redesign. In *Proceedings of the 27th Hawaii International Conference on System Sciences* (pp. 723-732). Los Alamitos, CA: IEEE Computer Society Press.

Deugo, D., Weiss, M., & Kendall, E. (2001). Reusable patterns for agent coordination. In A. Omicini et al. (Eds.), *Coordination of Internet Agents* (pp. 347-368). Springer.

Dietz, J.L.G. (2006). *Enterprise ontology: Theory and methodology.* Springer.

Ding, Y., Fensel, D., Klein, M., & Omelayenko, B. (2002). The Semantic Web: Yet another hip? *Data & Knowledge Engineering, 42*(2-3), 205-227.

Domingue, J., Stutt, A., Martins, M., Tan, J., Peterson, H., & Motta, E. (2003). Supporting online shopping through a combination of ontologies and interface metaphors. *International Journal of Human-Computer Studies, 59,* 699-723.

Domnitcheva, S. (2001). *Location modeling: State of the art and challenges.* Paper presented at the UbiComp Workshop on Location Modeling for Ubiquitous Computing, Atlanta, GA.

Doorenbos, R. B., Etzioni, O., & Weld, D. S. (1997). *A scalable comparison-shopping agent for the World Wide Web.* Paper presented at the International Conference on Autonomous Agents, Marina del Rey, CA.

Doran, J. E., & Palmer, M. (1995). The EOS Project: Integrating two models of palaeolithic social change. In N. Gilvert & R. Conte (Eds.), *Artificial Societies: The Computer Simulation of Social Life* (pp. 103-105). London: UCL Press.

Dourish, P. (1998). Using metalevel techniques in a flexible toolkit for CSCW applications. *ACM Transactions on Computer-Human Interaction (TOCHI), 5*(2), 109-155.

Downes, L., & Mui, C. (2000). *Unleashing the Killer App.* Harvard Business School Press.

Doz, L. Y. (2002). *Managing partnerships and strategic alliances.* Insead. Retrieved August 22, 2007, from http://www.insead.edu/executives

Dujmovic, J. J. (1975). Extended continuous logic and the theory of complex criteria. *Series on Mathematics and Physics, 537,* 197-216.

Dustdar, S., & Schreiner, W. (2005). A survey on Web services composition. *International Journal on Web and Grid Services, 1*(1), 1-30. Inderscience Enterprises Ltd.

Edwards, W. K., Newman, M. W., Sedivy, J. Z., & Smith, T. F. (2004). Supporting serendipitous integration in mobile computing environments. *International Journal of Human-Computer Studies, 60,* 666-700.

Ehrlich, I. (1975). The deterrent effect of capital punishment: A question of life and death. *American Economic Review, 65,* 397-417.

Ehrlich, I. (1996). Crime, punishment, and the market for offenses. *Journal of Economic Perspectives, 10*(1), 43-67.

Eischen, K. (2002). *The social impact of informational production: Software development as an informational practice* (CGIRS working paper 2002-1). Center for Global, International and Regional Studies, University of California, Santa Cruz.

Elammari, M., & Lalonde, W. (1999, June 14-15). An agent-oriented methodology: High-level and intermediate models. In *Proceedings of the International Workshop on Agent-Oriented Information Systems (AOIS '99)*, Heidelberg, Germany.

Elfoson, G. (1998). Developing trust with intelligent agents: An exploratory study. In *Proceedings of the First International Workshops on Trust* (pp. 125-138).

Ellman, J. (2004, January/February). Corporate ontologies as information interfaces. *IEEE Intelligent Systems*, pp. 79-80.

Engel, J., & Blackwell, R. (1982). *Consumer behavior* (4th ed.). CBS College Publishing.

Ephrati, E., & Rosenschein, J.S. (1992). *Reaching agreement through partial revelation of preferences.*

Ericsson Enterprise. (2002). *The Path to the mobile enterprise.* Retrieved August 22, 2007, from http://www.ericsson.com/products/whitepapers_pdf/whitepaper_mobile_enterprise_rc.pdf

Evans, P., & Wurster, T. (2000). *Blown to bits.* Harvard Business School Press.

Example for Reinforcement Learning: Playing Checkers. (2001). Retrieved August 22, 2007, from http://www.cs.wustl.edu/~sg/CS527_SP02/lecture2.html.

Faratin, P. (2000). *Automated service negotiation between autonomous computational agents.* Unpublished doctoral dissertation, University of London.

Faratin, P., Sierra, C., & Jennings, N. (1998). Negotiation decision functions for autonomous agents. *International Journal of Robotics and Autonomous Systems, 24*(34), 159-182.

Faratin, P., Sierra, C., & Jennings, N. R. (2002). Using similarity criteria to make issue trade-offs in automated negotiations. *Artificial Intelligence, 142*, 205-237.

Fedoruk, A., & Deters, R. (2002). Improving fault-tolerance by replicating agents. In *Proceedings of the First International Joint Conference on Autonomous Agents and Multi-Agent Systems*, Bologna, Italy (pp. 737-744).

Feigenbaum, J., & Shenker, S. (2002). Distributed algorithmic mechanism design: Recent results and future directions. In *Proceedings of the 6th International Workshop on Discrete Algorithms and Methods for Mobile Computing and Communications.*

Feigenbaum, J., Papadimitriou, C., & Shenker, S. (2001). Sharing the cost of multicast transmissions. *Journal of Computer and System Sciences, 63*, 2141.

Fender, J. (1999). A general equilibrium model of crime and punishment. *Journal of Economic Behavior & Organization, 39*, 437-453.

Feng, N. (1999). *S-module design for software hot swapping technology.* Unpublished master's thesis, Carleton University, System and Computer Engineering Department. Retrieved August 23, 2007, from http://citeseer.ist.psu.edu/feng99smodule.html

Fensel, D., Bussler, C., Ding, Y., Kartseva, V., Klein, M., Korotkiy, M., Omelayenko, B., & Siebes, R. (2002). Semantic Web application areas. In *Proceedings of the 7th International Workshop on Applications of Natural Language to Information Systems (NLDB 2002)*, Stockholm, Sweden.

Ferber, J. (1999). *Multi-agent systems: An introduction to distributed srtificial intelligence* (pp. 13-16). Addison-Wesley.

Fetnet. http://enterprise.fetnet.net/event /Special _02.htm.

Finnie, G., & Sun, Z. (2003). *A knowledge-based model of multiagent CBR systems.* Paper presented at the International Conference on Intelligent Agents, Web Technologies, and Internet Commerce (IAWTIC'2003), Vienna, Austria.

Finnie, G., Berker, J., & Sun, Z. (2004, August). *A multi-agent model for cooperation and negotiation in supply networks*. Paper presented at the Americas Conference on Information Systems, New York.

Firat, A., Madnick, S., & Siegel, M. (2000, December). *The camn Web wrapper engine*. Paper presented at the Workshop on Information Technology and Systems, Brisbane, Queensland, Australia.

Fischer, L. (2003). *The workflow handbook 2003*. Future Strategies Inc.

Fisher, M. (1997). What is the right supply chain for you? *Harvard Business Review,* March-April, 105-117.

Fisher, R. (1978). *International mediation: A working guide*. New York: International Peace Academy.

Fonseca, S., Griss, M., & Letsinger, R. (2001). *An agent-mediated e-commerce environment for the mobile shopper* (HP Labs Tech. Rep. No. HPL-2001-157).

Formica, A. (2006). Ontology-based concept similarity in formal concept analysis. *Information Sciences, 176*(18), 2624-2641.

Foroughi, A. (1998). Minimizing negotiation process losses with computerized negotiation support systems. *The Journal of Applied Business Research, 14*(4), 1526.

Forrester, J. W. (1958). Industrial dynamics. *Harvard Business Review,* July-August, 37-66.

Foster, H., Uchitel, S., Magee, J., & Kramer, J. (2003). Model-based verification of web service compositions. In *Proceedings of the IEEE International Conference on Automated, Software Engineering.*

Fox, M. S. (1981). An organizational view of distributed systems. *IEEE Transactions on Systems, Man and Cyvernetics, 11*(1), 70-79.

Franklin, S. (1996). *Coordination without communication.* Retrieved August 18, 2007 from http://www.msci.memphis.edu/~franklin/coord.html

Franklin, S., & Graesser, A. (1997). *Is it an agent, or just a program? A taxonomy for autonomous agents.* Paper presented at the Third International Workshop on Agent Theories, Architectures, and Languages.

Frias-Martinez, E., Magoulas, G., Chen, S., & Macredie, R. (2005). Modeling human behavior in user-adaptive systems: recent advances using soft computing techniques. *Expert Systems with Applications, 29,* 320-329.

Frohlich, M.T. (2002). E-integration in the supply chain: Barriers and performance, *Decision Sciences, 33*(4), 537-556.

Fudenberg, D., & Levine, D.K. (1998). *The theory of learning in games.* Cambridge: M.I.T. Press.

Fujii, K., & Suda, T. (2004). Dynamic service composition using semantic information. In *Proceedings of the 2nd ACM International Conference on Service Oriented Computing (ICSOC '04)* (pp. 39-48, ISBN: 1-58113-871-7).

Gaber, J. (2000). *New paradigms for ubiquitous and pervasive computing* (Research Rep. No. RR-09-00). Université de Technologies de Belfort-Montbéliard (UTBM), France.

Gaber, J. (2006). New paradigms for ubiquitous and pervasive applications. In *Proceedings of the 1st Workshop on Software Engineering Challenges for Ubiquitous Computing,* Lancaster, UK.

Gaber, J., & Bakhouya, M. (2006). An affinity-driven clustering approach for service discovery and composition for pervasive computing. In *Proceedings of the ACS/IEEE International Conference on Pervasive Services ICPS'06* (pp. 277-280, ISBN: 1-4244-0237-9).

Gaither, N. & Frazier, G. (2002). *Operations management,* 6th Edition, Cincinnati: Southwest.

Galstyan, A., Kolar, S., & Lerman, K. (2003). Resource allocation games with changing resource capacities. *Proceedings of the 2nd International Joint Conference on Autonomous Agents and Multiagent Systems Aamas-03,* 145-152.

Genesereth, M. R. & Ketchpel, S.P. (1994). Software agents. *Communications of the ACM, 37*(7), 48-53.

Gibson, J.J. (1968). *The senses considered as perceptual systems.* Allen & Unwin.

Gibson, J.J. (1979). *The ecological approach to visual perception.* Mifflin Company.

Gilder, G. (1993, September 13). Metcalfe's Law and legacy. *Forbes ASAP.*

Glaser, H., Alani, H., Carr, L., Chapman, S., Ciravegna, F., Dingli, A., Gibbins, N., Harris, S., schraefel, M., & Shadbolt, N. (2004). CS AKTive space: Building a Semantic Web application. In *Proceedings of 1ˢᵗ European Semantic Web Symposium (ESWS 2004)*, Crete, Greece (pp. 417-432).

Gmytrasiewicz, P., & Durfee, E. (1995). A rigorous operational formalization of recursive modelling. In *Proceedings of the 1ˢᵗ International Conference on Multiagent Systems* (pp. 125-132).

Gode, D. K., & Sunder S. (1993). Allocative efficiency of markets with zero-intelligence traders: market as a partial substitute for individual rationality. *Journal of Political Economy, 101*(1), 119-137, 1993.

Goh, K., Teo, H., Wu, H., & Wei, K. (2000). Computer supported negotiations: An experimental study of bargaining in electronic commerce. In *Proceedings of the 21ˢᵗ Annual International Conference on Information Systems*, Brisbane, Australia.

Goldkuhl, G. (1997). *The six phases of business processes: Further development of business action theory.* Linkoping University, Centre for Studies on Human, Technology and Organization (CMTO), Sweden.

Green, H. (1998, May 4). A cybershopper's best friend. *BusinessWeek.*

Greengrass, E., Sud, J., & Moore, D. (1999, June). Agents in the virtual marketplace: Component dtrategies. In *Proceedings of SIGS* (pp. 42-52).

Grimes, S. (2005). Location, location, location. *intelligent enterprise.* Retrieved August 22, 2007, from http://www.intelligententerprise.com/toc/?day=01&month=09&year=2005

Griss, M. (1999). *My agent will call your agent ... but will it respond?* (HP Labs Tech. Rep. No. HPL-1999-159).

Gruhl, D., Chavet, L., Gibson, D., Meyer, J., Pattanayak, P., Tomkins, A., & Zien, J. (2004). How to build a Webfountain: An architecture for very large-scale text analytics. *IBM Systems Journal, 43*(1), 64-76.

Gupta, A. (1998, February 23-27. *Junglee: Integrating data of all shapes and sizes.* Paper presented at the Fourteenth International Conference on Data Engineering, Orlando, FL.

Guttman, E. (1999). Service location protocol: Automatic discovery of IP network services. *IEEE Internet Computing, 3*(4), 71-80.

Guttman, R., & Maes, P. (1999). Agent-mediated integrative negotiation for retail electronic commerce. *Lecture Notes in Artificial Intelligence, 1571,* 70-90.

Guttman, R., Moukas, A., & Maes, P. (1998). Agent-mediated electronic commerce: A survey. *Knowledge Engineering Review, 13*(2), 147-159.

Guttman, R., Moukas, A., & Maes, P. (1999). Agents as mediators in electronic commerce. In M. Klusch (Ed.), *Intelligent information agents* (chap. 6). Berlin: Springer.

Habermas, J. (1984). *The theory of communicative action: Reason and rationalization of society.* Cambridge: Polity Press.

Haeckel, S.H. (1999). *Adaptive enterprise: Creating and leading sense-and-response organizations.* Boston: Harvard Business School Press.

Halpern, J. (2003). *Reasoning about uncertainty.* MIT Press.

Hamadi, R., & Benatallah, B. (2003). A Petri-net-based model for Web service composition. In *Proceedings of the 14ᵗʰ Australasian Database Conference on Database Technologies, 17,* 191-200. ISSN: 1445-1336.

Harris, S., & Gibbins, N. (2003). 3store: Efficient Bulk RDF storage. In *Proceedings of the ISWC'03 Practi-*

cal and Scalable Semantic Systems (PSSS-1), Sanibel Island, FL.

Hawking, E. (2006, June). Web search engines: Part 1. Computer, pp. 86-88. IEEE Computer Society.

He, M., Jennings, N., & Leung, H. F. (2003a). On agent-mediated electronic commerce. *IEEE Transactions on Knowledge and Data Engineering, 15*(4).

He, M., Leung H. F., & Jennings, N. (2003b). A fuzzy logic based bidding strategy for autonomous agents in continuous double auctions. *IEEE Transactions on Knowledge and Data Engineering, 15*(6), 1345-1363.

Hendler, J. (2001, March/April). Agents and the Semantic Web. *IEEE Intelligent Systems,* pp. 30-37.

Hennesy, M., & Riely, J (1999). Trust and partial typing in open systems of mobile agents. In *Proceedings of the 26th ACM Symposium on Principles of Programming Languages (POPL '99),* San Antonio, TX (pp. 93-104).

Hertz, J., Krogh, A., & Palmer, R. G. (1991). *Introduction to the theory of neural computation.* Addison-Wesley.

Hewitt, C. (1986). Offices are open systems. *ACM Transactions on Office Systems, 4*(3), 271-287.

Hiramatsu, H. (2001). A spatial hypermedia framework for position-aware information delivery systems. *Lecture Notes in Computer Science, 2113,* 754-763.

Hodkinson, C. S., & Kiel, G. C. (2003). Understanding WWW information search behaviour: An exploratory model. *Journal of End User Computing, 15*(4), 27-48.

Hogg, L. M. I., & Jennings, N. R. (2001). Socially intelligent reasoning for autonomous agents. *IEEE Transactions on Systems, Man, & Cybernetics Part A: Systems & Humans, 31*(5), 381-393.

Hohl, F. (1998). Time limited blackbox security: Protecting mobile agents from malicious hosts. In G. Vigna (Ed.), *Mobile Agents and Security* (LNCS 1419, pp. 92-113). Springer.

Holt, A. W. (1988). Diplans: A new language for the study and implementation of coordination. *ACM Transformations on Office Information Systems, 6*(2), 109-125.

Hostler, R.E., Yoon, V.Y., & Guimaraes, T. (2005). Assessing the impact of Internet agent on end users' performance. *Decision Support Systems, 41,* 313-323.

Howard, J., & Sheth, J. (1969). *The theory of buyer behavior.* John Wiley & Sons.

Hu, J., & Weliman, M. P. (2001). Learning about other agents in a dynamic multiagent system. *Cognitive Systems Research, 2*(1), 67-79.

Hu, M. (2003). Web services composition, partition, and quality of service in distributed system integration and re-engineering. In *Proceedings of the XML Conference,* Philadelphia, PA. IDEAlliance. Retrieved August 23, 2007, from http://www.idealliance.org/papers/dx_xml03/papers/05-05-04/05-05-04.pdf

Huberman, B. A. (1988). *The ecology of computation.* Amsterdam: North-Holland.

Huq, G. B. (2006, February 13-20). *Analysis, planning and practice of trading agent competition supply chain management (TAC/SCM).* Paper presented at the 2nd International Conference on Information Management Business, Sydney.

Hynes, N., & Mollenkopf, D. (1998). *Strategic alliance formation: Developing a framework for research.* Paper presented at the Australia New Zealand Academy of Marketing Conference, Otago, New Zealand.

İmrohoroğlu, A., Merlo, A., & Rupert, P. (1996). *On the political economy of income redistribution and crime* (Staff Rep. No. 216). Federal Reserve Bank of Minneapolis Research Department.

Itao, T., Nakamura, T., Matsuo, M., Suda, T., & Aoyama, T. (2002a). Adaptive creation of network applications in the jack-in-the-net architecture. *IFIP Networking,* pp. 129-140.

Itao, T., Nakamura, T., Matsuo, M., Suda, T., & Aoyama, T. (2002b). Service emergence based on cooperative interaction of self-organizing entities. In *Proceedings of the IEEE Symposium on Applications and the Internet* (pp. 194-203), Nara, Japan.

Jasper, R., & Uschold, M. (1999). *A framework for understanding and classifying ontology applications.* Paper presented at the 12th Workshop on Knowledge Acquisition, Modeling and Management, Banff, Canada.

Jaynes, E. (1957). Information theory and statistical mechanics: Part I. *Physical Review, 106,* 620-630.

Jennings, N. R. (1990). Coordination techniques for distributed artificial intelligence. In G.M.P. O'Hare & N.R. Jennings (Ed.), *Foundations of Distributed Artificial Intelligence* (pp. 187-210). London: Wiley.

Jennings, N. R. (2000) On agent-based software engineering. *Artifical Intelligence, 117*(2), 277-296.

Jennings, N. R., Faratin, P., Norman, T. J., O'Brien, P., Odgers, B., & Alty, J. L. (2000). Implementing a business process management system using ADEPT: A real-world case study. *International Journal of Applied Artificial Intelligence, 14*(5), 421-465.

Jerne, N. (1974). Towards a network theory of the immune system. *Ann. Immunol, 125,* 125-373.

Jini. (2000). *Jini network technology: Specifications* (Version 1.1 Beta). Sun Microsystems Inc. Retrieved August 23, 2007, from http://www.sun.com/software/jini/specs/.

Jonker, C. M., & Treur, J. (1999). Formal analysis of models for the dynamics of trust based on experiences. In *Proceedings of MAAMAW'99* (LNAI 1647, pp. 221-232). Springer.

Jung, J. J., & Jo, G. S. (2000). Brokerage between buyer and seller agents using constraint satisfaction problem models. *Decision Support Systems, 28*(4), 293-304.

Jureta, I., Faulkner, S., & Kolp, M. (2005). Best practices agent patterns for on-line auctions. In *Proceedings of the International Conference on Enterprise Information Systems* (pp. 814-822).

Kalfoglou, Y., Alani, H., Schorlemmer, M., & Walton, C. (2004). In On the emergent Semantic Web and overlooked issues: *Proceedings of the 3rd International Semantic Web Conference (ISWC 2004)*, Hiroshima, Japan.

Kavantzas, N., Budett, B., Ritzinger, G., Fletcher, T., Lafon, Y., & Bareto, C. (2005). Web services choreography description language Version 1.0. *World Wide Web Consortium.* Retrieved August 23, 2007, from http://www.w3.org/TR/2004/WD-ws-cdl-10-20041217/

Kelkoo.com. (2005). *Company information of Kelkoo.com.* Retrieved August 15, 2007, from http://www.kelkoo.co.uk/b/a/co_4293_128501_corporate_information_company.html

Kelly, K. (1998). *New rules for the new economy: 10 Radical strategies for a connected world.* Penguin Books.

Kendall, E. (1999). Role models: Patterns of agent system analysis and design. In *Proceedings of the Agent Systems and Applications/Mobile Agents.* ACM.

Kendall, E., Murali Krishna, P., Pathak, C., et al (1998). Patterns of intelligent and mobile agents. In *Proceedings of the Conference on Autonomous Agents.* ACM.

Kersten, G.E. (2001). Modeling distributive and integrative negotiations. *Group Decision and Negotiation, 10*(6), 493-514.

Kinny, D., Georgeff, M., & Rao, A. (1996). A methodology and modelling technique for systems of BDI agents. In W. van der Velde & J. Perram (Eds.), *Agents Breaking Away: Proceedings of the 7th European Workshop on Modelling Autonomous Agents in a Multi-Agent World (MAAMAW '96)* (LNAI 1038). Heidelberg, Germany: Springer-Verlag.

Kirche, E., Zalewski, J., & Tharp, T. (2005). Real-time sales and operations planning with CORBA: Linking demand management and production Planning. In C.S. Chen, J. Filipe, I. Seruca, J. Cordeiro (Eds.), *Proceedings of the 7th International Conference on Enterprise Information Systems* (pp. 122-129). Washington, DC: ICEIS, Setubal, Portugal.

Klein, M., & Fensel, D. (2001). Ontology versioning on the Semantic Web. In *Proceedings of the 1st International Semantic Web Working Symposium*, Stanford University, CA (pp. 75-91).

Klein, M., Faratin, P., & Sayama, H. (2003). Negotiating complex contracts. *Group Decision and Negotiation Journal: Special Issue on eNegotiations.*

Kolp, M., Giorgini, P., & Mylopoulos, J. (2001). A goal-based organizational perspective on multi-agent architectures. In *Proceedings of the Workshop on Agent Theories, Architectures, and Languages* (LNCS 2333, pp. 128-140). Springer.

Koshkina, M. (2003). *Verification of business processes for web services.* Unpublished master's thesis, York University, Departement of Computer Science, Toronto. Retrieved August 23, 2007, from http://www.cse.yorku.ca/~franck/research/students/maria.pdf

Kraus, S. (2001). *Strategic negotiation in multiagent environments.* MIT Press.

Kraus, S., & Lehmann, D. (1995). Designing and building an automated negotiation agent. *Computational Intelligence, 11*(1), 132-171.

Kraus, S., Wilkenfeld, J., & Zlotkin, G. (1995). Multiagent negotiation under time constraints. *Artificial Intelligence Journal, 75*(2), 297-345.

Kreps, D., & Wilson, R. (1982). Sequential equilibria. *Econometrica, 50,* 863-894.

Krulwich, B. (1996). The BargainFinder agent: Comparison price shopping on the Internet. In J. Williams (Ed.), *Bots, and Other Internet Beasties* (pp. 257-263). Indianapolis: Macmillan Computer Publishing.

Kushchu, I. (2005). Web-based evolutionary and adaptive information retrieval. *IEEE Transactions on Evolutionary Computation, 9*(2), 117-125.

Kuttner, R. (1998, May). The Net: A market too perfect for profits. *BusinessWeek.*

Kwan, I.S., Fong, J., & Wong, H.K. (2005). An e-customer behavior model with online analytical mining for internet marketing planning. *Decision Support Systems, 41,* 189-204.

Kwon, O. (2006). The potential roles of context-aware computing technology in optimization-based intelligent decision-making. *Expert Systems with Applications, 31,* 629-642.

Kwon, O. B., & Lee, K. C. (2002). MACE: Multi-agents coordination engine to resolve conflicts among functional units in an enterprise. *Expert Systems with Applications, 23*(1), 9-21.

Kwon, O.B., & Sadeh, N. (2004). Applying case-based reasoning and multi-agent intelligent system to context-aware comparative shopping. *Decision Support Systems, 37,* 199-213.

Lai, H., & Yang, T. C. (1988). A system architecture of intelligent-guided browsing on the Web. In *Proceedings of Thirty-First Hawaii International Conference on System Sciences* (pp. 423-432).

Lai, H., & Yang, T. C. (2000). A system architecture for intelligent browsing on the Web. *Decision Support Systems, 28*(3), 219-239.

Lang, K. (1995). Newsweeder: Learning to filter netnews. In *Proceedings of the 12th International Conference Machine Learning* (pp. 331-339). San Fransisco: Morgan Kaufmann.

Langendoerfer, P., Maye, O., Dyka, Z., Sorge, R., Winkler, R., & Kraemer, R. (2004). Middleware for location-based services, design and implementation issues. In Q. Mahmoud (Ed.), *Middleware for Communication.* Wiley.

LeBaron, B. (2000). Agent-based computational finance: suggested readings and early research. *Journal of Economic Dynamics and Control, 24,* 679-702.

Lee, H., Padmanabhan, V., & Whang, S. (1997). The bullwhip effect. *Sloan Management Review, 38*(3), 93-103.

Lee, H., Padmanabhan, V., & Whang, S. (2004). Information distortion in a supply chain: The bullwhip effect/comments on "information distortion in a supply chain: The bullwhip effect." *Management Science, 50*(12), 1875-1894.

Lee, W. J., & Lee, K. C. (1999). PROMISE: A distributed DSS approach to coordinating production and marketing decisions. *Computers and Operations Research, 26*(9), 901-920.

Lee, W. P., & Yang, T. H. (2003). Personalizing information appliances: A multi-agent framework for TV programme recommendations. *Expert Systems with Applications, 25*(3), 331-341.

Lenat, D.B. (1994). CYC: A large-scale investment in knowledge infrastructure. *Communications of the ACM, 38*(11), 33-38.

Leung, S. F. (1995). Dynamic deterrence theory. *Economica, 62,* 65-87.

Leymann, F. (2001, May). *Web services flow language* (Rep. No. WSFL10). IBM Software Group.

Liang, T.P., & Lai, H.J. (2002). Effect of store design on consumer purchases: An empirical study of on-line bookstores. *Information & Management, 39,* 431-444.

Lieberman, H. (1998). *Integrating user interface agents with conventional applications.* Paper presented at the Conference on Intelligent User Interfaces. ACM.

Lin, R.J. (2004). Bilateral multiissue contract negotiation for task redistribution using a mediation service. In *Proceedings of the Agent Mediated Electronic Commerce Workshop in AAMAS.*

Lin, R.J., & Chou, S.C.T. (2004). Mediating a bilateral multiissue negotiation. *Electronic Commerce Research and Applications, 3*(2).

Lin, W.S. (2004). *Framework for intelligent shopping support.* Unpublished doctoral thesis, University of Manchester, UK.

Lind, J. (2002). Patterns in agent-oriented software engineering. In *Proceedings of the Workshop on Agent-Oriented Software Engineering* (LNCS 2585, pp. 45-58). Springer.

Liu, J., & You, J. (2003, April). Smart shopper: An agent-based Web-mining approach to Internet shopping. *IEEE Transactions on Fuzzy Systems, 11*(2), 226-237.

Liu, J., Jin, X., & Tsui, K. (2005). *Autonomy oriented computing: From problem solving to complex systems modeling.* Kluwer Academic Publisher/Springer.

Liu, J., Jing, H., & Tang, Y.Y. (2002). Multi-agent oriented constraint satisfaction. *Artificial Intelligence, 136*(1), 101-144.

Liu, K. (2000). *Semiotics in information systems development.* Cambridge: Cambridge University Press.

Liu, K., Alderson, A., & Qureshi, Z. (1999a). Requirements recovery of legacy systems by analysing and modeling behavior. In *Proceedings of the International Conference on Software Maintenance* (pp. 3-12). Los Alamitos: IEEE Computer Society.

Liu, K., Sharp, B., Crum, G., & Zhao, L. (1995). Applying a semiotic framework to re-engineering legacy information systems. In *Proceedings of the 1st International Conference on Organizational Semiotics*, University of Twente, The Netherlands.

Lomuscio, M., & Jennings, N. (2001). A classification scheme for negotiation in electronic commerce. In F. Dignum & E.C. Sierra (Eds.), *Agent Mediated Electronic Commerce: A European Agentlink Perspective.* Springer-Verlag.

Lottaz, C., Smith, I. F. C., Robert-Nicoud, Y., & Faltings, B. V. (2000). Constraint-based support for negotiation in collaborative design. *Artificial Intelligence in Engineering, 14*(3), 261-280.

Lucas, J. H. C. (2001). Information technology and physical space. *Communications of the ACM, 44*(11), 89-96.

Luo, X., Zhang, C., & Leung, H. F. (2001). Information sharing between heterogeneous uncertain reasoning models in a multi-agent environment: A case study. *International Journal of Approximate Reasoning, 27*(1), 27-59.

Luo, Z., & Li, J. (2005a, October 18-21). A Web services provisioning optimization model in a Web services community. In *Proceedings of 2005 IEEE International Conference on E-Business Engineering* (pp. 689-696). Los Alamitos: IEEE Computer Society Press.

Luo, Z., Li, J., Tan, C.J., Tong, F., Kwok, A., Wong, E., & Wang, H. (2006b, June 21-23). Intelligent service middleware framework. In *Proceedings of the 2006 IEEE Conference on Service Operation, Logistics and Informatics* (pp. 1113-1118). Los Alamitos: IEEE Computer Society Press.

Luo, Z., Sheth, A., Miller J., & Kochut, K. (1998, November 14). Defeasible workflow, its computation, and exception handling. In *Proceedings of 1998 ACM Conference on Computer-Supported Cooperative Work, Towards Adaptive Workflow Systems Workshop.*

Luo, Z., Zhang, J., & Badia, R.M. (2006a). Service Grid for business computing. In *Grid Technologies, Emerging from Distributed Architecture to Virtual Organizations* (pp. 441-468). WIT Press.

Luo, Z., Zhang, J., Cai, D., & Kun, Q. (2005b, November 15-17). An integrated services framework for location discovery to support location based services. In *Proceedings of the 2nd International Conference on Mobile Technology, Applications, and Systems* (p. 7).

Maamar, Z., Mostefaoui, S. K., & Yahyaoui, H. (2005). Toward an agent-based and context-oriented approach for web services composition. *IEEE Transactions on Knowledge and Data Engineering, 17*(5), 686-697.

Machiraju, V. (2001). *Service-oriented research opportunities in the in the world of appliances (Tech. Rep.).* HP Software Technology Lab.

MacKay, D. (2003). *Information theory, inference and learning algorithms.* Cambridge University Press.

Madnick, S., & Siegel, M. (2001). Seizing the opportunity: Exploring Web aggregation. *MISQ Executive.*

Maes, P. (1994). Agents that reduce work and information overload. *Communications of the ACM, 42*(3).

Maes, R., & Moukas, A. (1999). Agents that buy and sell. *Communications of the ACM, 42*(3), 81-91.

Mahmoud, Q. (2004). *Middleware for communication.* Wiley.

Mahmoud, Q., & Zahreddine, W. (2005). A framework for adaptive and dynamic composition of web services. *Journal of Interconnection Networks, 6*(3), 209-228.

Majithia, S., Walker, D. W., & Gray, W. A. (2004). A framework for automated service composition in service-oriented architecture. *LNCS 3053*, 269-283.

Mandry, T., Pernul, G., & Rohm, A. W. (2000-2001). Mobile agents in electronic markets: Opportunities, risks, agent protection. *International Journal of Electronic Commerce, 5*(2), 47-60.

Mao, M. Z., Brewer, E. A., & Katz, R. H. (2001). *Fault-tolerant, scalable, wide-area Internet service composition* (Tech. Rep. No. UCB//CSD01-1129). University of California-Berkeley. Retrieved August 23, 2007, from http://www.cs.berkeley.edu/zmao/Papers/techreport.ps.gz

Marakas, G.M. (1999). *Decision support systems in the twenty-first century.* Upper Saddle River, NJ: Prentice Hall.

Marjit, S., & Shi, H. (1998). On controlling crime with corrupt officials. *Journal of Economic Behavior & Organization, 34*, 163-172.

Markillie, P. (2005, March 31). Crowned at last. *The Economist.*

Marsh, S. (1992). Trust and reliance in multi-agent systems: A preliminary report. In *Proceedings of the 4th European Workshop on Modeling Autonomous Agents in a Multi-Agent World (MAAMAW'92)*, Rome.

Marsh, S. (1994). *Formalizing trust as a computational concept.* Unpublished doctoral thesis, University of Stirling, Department of Mathematics and Computer Science, UK.

Marsh, S. (1994). Trust in distributed artificial intelligence. In C. Castelfranchi & E. Werner (Eds.), *Artificial Social Systems* (LNAI 830, pp. 94-112). Berlin: Springer-Verlag.

Marshall, C., & Shipman, F.M. (2003). Which Semantic Web? In *Proceedings of the 14th HyperText Conference (HT 2003)*, Nottingham, UK (pp. 57-66).

Matwin, S., Szapiro, T., & Haigh, K. (1991). Genetic algorithms approach to a negotiation support system.

IEEE Transactions on Systems, Man, and Cybernetics, 21(1), 102-114.

McCauley, J. L. (2005). Making dynamics modeling effective in economics. *Physica A.*

McGuinness, D.L. (2002). Ontologies Come of Age. In D. Fensel, J. Hendler, H. Lieberman, & W. Wahlster (Eds.), *Spinning the Semantic Web: Bringing the World Wide Web to Its Full Potential.* MIT Press.

McMullen, P. R. (2001). An ant colony optimization approach to addressing a JIT sequencing problem with multiple objectives. *Artificial Intelligence in Engineering, 15*(3), 309-317.

Medjahed, B. (2004). *Semantic Web enabled composition of Web services.* Unpublished doctoral dissertation. Virginia Tech, Department of Computer Science. Retrieved August 23, 2007, from http://www-personal.engin.umd. umich.edu/~brahim/

Meixell, M.J. & Gargeya, V.B. (2005). Global supply chain design: A literature review and critique. Transportation Research, *41*(6), 531- 550 Science Direct. Retrieved February 15, 2006 http://top25.sciencedirect.com/index. php?subject_area_id=4 .]

Meixell, M.J. (2006). *Collaborative manufacturing for mass customization.* George Manson University. Retrieved February 15,2006 http://www.som.gmu. edu/faculty/profiles/mmeixell/collaborative%20Planni ng%20&%20Mass%20Customization.pdf

Melone, T. W., & Crowston, K. (1990, October 7-10). *What is coordination theory and how can it help design cooperative work systems?* Paper presented at the ACM Conference on Computer Supported Cooperative Work (CSCW), Los Angeles.

Menasce, D.A. (2004). Composing Web services: A QoS view. *IEEE Internet Computing, 8*(6), 88-90. IEEE Computer Society Press.

Mennie, D., & Pagurek, B. (2000). An architecture to support dynamic composition of service components, systems and computer engineering, Carleton University. In *Proceedings of the 5ᵗʰ International Workshop*

on Component-Oriented Programming (WCOP 2000). Retrieved August 23, 2007, from http://citeseer.ist.psu. edu/mennie00architecture.html

Meyer, R. A. (1976). *Microeconomic decisions.* Houghton Mifflin Company.

Miah, T., & Bashir, O. (1997). Mobile workers: Access to information on the move. *Computing and Control Engineering, 8,* 215-223.

Miao, C.Y., Goh, A., Miao, Y., & Yang, Z.H. (2002). Agent that models, reasons and makes decisions. *Knowledge-Based Systems, 15,* 203-211.

Mika, P. (2005). Flink: Semantic Web technology for the extraction and analysis of social networks. *Journal of Web Semantics, 3*(2-3), 211-223.

Milanovic, N., & Malek, M. (2004). Current solutions for Web service composition. *IEEE Internet Computing, 8*(6), 51-59.

Miller, M. S., & Drexler, K. E. (1988). Markets and computation: Agoric open systems. Amsterdam: North-Holland. In B.A. Huberman (Eds.), *The Ecology of Computation* (pp. 133-176).

Mitchell, T. M. (1997). *Machine learning* (pp. 10-11). McGraw-Hill.

Mitchell, T., Caruana, R., Freitag, D., McDermott, J., & Zabowski, D. (1994). Experience with a learning personal assistant. *Communications of the ACM, 37*(7), 80-91.

Mokhtar, S. B., Georgantas, N., & Issarny, V. (2005). Ad hoc composition of user tasks in pervasive computing environments. *LNCS, 3628,* 31-46.

Mokhtar, S. B., Georgantas, N., & Issarny, V. (2006). COCOA: Conversation-based service composition in pervasive computing environments. In *Proceedings of the IEEE International Conference on Pervasive Services (ICPS'06)* (pp. 29-38).

Mouratidis, H., Weiss, M., & Giorgini, P. (2006). Modelling secure systems using an agent-oriented approach and security patterns. *International Journal on*

Software Engineering and Knowledge Engineering, 16(3), 471-498.

Murata, T. (1989, April). Petri Nets: Properties, analysis and applications. *Proceedings of the IEEE, 77*(4), 541-580.

Mussbacher, G., Amyot, D., & Weiss, M. (2007). Formalizing patterns with the user requirements notation. In T. Taibi (Ed.), *Design Pattern Formalization Techniques.* Hershey, PA: IGI Global.

Narayanan, S., & McIlraith, S. A. (2002). Simulation, verification and automated composition of Web services. In *Proceedings of the 11th International World Wide Web Conference (WWW-11)*, Honolulu, HI (pp. 77-88).

Nash, J. F. (1950). The bargaining problem. *Econometrica, 18*(2), 155-162.

Ngai, E. W. T., & Gunasekaran, A. (2005). A review for mobile commerce research and applications. *Decision Support Systems.* Retrieved August 20, 2007, from http://www.sciencedirect.com

Nicosia, F. (1996). *Consumer decision processes: Marketing and advertising implications.* Prentice Hall.

Nisan, N., & Ronen, A. (2001). Algorithmic mechanism design. *Games and Economic Behavior, 35*, 166-196.

Nixon, L. (2004, December). *Prototypical business use cases* (FP6 IST NoE Deliverable D1.1.2, KnowledgeWeb EU NoE FP6 507482). Retrieved August 20, 2007, from http://www.starlab.vub.ac.be/research/projects/knowledgeweb/KWebDel112v1.pdf

Noriega, P., & Sierra, C. (1999). Agent-mediated electronic commerce. *Lecture Notes in Artificial Intelligence, 1571.* Springer.

Noy, N.F., Sintek, M., Decker, S., Crubezy, M., Fergerson, R.W., & Musen, M.A. (2001, March/April). Creating Semantic Web contents with Protégé-2000. *IEEE Intelligent Systems*, pp. 60-71.

Nwana, H. S. (1994). *Negotiation strategies: An overview* (BT Laboratories internal report).

Nwana, H. S., Lee, L., & Jennings, N. (1996). Co-ordination in software agent systems. *British Telecom Technical Journal, 14*(4), 79-88.

Nwana, H.S., Rosenschein, J., Sandholm, T., Sierra, C., Maes, P., & Guttman, R. (1998). Agent-mediated electronic commerce: Issues, challenges and some viewpoints. In *Proceedings of the Second International Conference on Autonomous Agents* (pp. 189-196). ACM Press.

O'Hara, K. & Shadbolt, N. (2008). *The spy in the coffee machine: The end of privacy as we know it.* Oneworld Publications.

Oliver, J. (1997). A machine learning approach to automated negotiation and prospects for electronic commerce. *Journal of Management Information Systems, 13*(3), 83-112.

Omicini, A., & Zambonelli, F. (1988). Co-ordination of mobile information agents in TuCSoN. *Internet Research: Electronic Networking Applications and Policy, 8*(5), 400-413.

Oprescu, J. (2004). *Découverte et Composition de Services dans des Réseaux Ambiants.* Unpublished doctoral thesis, Ecole Doctorale Mathematiques, Sciences et Technologie de l'Information, laboratoire LSR–IMAG. Retrieved August 23, 2007, from http://drakkar.imag.fr/article.php3?id_article=175

OWL. Ontology Web Language. Retrieved August 23, 2007, from http://www.w3.org/TR/owl-ref/

Padhye, M. (2004). *Coordinating heterogeneous Web services through handhelds using SyD's wrapper framework.* Unpublished doctoral thesis, Georgia State University, College of Arts and Sciences. Retrieved August 23, 2007, from http://etd.gsu.edu/theses/

Paiva, A., Machado, I., & Prada, R. (2001). The child behind the character. *IEEE Transactions on Systems, Man, and Cybernetics - Part A: Systems and Humans, 31*(5), 361-368.

Paolucci, M., Kawmura, T., Payne, T., & Sycara, K. (2002). Semantic matching of Web services capabilities. In *Proceedings of the 1st International Semantic*

Web Conference on The Semantic Web (LNCS 2342, pp. 333-347, ISBN:3-540-43760-6).

Papazoglou, M. (2001, April). Agent-oriented technology in support of e-business. *Communications of the ACM*, pp. 71-77.

Papazoglou, M.P. (2001, April). Agent-oriented technology in support of e-business. *Communications of the ACM, 44*(4), 35-41.

Parusha, A., & Yuviler-Gavishb, N. (2004). Web navigation structures in cellular phones: The depth/beadth trade-off issue. *International Journal of Human-Computer Studies, 60*, 753-770.

Peltz, C. (2003a). Web services orchestration and choreography. *Computer, 36*(10), 46-52.

Peltz, C. (2003b). Web services orchestration: A review of emerging technologies, tools, and standards. *Hewlett Packard Co.* Retrieved August 23, 2007, from devresource.hp.com/drc/technical_white_papers/WSOrch/WSOrchestration.pdf

Persson, P., Laaksolahti, J., & Lonnqvist, P. (2001). Understanding socially intelligent agents: A multilayered phenomenon. *IEEE Transactions on Systems, Man, and Cybernetics - Part A: Systems and Humans, 31*(5), 349-360.

Peterson, J.L. (1981). *Petri net theory and the modeling of systems.* Englewood Cliffs, NJ: Prentice Hall.

Peyman, F., Jennings, N.R., Lomuscio, A.R., Parsons, S., Sierra, C., & Wooldridge, M. (2001). Automated negotiation: Prospects, methods and challenges. *International Journal of Group Decision and Negotiation, 10*(2), 199-215.

Pham, T., Schneider, G., & Goose, S. (2000). A situated computing framework for mobile and ubiquitous multimedia access using small screen and composite devices. In *Proceedings of the Eighth ACM International Conference on Multimedia* (pp. 323-331).

Pietra, S. D., Pietra, V. D., & Lafferty, J. (1997). Inducing features of random fields. *IEEE Transactions on Pattern Analysis and Machine Intelligence, 19*(2), 380-393.

Pietrula, M.J., & Weingart, L.R. (1994). Negotiation as problem solving. In *Advances in Managerial Cognition and Organizational Information Processing.* JAI Press.

Plitch, P. (2002). Are bots legal? *Wall Street Journal, 240*(54), R13.

Pontelli, E., & Son, T.C. (2003). Designing intelligent agents to support universal accessibility of e-commerce services. *Electronic Commerce Research and Applications, 2*, 147-161.

Porn, L. M., & Patrick, K. (2002). Mobile computing acceptance grows as applications evolve. *Healthcare Financial Management, 56*(1), 66-70.

Pradhan, S. (2002). *Semantic location.* Retrieved August 22, 2007, from http://cooltown.hp.com/dev/wpapers/semantic/semantic.asp

Pruitt, D. G. (1981). *Negotiation behavior.* Academic Press.

PRWeb.com. (1998, February 5). Anneneberg, Bloomberg, and now David Cost and Jeff Trester's PriceScan, Proves that Price Information is POWER! Retrieved August 15, 2007, from http://ww1.prweb.com/releases/1998/2/prweb3455.php

Pyle, D. J. (1983). *The Economics of Crime and Law Enforcement.* New York: St. Martin Press.

Qin, Z. (2002). *Evolving marketplace designs by artificial agents.* Unpublished master's dissertation, University of Bristol, Computer Science.

Qin, Z., & Kovacs, T. (2004). Evolution of realistic auctions. In M. Withall & C. Hinde (Eds.), *Proceedings of the 2004 UK Workshop on Computational Intelligence*, Loughborough, UK (pp. 43-50).

Rahwan, I., Kowalczyk, R., & Pham, H. H. (2002). Intelligent agents for automated one-to-many e-commerce negotiation. In *Proceedings of the 25th Australasian Conference on Computer Science* (pp. 197-204).

Raiffa, H. (1982). *The art and science of negotiation.* Cambridge: Harvard University Press.

Ramchurn, S., Jennings, N., Sierra, C., & Godo, L. (2003). A computational trust model for multi-agent interactions based on confidence and reputation. *Proceedings of the 5th International Workshop on Deception, Fraud and Trust in Agent Societies.*

Ran, S. (2003, March). A model for web services discovery with QoS. *ACM SIGecom Exchange, 4*(1).

Reich, J. (2000). *Ontological design patterns: Modelling the metadata of molecular biological ontologies, information and knowledge.* Paper presented at the Conference on Database and Expert System Applications.

Reisig, W. (1985). Petri Nets: An introduction (eatcs monographs on theoretical computer science). In W. Brauer, G. Rozenberg, & A. Salomaa (Eds.). Berlin: Springer Verlag.

Riehle, D., & Gross, T. (1998). Role model based framework design and integration. In *Proceedings of the Conference on Object-Oriented Programs, Systems, Languages, and Applications* (pp. 117-133). ACM.

Robert, E. M. (2000). *Discovery and its discontents: Discovery protocols for ubiquitous computing* (Research Rep. No. UIUCDCS-R-2000-2154). Department of Computer Science, University of Illinois Urbana-Champaign, Urbana. Retrieved August 23, 2007, from http://citeseer.ist.psu.edu/mcgrath00discovery.html

Rodgera, J. A., & Pendharkarb, P. C. (2004). A field study of the impact of gender and user's technical experience on the performance of voice-activated medical tracking application. *International Journal of Human-Computer Studies, 60*, 529-544.

Rogers, E. M. (2003). *Diffusion of Innovations* (5th ed.). New York: Free Press.

Rohs, M., & Roduner, C. (2006). Towards an enterprise location service. In *Proceedings of the International Symposium on Applications and the Internet Workshops*, Phoenix, AZ.

Rosenschein, J. S., & Zlotkin, G. (1994). *Rules of encounter: Designing conventions for automated negotiation among computers.* Cambridge: MIT Press.

Rosenschein, J.S., & Zlotkin, G. (1994). *Rules of encounter.* The MIT Press.

Rothkopf, M., & Harstad, R. (1995). Two model of bidtaker cheating in vickrey auctions. *Journal of Business, 68*, 257-267.

Rothkopf, M.H., Teisberg, T.J., & Kahn, E.P. (1990). Why are vickrey auctions rare? *Journal of Political Economy, 98*(1), 94-109.

Rubinstein, A. (1982). Perfect equilibrium in a bargaining model. *Econometrica, 50* (1), 155-162.

Sah, R. K. (1991). Social Osmosis and Patterns of Crime. *The Journal of Political Economy, 99*(6), 1272-1295.

Sahin, F. & Powell Robinson, E.P. (2002). Flow coordination and information sharing in supply chains: Review, implications, and directions for future research. *Decision Sciences, 33*(4), 505-536.

Sandholm, T. (2000). Agents in electronic commerce: Component technologies for automated negotiation and coalition formation. *Autonomous Agents and MultiAgent Systems, 3*(1), 73-96.

Sathi, A., & Fox, M.S. (1989). Constraint directed negotiation of resource allocation. In L. Gasser & M. Huhns (Eds.), *Distributed artificial intelligence II* (pp. 163-195). Morgan Kaufmann.

Satish, T. (2001). *XLANG Web services for business process design.* Retrieved August 23, 2007, from http://www.gotdotnet.com/team/xml_wsspecs/xlang-c/default.htm

Schaeffer, J., Plaat, A., & Junghanns, A. (2001). Unifying single-agent and two-player search. *Information Sciences, 135*(3-4), 151-175.

Schelfthout, K., Coninx, T., et al (2002). *Agent implementation patterns.* Paper presented at the OOPSLA Workshop on Agent-Oriented Methodologies.

Schilit, B. N., Adams, N. I., & Want, R. (1994). Context-aware computing applications. In *Proceedings of the First International Workshop on Mobile Computing Systems and Applications* (pp. 85-90).

Schilit, W. N. (1995). *System architecture for context aware mobile computing.* Unpublished doctoral thesis, Columbia University.

Schillo, M., Funk, P., & Rovatsos, M. (2000). Using trust for detecting deceitful agents in artificial societies [Special issue: Deception, Fraud, and Trust in Agent Societies]. *Applied Artificial Intelligence Journal, 14,* 825-848.

Schneider, F. B. (1997). Towards Fault-Tolerant and Secure Agentry. In M. Mavronicolas & P. Tsigas (Eds.), *Proceedings of the 11ᵗʰ International Workshop on Distributed Algorithms* (LNCS 1320, pp. 1-14). Berlin: Springer.

Schneider, G.P., & Perry, J.T. (2000). *Electronic Commerce.* Cambridge, MA: Course Technology.

Schneider, P. G., & Perry, J. T. (2001). Electronic Commerce. In *Course Technology.* Canada.

schraefel, M.C., Preece, A., Gibbins, N., Harris, S., & Millard, I. (2004). Ghosts in the Semantic Web machine? In *Proceedings of the 1ˢᵗ Workshop on friend of a friend, social networking and the Semantic Web,* Galway, Ireland.

Schreiber, G., Akkermans, H., Anjewierden, A., de Hoog, R., Shadbolt, N, de Velde, W.V., & Wielinga, B. (1999). *Knowledge engineering and management: The CommonKADS approach.* MIT Press.

Seager, A. (2003). M-commerce: An integrated approach. *Telecommunications International, 37*(2), 36.

Searle, J.R. (1969). *Speech acts: An essay in the philosophy of language.* London: Cambridge University Press.

Sebenius, J.K. (1992). Negotiation analysis: A characterization and review. *Management Science, 38*(1), 18-38.

Selen, W., & Soliman, F. (2002). Operations in today's demand chain management framework. *Journal of Operations Management, 20*(6), 667-673.

Senecal, S., Kalczynski, P.J., & Nantel, J. (2005). "Consumers" decision-making process and their online shopping behavior: A clickstream analysis. *Journal of Business Research, 58,* 1599-1608.

Senn, J. A. (2000). The emergence of m-commerce. *Computer, 33*(12), 148-150.

Shadbolt, N., O'Hara, K., & Crow, L. (1999). The experimental evaluation of knowledge acquisition techniques and methods: History, problems, and new directions. *International Journal of Human-Computer Studies (IJHCS), 51,* 729-755.

Shadbolt, N., schraefel, M., Gibbins, N., & Harris, S. (2003). CS AKTive Space: Or how we stopped worrying and learned to love the Semantic Web. In *Proceedings of the 2ⁿᵈ International Semantic Web Conference,* Florida.

Shapiro, C., & Varian, H. R. (1998). *Information rules: A strategic guide to the network economy.* Cambridge, MA: Harvard Business School Press.

Sheng, Q. Z., Benatallah, B., Dumas, M., & Mak, E. (2002). SELFSERV: A platform for rapid composition of web services in a peer-to-peer environment. In *Proceedings of the 28ᵗʰ Very Large DataBase Conference (VLDB'2002),* Hong Kong, China. Retrieved August 23, 2007, from http://sky.fit.qut.edu.au/~dumas/

Shepard, S. (2005). *RFID: Radio frequency identification.* New York: McGraw-Hill.

Shu, S., & Norrie, D. (1999). *Patterns for adaptive multi-agent systems in intelligent manufacturing.* Paper presented at the International Workshop on Intelligent Manufacturing Systems.

Shugan, S. M. (1980). The cost of thinking. *Journal of Consumer Research, 7,* 99-111.

Sichman, S. J. (1994). A social reasoning mechanism based on dependence networks. In *Proceedings of the 11ᵗʰ European Conference on Artificial Intelligence (ECAI-94),* Amsterdam.

Sichman, S. J., & Demazeau, Y. (1995). *Exploiting social reasoning to deal with agency level inconsistency.* Paper presented at the 1ˢᵗ International Conference on Multi-Agent Systems (ICMAS-95), San Francisco.

Sierra, C., Faratin, P., & Jennings, N. R. (2000). Deliberative automated negotiators using fuzzy similarities. In *Proceedings of EUSFLAT* (pp. 155-158).

Sikora, R., & Shaw, M. J. (1998). A multi-agent framework for the coordination and integration of information systems. *Management Science, 44*(11), 65-78.

Sillince, J. A. A. (1998). Extending electronic coordination mechanisms using argumentation: The case of task allocation. *Knowledge-Based Systems, 10*(6), 325-336.

Sillince, J. A. A., & Saeddi, M. H. (1999). Computer-mediated communication: Problems and potentials of argumentation support systems. *Decision Support Systems, 26*(4), 287-306.

Simch-Levy, D., Kaminsky, P., & Simchi-Levy, E. (2003). *Designing and managing the supply chain— concepts, strategies and case studies, Second Edition.* New York: McGraw-Hill.

Singh, M. (2004). Business process management: A killer ap for agents? *Proceedings of the Third International Conference on Autonomous Agents and Multi Agent Systems AAMAS-2004,* (pp. 26-27).

Sinha, I. (2000, March-April). Cost transparency: The net's real threat to prices and brands. *Harvard Business Review*, pp. 43-50.

Sirin, E., Hendler, J., & Parsia, B. (2003). Semi-automatic composition of web services using semantic descriptions. In *Proceedings of the Web Services, Modeling, Architecture and Infrastructure Workshop in conjunction with ICEIS 2003.* Retrieved August 23, 2007, from http://citeseer.ist.psu.edu/sirin02semiautomatic.html

Slack, F., & Rowley, J. (2002). Online kiosks: The alternative to mobile technologies for mobile users. *Internet Research: Electronic Networking Applications and Policy, 12*(3), 248-257.

Smith, H., & Fingar, P. (2003). *Business process management (bpm): The third wave.* Meghan-Kiffer Press.

Smith, R. G., & Davis, R. (1981). Frameworks for cooperation in distributed problem solving. *IEEE Transactions on Systems, Man and Cyvernetics, 11*(1), 61-70.

Smith, V. L. (1962). An experimental study of competitive market behavior. *Journal of Political Economy, 70,* 111-137.

SOAP. Simple Object Access Protocol. Retrieved August 23, 2007, from http://www.w3.org/TR/soap12-part0/

Sokoloff, L. (2004). *Applications in LabVIEW.* New Jersey: Prentice Hall.

Solomon, M.R. (2007). *Consumer behavior: Buying, having, being* (7th ed.). Pearson/Prentice Hall.

Srivastava, B., & Koehler, J. (2003). Web service composition: Current solutions and open problems. In *Proceedings of the ICAPS 2003 Workshop on Planning for Web Services* (pp. 28-35). Retrieved August 23, 2007, from http://citeseer.ist.psu.edu/srivastava03Web.html

Staab, S., & Stuckenschmidt, H. (Eds.). (2006). *Semantic Web and peer-to-peer.* Springer.

Stamper, R.K. (1988). *MEASUR.* Enschede, The Netherlands: University of Twente.

Stamper, R.K. (1992). Language and computing in organised behavior. In R.P. van de Riet & R.A. Meersman (Eds.), *Linguistic Instruments in Knowledge Engineering* (pp. 143-163). Elsevier Science Publishers B.V.

Stamper, R.K. (1997). Organizational semiotics. In J. Mingers & F. Stowell (Eds.), *Information Systems: An Emerging Discipline.* London: McGraw-Hill.

Steinmetz, E., Collins, J., Jamieson, S., et al. (1999). Bid evaluation and selection in the MAGNET automated contracting system. *Lecture Notes in Artificial Intelligence, 1571,* 105-125.

Stiglitz, J. E., & Driffill, J. (2000). *Economics* (chap. 4: demand, supply and price). W. W. Norton & Company, Inc.

Strader, T. J., Lim, F. R., & Shaw, M. J. (1998). Information infrastructure for electronic virtual organization management. *Decision Support Systems, 23*(1), 75-94.

Su, X., & Rao, J. (2004). A survey of automated Web service composition methods. In *Proceedings of the 1st International Workshop on Semantic Web services and*

Web process composition. Retrieved August 23, 2007, from http://www.cs.cmu.edu/~jinghai/

Su, Y. W., Huang, C., Hammer, J., Huang, Y., Li, H., Wang, L., Liu, Y., Pluempitiwiriyawej, C., Lee, M., & Lam, H. (2001). An internet-based negotiation server for e-commerce. *The VLDB Journal, 10,* 72-90.

Sugumaran, V., & Storey, V.C. (2002). Ontologies for conceptual modeling: Their creation, use, and management. *Data and Knowledge Engineering, 42*(3), 251-271.

Sycara, K. (1989). Multiagent compromise via negotiation. In L. Gasser & M. Huhns (Eds.), *Distributed artificial intelligence II* (pp. 119-139). Morgan Kaufmann.

Sycara, K. (1990). Negotiation planning: An AI Approach. *European Journal of Operational Research, 46,* 216-234.

Sycara, K. P. (1989). Multiagent compromise via negotiation. In L. Gasser & M. Huhns (Eds.), *Distributed Artificial Intelligence* (Vol. II, , pp. 119-138). London: Morgan Kaufmann/San Mateo, CA: Pitman Publishing.

Sycara, K., Paolucci, M., Ankolekar, A., & Srinivasan, A. (2003). Automated discovery, interaction and composition of Semantic Web services. *Journal of Web Semantics, 1*(1), 27-46.

Tahara, Y., Oshuga, A., & Hiniden, S. (1999). *Agent system development method based on agent patterns.* Paper presented at the International Conference on Software Engineering. ACM.

Tang, T. Y., Winoto, P., & Niu, X. (2003). I-TRUST: Investigating Trust between Users and Agents in a Multi Agent Portfolio Management System. *Electronic Commerce Research and Applications, 2*(4), 302-314. Elsevier Science B.V.

Tangmunarunkit, H., Decker, S., & Kesselman, C. (2003, May 20). Ontology-based resource matching in the Grid: The Grid meets the Semantic Web. In *Proceedings of the 1st Workshop on Semantics in Peer-to-Peer and Grid Computing, in conjunction with the 12th International World Wide Web Conference.*

Taveter, K., & Wagner, G. (2000). Combining AOR diagrams and Ross Rusiness rules' diagram for enterprise modelling. In *Proceedings of the International Bi-Conference workshop on Agent-Oriented Information Systems 00' (AOIS00),* Stockholm, Sweden and Texas.

Thompson, L., & Hastie, R. (1990). Social perception in negotiation. *Organizational Behavior and Human Decision Processes, 47,* 98-123.

Tung, B., & Lee, J. (1999). An agent-based framework for building decision support systems. *Decision Support Systems, 25*(3), 225-237.

Turban, E., & Aronson, J.E. (2001). Decision support systems and intelligent systems (6th international ed.). Upper Saddle River, NJ: Prentice Hall.

Turisco, F. (2000). Mobile computing is next technology frontier for healthcare providers. *Health Care Financial Management, 54*(11), 78-80.

UDDI. Universal Discovery Description and Integration Protocol. Retrieved August 23, 2007, from http://www.uddi.org/

Ulieru, M., Norrie, D., Kremer, R., & Shen, W. (2000). A multi-resolution collaborative architecture for Web-centric global manufacturing. *Information Sciences, 127*(1-2), 3-21.

Ulmer, D. (2004). *Architectural solutions to agent-enabling e-commerce portals with pull/push abilities.* Unpublished doctoral thesis, Pace University.

Uschold, M. (2003). Where are the semantics in the Semantic Web? *AI Magazine, 24*(3), 25-36.

Uschold, M., & Gruninger, M. (2004). Ontologies and semantics for seamless connectivity. *SIGMOD Record, 33*(4).

Vahidov, R. (2005). Intermediating user-DSS interaction with autonomous agents. *IEEE Transactions on Systems, Man, and Cybernetics-Part A: Systems and Humans, 35*(6), 964-970.

Vakharia, A.J. (2002). E-business and supply chain management. *Decision Sciences, 33*(4), 495-504.

Van der Aalst, W. M. P., Dumas, M., & ter Hofstede, A. H. M. (2003). Web service composition languages: Old wine in new bottles? In *Proceedings of the 29th EURO-MICRO Conference* (pp. 298-305).

van der Aalst, W., & van Hee, K. (2002). *Workflow management: Models, methods, and systems.* MIT Press.

Varian, H.R. (2004). Competition and Market Power. In H.R. Varian, J. Farrell, & C. Shapiro (Eds.), *The economics of information technology: An introduction* (pp. 1-47). Cambridge University Press: Cambridge.

Varshney, U. (1999). Networking support for mobile computing. *Communications of AIS, 1*(1), 1-30.

Varshney, U. (2000). Recent advances in wireless networking. *IEEE Computer, 33*(6), 100-103.

Vassileva, J. (2002). Bilateral negotiation with incomplete and uncertain information. In P.G.S. Parsons & M. Wooldridge (Eds.), *Game theory and decision theory in agentbased systems.* Kluwer Academic Publishers.

Ververidis, C., & Polyzos, G. (2002). Mobile marketing using location based services. In *Proceedings of the 1st International Conference on Mobile Business*, Athens, Greece.

Vickrey, W. (1961). Counterspeculation, auctions, and competitive sealed tenders. *Journal of Finance, 16,* 8-37.

Vigna, G. (1998). Cryptographic Traces for Mobile Agents. In G. Vigna (Ed.), *Mobile Agents and Security* (LNCS 1419, pp. 137-153). Springer.

Vincent, D. (1989). Bargaining with common values. *Journal of Economic Theory, 48,* 47-62.

Vinoski, S. (2002). Where is middleware? *IEEE Internet Computing, 6*(2), 83-85. IEEE Computer Society Press.

vLeonhardt, U. (1998). *Supporting location: awareness in open distributed systems.* Unpublished doctoral thesis, Imperial College, Department of Computing, London.

Voss, A., & Kreifelts, T. (1997). *SOaP: Social agents providing people with useful information.* Paper presented at the Conference on Supporting Groupwork. ACM.

W3C. (1997). *Open profiling standard* (version 1.0, note 1997/6).

Wang, F. H., & Shao, H. M. (2004). Effective personalized recommendation based on time-framed navigation clustering and association mining. *Expert Systems with Applications, 27*(3), 365-377.

Wang, M., Cheung, W.K., Liu, J., & Luo, Z. (2006a, June 26-29). Agent-based Web service composition for supply chain management. In *Proceedings of IEEE Joint Conference on E-Commerce Technology and Enterprise Computing, E-Commerce and E-Services* (pp. 328-332). Los Alamitos: IEEE Computer Society Press.

Wang, M., Cheung, W.K., Liu, J., Xie, X., & Luo, Z. (2006b, September 5-7). E-service/process composition through multi-agent constraint management. In *Proceedings of the 4th International Conference on Business Process Management* (LNCS 4102, pp. 274-289). Springer-Verlag.

Wang, M., Wang, H., & Liu, J. (2007, January 7-10). Dynamic supply chain integration through intelligent agents. In *Proceedings of 40th Hawaii International Conference on System Sciences* (p. 46). Los Alamitos: IEEE Computer Society Press.

Wang, Y., Tan, K. L., & Ren, J. (2002). A study of building Internet marketplaces on the basis of mobile agents for parallel processing. *World Wide Web, 5*(1), 41-66.

Want, R., Hopper, A., Falcao, V., & Gibbons, J. (1992). The active badge location system. *ACM Transactions on Information Systems, 10*(1), 91-102.

Want, R., Schilit, B., Adams, N., Gold, R., Petersen, K., Ellis, J., et al. (1995). *The PARCTAB ubiquitous computing experiment* (Tech. Rep. No. CSL-95-1). Xerox Palo Alto Research Center.

Watanabe, Y., Ishiguro, A., & Uchkawa, Y. (1999). Decentralized behavior arbitration mechanism for autonomous mobile robot using immune system. In *Artificial Immune Systems and Their Applications*. Springer-Verlag.

Weiser, M. (1996). *Hot topics: Ubiquitous computing.* IEEE Computer.

Weiss, M. (2001). *Patterns for e-commerce agent architectures: Agents as delegates.* Paper presented at the Conference on Pattern Languages of Programs.

Weiss, M. (2003). Pattern-driven design of agent systems: Approach and case study. In *Proceedings of the Conference on Advanced Information System Engineering* (pp. 711-723). Springer.

Weiss, M. (2004). A pattern language for motivating the use of agents. In *Agent-Oriented Information Systems: Revised Selected Papers* (LNAI 3030, pp. 142-157). Springer

Weiss, M., Gray, T., & Diaz, A. (1997). Experiences with a service environment for distributed multimedia applications. feature interactions in telecommunications and distributed systems, IOS, 242-253.

Weitzner, D.J., Hendler, J., Berners-Lee, T., & Connolly, D. (2005). Creating a policy-aware Web: Discretionary, rule-based access for the World Wide Web. In E. Ferrari & B. Thuraisingham (Eds.), *Web and Information Security*. Idea Group Inc.

Wenger, E. (1998). *Communities of practice: learning, meaning and identity.* Cambridge: Cambridge University Press.

White, E. (2000, October 23). No comparison. *Wall Street Journal*.

Wilks, Y. (2004). Are ontologies distinctive enough for computations over knowledge? *IEEE Intelligent Systems, 19*(1), 74-75.

Williamson, O.E. (1975). *Markets and hierarchies.* New York: Free Press.

Williamson, O.E. (1991). *The nature of the firm: origins, evolution and development.* Oxford: Oxford University Press.

Winogard, T., & Flores, F. (1986). *Understanding computers and cognition: A new foundation for design.* Noorwood, NJ: Ablex.

Winograd, T., & Flores, F. (1986). *Understanding computers and cognition: A new foundation for design.* Norwood: Ablex.

Winoto, P. (2003a). A Simulation of the Market for Offenses in Multiagent Systems: Is Zero Crime Rates Attainable? In J. S. Sichman, F. Bousquet, & P. Davidsson (Eds.), *Multi-Agent-Based Simulation II* (LNCS 2581, pp. 181-193). Springer.

Winoto, P. (2003b). Controlling Malevolent Behavior in Open Multi-Agent Systems by Means of Deterrence Theory. In *Proceedings of 2003 IEEE/WIC International Conference on Intelligent Agent Technology*, Halifax (pp. 268-274).

Wong, W. Y., Zhang, D. M., & Kara-Ali, M. (2000). Negotiating with experience. In *Proceedings of KBEM-2001*, Austin, TX.

Wooldridge, M. (1997). Agent based software engineering. *IEEE Proceedings of Software Engineering, 144*(1), 26-37.

Wooldridge, M. (2002). *An introduction to multiagent systems.*: John Wiley & Sons.

Wooldridge, M., & Jennings, N. (1995). Intelligent agents: Theory and practice. *The Knowledge Engineering Review, 10*(2), 115-152.

Wooldridge, M., & Jennings, N. R. (1999). The cooperative problem-solving process. *Journal of Logic Computation, 9*(4), 563-592.

Wooldridge, M., Jennings, N.R., & Kinny, D. (2000). The Gaia methodology for agent-oriented analysis and design. *Journal of Autonomous Agents and Multi-Agent Systems, 3*(3), 285-312.

Wooldridge., M. & Jennings, N.R. (1995). Intelligent agents: Theory and practice. GRACO. Retrieved on February 15, 2006 at http://www.graco.unb.br/alvares/DOUTORADO/disciplinas/feature/agente_definicao.pdf .]

WSCI. *Web Service Choreography Interface 1.0 Specification.* Intalio, Sun Microsystems, BEA Systems, SAP.

WSDL. Web Services Description Language. Retrieved August 23, 2007, from http://www.w3.org/TR/wsdl

Wu, C., & Chang, E. (2005). State-of-the-art Web services architectural styles. In *Proceedings of the 3rd IEEE European Conference on Web Services (ECOWS),* Vaxjo, Sweden. Retrieved August 23, 2007, from http://wscc.info/p51561/files/paper54.pdf

Wu, D. J. (2001). Software agents for knowledge management: Coordination in multi-agent supply chains and auctions. *Expert Systems with Applications, 20*(1), 51-64.

Wu, G., Yuan, H., Tseng, S. S., & Fuyan, Z. (1999). A knowledge sharing and collaboration system model based on Internet. In *Proceedings of IEEE International Conference on Systems, Man, and Cybernetics* (pp.148-152).

Wurman, P. R. (2001). A parameterization of the auction design space. *Games and Economic Behavior, 35,* 304-338.0

Wurman, P., Walsh, W., & Wellman, M. (1998). Flexible double auction for electronic commerce: Theory and implementation. *Decision Support System, 24,* 17-27.

Wurman, P., Wellman, M., & Walsh, W. (2001). A parameterization of the auction design space. *Games and Economic Behavior, 35,* 304-338.

Xiaohui, G., Nahrstedt, K., & Yu, B. (2004). SpiderNet: An Integrated peer-to-peer service composition framework. In *Proceedings of the 13th IEEE International Symposium on High Performance Distributed Computing (HPDC)* (pp. 110-119).

Xu, D., Nahrstedt, K., & Wichadakul, D. (2001). Qos-aware discovery of wide-area distributed services. In *Proceedings of the 1st IEEE/ACM International Symposium on Cluster Computing and the Grid (CCGrid),* Brisbane, Australia (pp. 92-99).

Yang, Y., Tan, Q., & Xiao, Y. (2005). Verifying Web services composition based on hierarchical colored petri nets. In *Proceedings of the 1st International Workshop on Interoperability of Heterogeneous Information Systems* (pp. 47-54, ISBN: 1-59593-184-5).

Yokoo, M. (2001). *Distributed constraint satisfaction: Foundation of cooperation in multi-agent systems.* Berlin/New York: Springer.

Yuan, S. T., & Peng, K. H. (2004). Location based and customized voice information service for mobile community. *Information Systems Frontiers, 6*(4), 297-311.

Yuan, S. T., & Tsao, E. (2003). A recommendation mechanism for contextualized mobile advertising. *Expert Systems with Applications, 24*(4), 399-414.

Yuan, S.T. (2002). A personalized and integrative comparison-shopping engine and its applications. *Decision Support Systems, 34,* 139-156.

Zacharia, G., Moukas, A., & Maes, P. (1999). *Collaborative reputation mechanisms in electronic marketplaces.* Paper presented at the Hawaii International Conference On System Science. IEEE.

Zambonelli, F., Jennings, N., et al. (2001). Agent-oriented software engineering for Internet applications. In A. Omicini, et al. (Eds.), *Coordination of Internet Agents* (pp. 326-346). Springer.

Zeng, D., & Sycara, K. (1998). Bayesian learning in negotiation. *International Journal of Human Computer Studies, 48,* 125-141.

Zeng, L., Benatallah, B., Ngu, A., Dumas, M., Kalagnanam, J., & Chang, H. (2004). QoS-aware middleware for Web services composition. *IEEE Transactions on Software Engineering, 30*(5), 311-327.

Zheng, D., & Sycara, K. (1997). Benefits of learning in negotiation. In *Proceedings of the 14th National Conference on Artificial Intelligence* (pp. 36-41).

Zlokin, G., & Rosenschein, J.S. (1989, August). Negotiation and task sharing among autonomous agents in cooperative domains. In *Proceedings of the 11th International Joint Conference on Artificial Intelligence,* Detroit, MI (pp. 912-917).

About the Contributors

Eldon Y. Li is university chair professor of the College of Commerce at National Chengchi University, Taiwan. He was professor and dean of the College of Informatics at Yuan Ze University in Taiwan from 2003 to 2005. He was a professor and coordinator of the MIS Program at the College of Business, California Polytechnic State University, San Luis Obispo, California, USA. He visited the Department of Decision Sciences and Managerial Economics at the Chinese University of Hong Kong from 1999 to 2000. He was the professor and founding director of the Graduate Institute of Information Management at National Chung Cheng University in Chia-Yi, Taiwan. He holds a PhD from Texas Tech University. His current research interests are in human factors in information technology (IT), strategic IT planning, software engineering, quality assurance, and information and systems management. He is editor-in-chief of *International Journal of Electronic Business, International Journal of Information and Computer Security, International Journal of Information Policy and Law, International Journal of Internet and Enterprise Management,* and *International Journal of Internet Marketing and Advertising.*

Soe-Tsyr Yuan is a professor of information management of National Chengchi University, Taiwan. She received a PhD in computer science from Oregon State University, USA. Her research interests include service science, M(U-) commerce, service-oriented computing, business intelligence management, intelligent agents, knowledge discovery and data mining, and so forth. She has served as a member of various program and editorial committees for international conferences and journals such as the *International Journal of Web Services Research, International Journal of E-Business Research, Service Oriented Computing and Applications,* and so forth.

* * * * *

Harith Alani is a doctor of computer science. He is currently working as a senior research fellow with the Intelligence, Agents, Multimedia Group in the School of Electronics and Computer Science, University of Southampton. For the past six years, Dr. Alani has been affiliated with the AKT IRC. His current research activities include semantic representation and analysis of communities of practice, ontology change management, and ontology searching, ranking, evaluation, and minimization. He has organized a number of knowledge management and Semantic Web related workshops in various international conferences.

Mohamed Bakhouya received a PhD in 2005 in computer science and computer engineering from the University of Technology of Belfort Montbeliard, France. From September 2001 to August 2006, he was a lecturer at the Computer Engineering Department of UTBM. He is currently a postdoctoral researcher at the High Performance Computing Laboratory at The George Washington University (WDC, USA). His research interests include ubiquitous and pervasive computing, distributed and parallel algorithms, mobile computing, artificial intelligence, grid computing, and reconfigurable computing.

Joseph Barjis is a member of the Faculty, Department of IT at Georgia Southern University (USA). Prior to that, he worked in The Netherlands, UK, and Russia, where he had teaching and research appointments. Dr. Barjis has an MSc summa cum laude and PhD summa cum laude in computer science and a postdoctorate in information systems. His research interest is focused on areas such as e-business, business process analysis, design, modeling and simulation, IS design, and application of Petri nets. He is author/editor of over 80 works: 3 edited books, 2 college handbooks, 7 book chapters, 5 journal articles, and over 60 conference papers.

Yee-Ming Chen is associate professor in the Department of Industrial Engineering and Management at Yuan Ze University, where he carries out basic and applied research in agent-based computing and soft computing. His current research interests include information fusion and data mining, supply chain management, and system diagnosis/prognosis. He is also actively engaged in applied projects with industry that involve product development or the development of product realization infrastructure. For more information, visit http://kem.iem.yzu.edu.tw/.

William K. Cheung is an associate professor in the Department of Computer Science at Hong Kong Baptist University. He received a PhD in computer science from the Hong Kong University of Science and Technology (1999). He served as the co-chairs and program committee members of a number of international conferences as well as journals' guest editors on areas including artificial intelligence, Web intelligence, data mining, Web services, and e-commerce technology. Also, he has been on the editorial board of *IEEE Intelligent Informatics Bulletin* since 2002. His research interest includes artificial intelligence and machine learning and their applications to collaborative filtering, Web mining, distributed data mining, planning under uncertainty as well as dynamic Web/Grid service management.

Samuel Chong is an experienced consultant with Accenture Ltd, who has led functional and technical teams to deliver business solutions to blue chip clients across different industries. He currently leads an integration team on a cross European project for a major global client. Dr. Chong graduated with a BSc (Hons) in computing science and MSc in information systems (Distinction). He subsequently earned a PhD in computing science specializing in the field of organizational semiotics and agent-based system design. His PhD thesis was the first treatment of its kind to agent-based system design in the UK. During his PhD tenure, he published over 15 conference papers, journal articles, and book chapters.

Seng-cho T. Chou is professor of information management at National Taiwan University. He received his BSc from the Chinese University of Hong Kong, MS from the University of California, and PhD in computer science from the University of Illinois at Urbana-Champaign. His current research interests are Web technologies and services, e-business and e-commerce, knowledge management, data mining, and ubiquitous and mobile computing.

Jaafar Gaber received a PhD in 1998 from University of Science and Technology of Lille (France) in computer science and computer engineering. He is currently an associate professor of computational sciences and computer engineering at the University of Technology of Belfort-Montbéliard UTBM (France). Prior to joining UTBM, he was a research scientist at the Institute of Computational Sciences and Informatics (CSI) in George Mason University in Fairfax (Virginia, USA). His research interests include ubiquitous and pervasive computing, high-performance computing, distributed data mining, bio-computing, distributed algorithms and mobile computing, computer networks, and communications.

Pei-Hung Hsieh received his master's degree from the Information Management Department of Fu-Jen University, Taiwan. His research interests include mobile commerce, security policy, technology map of advanced ICT technologies, and so forth. He is now working under the Science & Technology Policy Research and Information Center of National Science Council and actively engages in the collecting, organizing, constructing, analyzing, judging, and researching of science and technology policy information for the purpose of providing the relevant information for the government in drawing up science and technology policies.

Pei-Ni Huang received a master's degree in industrial engineering and management from Yuan Ze University in July 2006. Since 2003 she has been working as a research assistant within the Laboratory for Dr. Yee Ming Chen. She is an active member in the international conference and received the Excellent Student Paper Award at the 36th International Conference on Computers & Industrial Engineering on June 23, 2006. She is currently a logistic planner at IC Design House in Taipei.

Golenur Begum Huq is currently a PhD student in the School of Computing and Mathematics of the University of Western Sydney. Her research interests are agent technology including negotiation, cooperation, and coordination, conducting e-business, challenges of e-business, e-supply chain management, and e-learning. Golenur has publications of her research presented at international conferences. Until recently, she was a teaching fellow at UWS for four years, and is currently involved in teaching at the same university. Golenur has 12 years industry experience where she was involved in management activities.

Yannis Kalfoglou is a senior research fellow. He received his PhD in AI from the University of Edinburgh. He is working on the AKT project. He was the principle investigator of an industrial project funded by Hewlett Packard, CROSI, which explored applications of ontology mapping. He is working on Semantic Web technologies, in particular, semantic interoperability and integration. Dr. Kalfoglou has published over 48 works in leading journals and conferences in the areas of the Semantic Web, artificial intelligence, and knowledge engineering. He has served as a member of various program and editorial committees for international journals and conferences.

Robyn Lawson is a senior lecturer with the School of Computing and Mathematics at the University of Western Sydney. She is also associate director of the Advanced Enterprise Information Management Systems (AcIMS) Research Group, which focuses on research into e-business solutions for small and medium enterprises (SMEs). Robyn's PhD investigated electronic commerce adoption in manufacturing. In addition, Robyn has published papers in the areas of e-business and e-collaboration, and is also on the review board for the *Journal of Information Systems and Small Business.*

Kun Chang Lee is a full professor of MIS at Sungkyunkwan University in Seoul, Korea. He received a PhD in MIS from the Korea Advanced Institute of Science and Technology (KAIST), a Master of Science in MIS from KAIST, and a BA in business administration from Sungkyunkwan University, Seoul, Korea. His research focuses on decision analysis involved in electronic commerce management. His main research interests lie in electronic commerce, knowledge management, and artificial intelligence-based decision making. His research findings are published in the *Journal of Management Information Systems, IEEE Transactions on Engineering Management, Decision Support Systems, International Journal of Production Research, Journal of Artificial Societies and Social Simulation, Expert Systems with Applications, Fuzzy Sets and Systems, Intelligent Systems in Accounting Finance and Management, Computers & OR, Simulation, Expert Systems,* among others.

Namho Lee is a senior principal consultant at Oracle Consulting Korea. He received a PhD in MIS from Sungkyunkwan University. His research focuses on a multi-agent approach to business problem-solving. His publication paper is to appear in *Journal of Artificial Societies and Social Simulation.* Besides, part of his research work is to be published in a prestigious book series, Advances in Management Information Systems (Publisher: M.E. Sharpe, editor-in-chief: Vladimir Zwass). His work experience is focused on IS design and implementation using the ERP system.

Wen-Shan Lin is an assistant professor in the Department of Management Information Systems at National Chiayi University, Taiwan. She received an MSc from Manchester Metropolitan University and a PhD from the University of Manchester. Her research interests include Internet marketing, intelligent agents, customer behavior, and technology management. She has published several research papers in IEEE conferences, *Journal of Expert System with Applications,* and others. She is a member of IEEE.

Raymund J. Lin is section manager, service science at the Service Innovation Center of the Institute for Information Industry (III). Before joining III, he served as project manager, innovative applications at Samar Techtronics. He received his PhD in information management from National Taiwan University.

Jiming Liu is presently a professor and head of the Computer Science Department at Hong Kong Baptist University. He holds a BSc in physics from East China Normal University, an MA in educational technology from Concordia University, and an MEng and PhD both in electrical engineering from McGill University. Prof. Liu has published over 200 scientific articles in refereed international journals and conferences. In addition, he has published numerous books, including seven research monographs. His present research interests include: autonomy oriented computing (AOC), self-organizing systems and networks, multi-agent systems and distributed problem solving, and Web intelligence (WI). Prof. Liu has served as editor-in-chief of *Web Intelligence and Agent Systems, Annual Review of Intelligent Informatics,* and *The IEEE Intelligent Informatics Bulletin.* He is an associate editor of *IEEE Transactions on Data and Knowledge Engineering* as well as an editor for seven other international journals and book series.

Zongwei Luo (zwluo@eti.hku.hk.) is a senior researcher and managing the emerging technologies group in the E-business Technology Institute, University of Hong Kong. His current interests include enabling technologies for e-logistics and supply chain management. Dr. Luo sits on the editorial board for several international journals and serves as a panelist for many events.

Kieron O'Hara is a senior research fellow in memories for life at the University of Southampton. He researches in politics and technology, and is the author of six books. He is a fellow of the Web Science Research Initiative, and co-author of *A Framework for Web Science* with Sir Tim Berners-Lee and others. He is currently editing a special issue of the *International Journal of Human-Computer Studies* on ontologies and knowledge representation, and writing a book on digital privacy. He is a trustee of the higher education think tank Agora.

Zengchang Qin received a BSc in control engineering from Heilongjiang University, Harbin, China (2001), and an MSc in machine learning and data mining from the University of Bristol, UK (2002). He also obtained his PhD in soft computing and data mining from the same university in 2005. He has been working as a temporary lecturer in the Department of Engineering Mathematics of Bristol University since October 2005, before taking the BT/BISC research fellowship. He is currently at BT (British Telecommunications) as research fellow working on automated question answering systems. His research interests focus on machine learning and data mining, soft computing and computational economics/finance. He is one of the Best Student Paper Award winners in IFSA World Congress 2005. He was also a visiting fellow of Intelligent Systems Lab, BT Research from February to March 2006.

Nigel Shadbolt is a professor of AI. He was the director of the Advanced Knowledge Technologies (AKT) project. He is a member of the IEEE Computer Societies Publications board. He was an associate editor of the *International Journal for Human Computer Systems* and is currently on the editorial board of the *Knowledge Engineering Review* and the *Computer Journal*. He is a member of various national committees, including the UK e-Science Technical Advisory Committee (TAG), the Strategic Advisory Team (SAT) of the ICT program of the EPSRC, and the Defence Scientific advisory board's human sciences sub-committee. He is currently the president of the British Computer Society.

Chung-Jen (CJ) Tan is chief executive officer of the Hong Kong R&D Center for Logistics and Supply Chain Management Enabling Technologies launched by the HKSAR government. The Center has the mission to conduct research and develop relevant technical solutions to serve industries in Hong Kong and the Pearl River Delta region. He is also the founding director of the E-Business Technology Institute (ETI) and holds the positions of visiting IBM chair professorship at the Department of Computer Science and the School of Business, University of Hong Kong. Before taking up the helm at ETI in September 1999, Prof. Tan served as senior manager at IBM Research. In 1997, he led the IBM Deep Blue team in an historic match against the world chess champion Garry Kasparov. Prof. Tan serves as adjunct professor at Chongqing University, China Central University of Science and Technology, South China University of Science and Technology, and Harbin Institute of Science and Technology, guest professor at Zhejiang University. He serves on the management board of the E-technology Research Center of Shanghai Jiaotong University. Prof. Tan is a fellow of the Association for Computing Machinery (ACM) and the Hong Kong Academy of Engineering Sciences. He was a founding council member of the IBM Deep Computing Institute. Prof. Tan received a BSE and DEng from Seattle University in 1963 and from the Columbia University in 1969, respectively.

Tiffany Y. Tang received a MS in computer science from Guangzhou Jinan University. She is an instructor in the Department of Computing at Hong Kong Polytechnic University. Her research interests include artificial intelligence in education, reputation, and recommendation systems, and human-computer interaction. She is currently a PhD candidate at the University of Saskatchewan.

Frank Tong is the senior assistant director of the E-Business Technology Institute of the University of Hong Kong. Currently, he is also appointed as the director of research and technology development with the Hong Kong R&D Centre for Logistics and Supply Chain Management Enabling Technologies. Frank's areas of interest include logistics and supply chain information management, RFID technologies, wireless security, mobile computing in e-business applications as well as the Web-based and Grid computing technologies for service provision and integration. Tong has a PhD in computer science from the Simon Fraser University, Canada. He has earned major government R&D grants, published in international journals and conferences, produced industrial solutions, and developed products.

Yun Wan is an assistant professor of computer science and information systems in the University of Houston, Victoria. His research interests include electronic commerce, information system analysis and design, and decision support systems. He has published papers about comparison-shopping agents in journals like *Communications of the ACM*. He presented his research about comparison-shopping agents and other topics of electronic commerce in several international conferences like the *International Conference on Electronic Commerce*. He is a fellow of the 2003 Doctoral Consortium of International Conference on Information Systems. He obtained his PhD in MIS from Liautaud Graduate School of Business, University of Illinois.

Minhong Wang (magwang@hkucc.hku.hk) is an assistant professor in the Division of Information & Technology Studies at the University of Hong Kong. She received her PhD in information systems from City University of Hong Kong. Her areas of research interest include agent-oriented software engineering, knowledge-based decision and coordination, business process management, and Web intelligence.

Michael Weiss is an associate professor at Carleton University, which he joined in 2000 after a five year stint in industry following his PhD in computer science in 1993 (University of Mannheim, Germany). In particular, he led the Advanced Applications Group within the Strategic Technology Group of Mitel Corporation. His research interests include service-oriented architectures, software architecture and patterns, business model design and evolution, and open source development.

Pinata Winoto received an MS in computer science from Guangzhou Jinan University, an MA in economics from the University of Mississippi, and a PhD in computer science from the University of Saskatchewan. He is an assistant lecturer in the Department of Computer Science at Hong Kong Baptist University. His research interests include multi-agent systems, Web intelligence, software usability engineering, and applied artificial intelligence.

Index

A

active-user scheme (AU) 251
active mobile group 228
active objects 8
adaptive intelligent 42
AdSense technology 60
Advanced Knowledge Technologies (AKT) 285
affordance 101
agent 25, 26, 28, 30, 239, 240, 241, 242, 243
  obligations 101
  society 239, 240, 261
agent-based e-business systems 1
agent-based shopping supports (ASS) 38, 39
agent-oriented-relationship (AOR) 121
agents 38
  interaction 96, 114
algorithmic mechanism design (AMD) 65
Andreasen model 59
antecedent 101
artificial-intelligence technology 42
artificial intelligence (AI) 63, 97, 126, 218, 269
Assembling Company (AC) 185
auction 3, 4, 9, 58, 61, 62, 77
  space 79, 85, 86
    model 79, 84, 92
automatic
  deterrence
    policy 260
  path
    creation (APC) 331

autonomy oriented computing (AOC) paradigm 193

B

behavioral
  analysis 109
  norms 103, 107, 117
Bettman information processing model 59
business
  oriented
    location
      model (BOLM) 294, 297
      models (BOLM) 320
  process
    management 147
  unit (BU) 294
business-to-business (B2B) 39, 58, 126, 132, 220, 326
business-to-consumer (B2C) 25, 39, 58, 126, 220

C

case based reasoning (CBR) 39, 132, 218, 219
CBB model 74
central processing unit (CPU) 325
choreography
  language 334
  methods 330
client to server paradigm (CSP) 340, 344
client unit (CU) 294